Monika Sethi, Shivani Sood, Saravjeet Singh, Arfat Ahmad Khan (Eds.)

Applied Mathematics for Healthcare Intelligent Systems

Mathematical Methods in the Digital Age

Computational Intelligence & Advancements

Edited by
Abhishek Kumar, Satya Prakash Yadav,
Ashutosh Kumar Dubey, Andrei Chtcheprov,
Sayavur Bakhtiyarov

Volume 9

Applied Mathematics for Healthcare Intelligent Systems

Data Representation, Smart Healthcare, Deep Learning and Medical Imaging

Edited by
Monika Sethi, Shivani Sood, Saravjeet Singh,
Arfat Ahmad Khan

DE GRUYTER

Editors

Monika Sethi
Department of Computer Science and Engineering
Chitkara University Institute of Engineering and Technology
140401 Rajpura
Punjab
India
sethi.monika4@gmail.com

Saravjeet Singh
Department of Computer Science and Engineering
Chitkara University Institute of Engineering and Technology
140401 Rajpura
Punjab
India
saravjeet.2009@gmail.com

Shivani Sood
School of Computer Applications
Lovely Professional University
144021 Phagwara
Punjab
India
shivanisd6@gmail.com

Arfat Ahmad Khan
Department of Computer Science
College of Computing
Khon Kaen University
40002 Khon Kaen
Thailand
abhishekkmr812@gmail.com

ISSN 2943-2731
ISBN 978-3-11-222621-6
ISBN 978-3-11-222622-3 (PDF)
ISBN 978-3-11-222623-0 (EPUB)
DOI https://doi.org/10.1515/9783112226223

Library of Congress Control Number: 2026932514

Bibliographic information published by the Deutsche Nationalbibliothek
The Deutsche Nationalbibliothek lists this publication in the Deutsche Nationalbibliografie; detailed bibliographic data are available on the Internet at http://dnb.dnb.de.

De Gruyter and Walter de Gruyter GmbH are part of De Gruyter Brill.
www.degruyterbrill.com

Questions about General Product Safety Regulation:
productsafety@degruyterbrill.com

Printing: CPI books GmbH, Leck

MIX
Papier | Fördert gute Waldnutzung
FSC® C083411

Preface

The rapid evolution of healthcare technologies has transformed the way medical data is generated, interpreted, and applied in clinical practice. Advances in intelligent systems, data-driven methodologies, and artificial intelligence (AI) have significantly influenced healthcare diagnostics, medical imaging, and decision-support systems. Recognizing this transformative shift, this edited volume has been carefully curated to provide a comprehensive and interdisciplinary perspective on intelligent healthcare solutions.

As editors of this volume, our primary motivation has been to bring together diverse scholarly contributions that reflect both theoretical insights and practical applications in the domain of smart healthcare. While applied mathematics forms an underlying foundation of many intelligent systems, this book does not focus on mathematical derivations or formal proofs. Instead, it emphasizes conceptual understanding, modeling perspectives, and application-driven approaches that demonstrate how intelligent systems and data representations are shaping modern healthcare. This approach allows the book to remain accessible to a broad audience while still maintaining academic depth and relevance.

This volume is the result of a collaborative effort involving contributions from researchers, academicians, and professionals across different institutions and countries. Each chapter has been independently authored, yet collectively aligned with the central theme of healthcare intelligent systems. As editors, we have carefully reviewed and organized these contributions to ensure coherence, quality, and relevance. The chapters address a wide spectrum of topics, including medical imaging modalities, data handling and representation, deep learning techniques, explainable AI, generative AI, three-dimensional reconstruction, performance evaluation metrics, ethical considerations, and emerging trends in healthcare technologies.

The organization of the book has been designed to guide readers from foundational concepts toward advanced applications. Initial chapters introduce the role of intelligent systems and data-driven approaches in healthcare, followed by discussions on medical imaging and deep learning frameworks. Subsequent sections focus on explainable and trustworthy AI, real-world case studies, and challenges associated with deploying intelligent systems in clinical environments. The concluding chapters explore future directions and technological advancements that are expected to further influence medical diagnostics and healthcare delivery. Readers may approach the book sequentially or select chapters according to their specific research or professional interests.

This edited volume is intended for a diverse readership. Scholars and researchers in computer science, biomedical engineering, applied sciences, and healthcare technology will find it useful as a reference for current research directions and interdisciplinary methodologies. Graduate and postgraduate students can use this book as a learning resource to understand how intelligent systems are designed and applied in

 | https://doi.org/10.1515/9783112226223-202

healthcare contexts. Industry professionals, engineers, and healthcare practitioners may also benefit from the practical insights and case-based discussions presented throughout the chapters. By addressing these varied audiences, the book aims to foster knowledge exchange and collaboration across disciplines.

The successful completion of this volume would not have been possible without the collective efforts of many individuals. We sincerely acknowledge all contributing authors for their valuable chapters, timely cooperation, and commitment to academic quality. Their contributions form the intellectual core of this book and reflect the diversity and depth of current research in healthcare intelligent systems.

We would like to express our special gratitude to **Dr Abhishek Kumar**, whose encouragement, guidance, and support played a crucial role in the acceptance of this book proposal. His confidence in this project and constructive suggestions significantly strengthened our vision and direction as editors, and we are deeply thankful for his contribution to this achievement.

We also extend our heartfelt thanks to our **family members**, whose patience, understanding, and continuous encouragement made it possible for us to undertake and complete this editorial responsibility. Editing an academic volume of this scale requires sustained effort and dedication, and the support of our families has been invaluable throughout this journey.

Finally, we appreciate the support and cooperation of the publishing team at De Gruyter for providing the opportunity to bring this edited volume to fruition. We hope that this volume serves as a valuable resource for the academic and professional community and contributes meaningfully to the advancement of intelligent healthcare technologies. Our aspiration is that this book will inspire further research, innovation, and interdisciplinary collaboration aimed at improving healthcare outcomes through intelligent systems.

Contents

List of Contributors

Mahak
Department of Optometry, SOAHS
Alakh Prakash Goyal Shimla University
Shimla 171009
Himachal Pradesh, India
Chapter 1

Sachitanand Singh
Department of Optometry, UIAHS
Chandigarh University
Mohali 140413
Punjab, India
Chapter 1

Nakul Sankhayadhar
Department of Optometry, UIAHS
Chandigarh University
Mohali 140413
Punjab, India
Chapter 1

Iqra Jamaal Khan
Department of Optometry, UIAHS
Chandigarh University
Mohali 140413
Punjab, India
Chapter 1

Ashith Tripathi
Department of Optometry, UIAHS
Chandigarh University
Mohali 140413
Punjab, India
Chapter 1

Rashi Rastogi
Sir Chottu Ram Institute of Engineering and Technology
Ch. Charan Singh University
Meerut
Uttar Pradesh
Indiarastogi.rashi4@gmail.com
Chapter 2

Chander Prabha
Chitkara University Institute of Engineering and Technology
Rajpura 140401
Punjab, India
prabhanice@gmail.com
Chapter 3

Shalini Kumari
Chitkara University Institute of Engineering and Technology
Rajpura 140401
Punjab, India
k.shalini@chitkara.edu.in
Chapter 3

Monika Sethi
Department of Computer Science and Engineering
Chitkara University Institute of Engineering and Technology
140401 Rajpura
Punjab, India
sethi.monika4@gmail.com
Chapter 4

Shivani Sood
School of Computer Applications
Lovely Professional University
144021 Phagwara
Punjab, India
shivanisd6@gmail.com
Chapters 4, 8

Saravjeet Singh
Department of Computer Science and Engineering
Chitkara University Institute of Engineering and Technology
140401 Rajpura
Punjab, India
saravjeet.2009@gmail.com
Chapter 4

 | https://doi.org/10.1515/9783112226223-204

Jatin Arora
Department of Computer Science and Engineering
Chitkara University Institute of Engineering and Technology
140401 Rajpura
Punjab, India
jatin.arora@chitkara.edu.in
Chapter 4

Neha Garg
Chitkara University Institute of Engineering and Technology
Rajpura 140401
Punjab, India
Chapter 5

Priyanka Datta
Laxmi Institute of technology
Sarigam
Gujarat, India
Chapter 5

Amitesh Aggarwal
Chitkara University Institute of Engineering and Technology
Rajpura 140401
Punjab, India
Chapter 5

Amanpreet Kaur
Chitkara University Institute of Engineering and Technology
Rajpura 140401
Punjab, India
Chapter 5

Yonis Gulzar
Department of Management Information Systems
College of Business Administration
King Faisal University
Al Ahsa, Saudi Arabia
Chapter 5

Ashith Tripathi
Department of Optometry
UIAHS
Chandigarh University
Punjab, India
Chapter 6

Sachitanand Singh
Department of Optometry
UIAHS
Chandigarh University
Punjab, India
Chapter 6

Iqra Jamal Khan
Department of Optometry
UIAHS
Chandigarh University
Punjab, India
Chapter 6

Padamandala Krishnasri
Department of Optometry
SOAHS
Malla Reddy University
Hyderabad
Telangana, India
Chapter 6

Sanskriti Singh
Department of Nursing
UIN
Chandigarh University
Punjab, India
Chapter 6

Hema Umapathi
Department of Biotechnology and Bioinformatics
JSS Academy of Higher Education and Research
Mysore
Karnataka, India
Chapter 7

Anunya Pattjoshi
Department of Biotechnology and Bioinformatics
JSS Academy of Higher Education and Research
Mysore
Karnataka, India
Chapter 7

Khushi S. Udasi
Department of Biotechnology and Bioinformatics
JSS Academy of Higher Education and Research
Mysore
Karnataka, India
Chapter 7

Tushar Tejonidhi
Department of Biotechnology and Bioinformatics
JSS Academy of Higher Education and Research
Mysore
Karnataka, India
Chapter 7

Yuvaraj Sivamani
Crescent School of Pharmacy
B.S. Abdur Rahman Crescent Institute of Science and Technology
Chennai
Tamil Nadu, India
Chapter 7

Sumitha Elayaperumal
Department of Biotechnology and Bioinformatics
JSS Academy of Higher Education and Research
Mysore
Karnataka, India
Chapter 7

Symah Bashir
School of Computer Applications
Lovely Professional University
144021 Phagwara
Punjab, India
symahbashir123@gmail.com
Chapter 8

Amanpreet Singh
School of Computer Applications
Lovely Professional University
144021 Phagwara
Punjab, India
Chapter 8

Rishu Chhabra
Chitkara University Institute of Engineering and Technology
Rajpura 140401
Punjab, India
rishu.miglani@chitkara.edu.in
Chapter 9

Shiwangi
Chitkara University Institute of Engineering and Technology
Rajpura 140401
Punjab, India
shiwangi2210.be23@chitkara.edu.in
Chapter 9

Saravjeet Singh
Chitkara University Institute of Engineering and Technology
Rajpura 140401
Punjab, India
saravjeet.singh@chitkara.edu.in
Chapter 9

Chetna Sharma
Chitkara University Institute of Engineering and Technology
Rajpura 140401
Punjab, India
chetna.kaushal@chitkara.edu.in
Chapter 10

Astha
Chitkara University Institute of Engineering and Technology
Rajpura 140401
Punjab, India
astha0392.be23@chitkara.edu.in
Chapter 10

Manya
Chitkara University Institute of Engineering and Technology
Rajpura 140401
Punjab, India
manya2146.be23@chitkara.edu.in
Chapter 10

Ria Khurana
Chitkara University Institute of Engineering and Technology
Rajpura 140401
Punjab, India
ria2192.be23@chitkara.edu.in
Chapter 10

Gagandeep Kaur
Chitkara University Institute of Engineering and Technology
Rajpura 140401
Punjab, India
gaganmalhotra1791@gmail.com
Chapter 11

Ashima Kukkar
Chitkara University Institute of Engineering and Technology
Rajpura 140401
Punjab, India
ashima@chitkara.edu.in
Chapter 11

Ramanjot Kaur
Chitkara University Institute of Engineering and Technology
Rajpura 140401
Punjab, India
ramanjot2189.be23@chitkara.edu.in
Chapter 11

Sartaj Singh
Lovely Professional University
144021 Phagwara
Punjab, India
sartaj.singh@lpu.co.in
Chapter 12

Jasleen Kaur Paintal
Lovely Professional University
144021 Phagwara
Punjab, India
jasleen.30857@lpu.co.in
Chapter 12

Namarta Vij
University of Windsor
Windsor, ON, Canada
vijn@uwindsor.ca
Chapter 12

Ravneet Kaur
Department of Computer Science and Engineering-APEX
Engineering College
Chandigarh Group of Colleges Jhanjeri
Mohali 140307
Punjab, India
ravneet.j3020@cgc.ac.in
Chapter 13

Chaitanya Singla
Department of Computer Science and Engineering-APEX
Engineering College
Chandigarh Group of Colleges Jhanjeri
Mohali 140307
Punjab, India
Chapter 13

Gurpreet Singh
Department of Computer Science and Engineering
Lovely Professional University
144021 Phagwara
Punjab, India
Chapter 13

Manjot Kaur
Department of Computer Science and Engineering-APEX
Engineering College
Chandigarh Group of Colleges Jhanjeri
Mohali 140307
Punjab, India
Chapter 13

Thuseethan Salavia
CSE, Darwin University
Melbourne, Australia
Chapter 13

Kanika Sharma
School of Computer Applications
Lovely Professional University
144021 Phagwara
Punjab, India
kanusharma0503@gmail.com
Chapter 14

Parul Khurana
School of Computer Applications
Lovely Professional University
144021 Phagwara
Punjab, India
Chapter 14

Balraj Kumar
School of Computer Applications
Lovely Professional University
144021 Phagwara
Punjab, India
Chapter 14

Himani Bhatt
Department of Mathematics
Bennett University
Greater Noida 201310
Uttar Pradesh, India
e23soep0031@bennett.edu.in
Chapter 15

Saurabh Rana
Department of Mathematics
Bennett University
Greater Noida 201310
Uttar Pradesh
India
2saurabh.rana@bennett.edu.in
Chapter 15

Mandeep Mittal
Department of Mathematics
Bennett University
Greater Noida 201310
Uttar Pradesh, India
Mandeep.Mittal@bennett.edu.in
Chapter 15

Mahak, Sachitanand Singh, Nakul Sankhayadhar, Iqra Jamaal Khan, Ashith Tripathi

1 An In-Depth Analysis of Medical Imaging Modalities and the Transformative Potential of Deep Learning and Enhancing Diagnostic Accuracy in Healthcare

Abstract: Medical imaging serves an important role in the diagnosis, treatment, and management of ocular and systemic disease. Advances in deep learning in artificial intelligence (AI) have transformed medical imaging through increased automation, predictive effects, and natural language processing. In this chapter we describe the various imaging modalities of optical coherence tomography, fundus photography, fluorescein angiography, magnetic resonance imaging (MRI), computed tomography (CT), ultrasound, and positron emission tomography (PET), and their clinical applications in ophthalmology and global health.

In ophthalmology, AI algorithms, including convolutional neural networks and generative adversarial networks, enhance the identification of associated retinal diseases (i.e., glaucoma, diabetic retinopathy, and age-related macular degeneration). The productivity of an automated screen decreases the labor time of the clinician and provides a level of accuracy that often exceeds human review, and it may allow for early identification of disease. Imaging, in addition to its application in ophthalmology, continues to play an important role in the identification of systemic illnesses such as cardiovascular diseases, neurological diseases, and oncological diseases. The use of AI in MRI, CT, and PET has made it easier to identify strokes, tumors, and other neurologic disorders, while also providing more accurate information to guide treatment decision-making. Deep learning facilitates quick classification, segmentation, and anomaly detection in image analysis. Computer-assisted diagnosis and telemedicine that utilize AI increase access to medical services in underserved and low-wealth areas. However, many challenges still exist in terms of privacy, legal challenges, and specific clinical validation. Ultimately, there needs to be underlying assurances that AI is interpretable and human in the loop before it becomes widespread. This chapter discusses the possible impacts of AI on medical imaging, future opportunities for use, and challenges that will arise in order to help advance precision medicine.

Keywords: medical imaging, deep learning, ocular diagnostics, computer-aided diagnosis (CAD), healthcare diagnostics, magnetic resonance imaging (MRI)

 | https://doi.org/10.1515/9783112226223-001

1.1 Introduction

Mathematics has always been essential in medical sciences, providing tools for modeling biological systems, analyzing medical data, and making a better diagnosis [1]. Intelligent systems utilize mathematical models, such as statistical inference, differential equations, and machine learning (ML) methods to interpret massive medical datasets to derive useful information [2]. The discipline of medical imaging is changing significantly by the applications of applied mathematics and artificial intelligence (AI) [3]. Imaging modalities such as X-ray, computed tomography (CT), magnetic resonance imaging (MRI) and ultrasound, all of which provide complicated datasets, rely on numerous applied mathematics methods to accomplish tasks such as noise reduction, image reconstruction automation, and pattern recognition techniques [4]. Deep learning (DL), an AI technology, revolutionized medical imaging with image analysis automation, detection of anomalies in disease, and human error reduction [5]. In the combination of mathematical reasoning and AI technology, medical discipline can achieve better accuracy, efficiency, and diagnostic automations [6]. In this chapter, you may expect to find a detailed examination of the evolution of medical imaging modalities, for example, X-ray, CT, MRI, positron emission tomography (PET), and optical coherence tomography (OCT), in terms of DL methods and approaches. The primary topics discussed are the principles and concepts of imaging, as well as the unique ability of AI to improve accuracy in diagnostic exams, automate diagnostics, and improve predictive analytics [7]. The ability of new and emerging processing technologies, like convolutional neural networks (CNNs), generative adversarial networks (GANs), and new transformer models, are examined with respect to medical image processing [8]. In addition to a discussion on how DL is aiding medical diagnostics through case studies, benchmarks, and advancements in service to provide better efficiency, accuracy, and information globally, this chapter provides a thorough investigation of DL's influence on medical diagnostic processes. Also, it demonstrates how DL has the potential to enhance diagnosis accuracy in healthcare using a thorough discussion of AI techniques and methods applicable to applied mathematics [9]. Firstly, it is important to understand the basic imaging technologies that provide the diagnostics that we discussed in this chapter.

1.1.1 Overview of Medical Imaging in Healthcare

Medical imaging allows us to observe interior anatomical and physiological processes in real time and noninvasively, a function that has progressed to include AI-based support for high-throughput screening, predictive diagnostics, and remote triage decision-making in digital health environments [10]. Advances in imaging technologies – from structural images (CT) to functional images (PET) – through DL-generated integration have vastly improved our ability to perform both anatomical and physiological assessments in tandem and established a pathway to precision medicine [11]. The advancing technology in medical imaging has also progressed rapidly, leading to improvements in clarity, speed of acquisition, and diagnostic accuracy [12]. In the early days of medical imaging, X-ray technology was the first specific form of medical imaging. This revolutionary advancement allowed clinicians to observe fractures of the bones and lung infections [13]. As technology became more advanced, cross-sectional and functional images became available, which allowed clinicians to detect and stage an illness earlier in its progression and assess treatment efficacy [14]. Medical imaging is central to contemporary medicine. It allows physicians to visualize internal organs and other tissues noninvasively and reproducibly, providing accurate diagnosis and treatment of disease [15]. Standard imaging methodologies, including radiography, MRI, and ultrasound, are all based on mathematical models to obtain, enhance, and interpret images [16]. Recent advancements in AI have further improved medical imaging by the use of automation to perform tasks of image segmentation, classification, and anomaly detection, increasing diagnostic efficiency [17]. This chapter encompasses a thorough review of key medical imaging modalities in terms of their principles, applications, and limitations, including X-ray, CT, MRI, PET, OCT, and ultrasound and includes the use of AI and its impact on imaging diagnosis through AI improvements in image reconstruction, segmentation, classification, and prediction as we look to define our roles. The incorporation of AI models into radiological practice together with advancements in transformer models, quantum computing, and wearable technology forms another key theme. The chapter also provides an overview of specific DL architectures, performance results from real-world datasets, and the future potential of AI to enable personalized healthcare diagnostics and accessible healthcare for all [18]. Table 1.1 illustrates the comparative analysis of commonly used medical imaging modalities based on resolution, cost, clinical applications, and limitations.

Table 1.1: Comparative overview of major medical imaging modalities.

Modality	Resolution	Cost	Use cases	Limitations
X-ray	Low to moderate	Low	Bone fractures, chest infections, and dental imaging	Radiation exposure and limited soft tissue visualization
CT	Moderate to high	Moderate to high	Trauma, head injuries, and lung pathologies	Higher radiation dose and expensive
MRI	High	High	Brain, spine, and musculoskeletal disorders	Contraindicated in metallic implants, expensive, and long duration
PET	Low (functional data)	Very high	Oncology and metabolic imaging	Radiation, cost, and low spatial resolution

Table 1.1 (continued)

Modality	Resolution	Cost	Use cases	Limitations
Ultrasound	Moderate	Low	Obstetrics, soft tissue, and cardiology	Operator-dependent and poor penetration through bone/gas
OCT	Very high (micron-level)	Moderate	Retinal disorders, glaucoma, and AMD	Limited to transparent tissues and short imaging depth

1.2 Progression of Medical Imaging Procedures

Advancements in medical imaging technology are conceptually based on fundamental tenets of mathematics that can be applied to capture, improve, reconstruct, and interpret images accurately. Combined with applied mathematics, medical imaging has raised the accuracy of clinical decision-making, as well as automation and predictive analytics in healthcare [19].

1.2.1 Early Developments

X-Ray Imaging and the Birth of Radiology: Modern medical imaging began with Röntgen's discovery of X-rays in 1895, but the biggest change has occurred with AI-assisted systems for interpreting X-rays; these systems have demonstrated accuracy comparable to radiology level in large validation studies, which transformed healthcare by allowing the noninvasive visualization of internal structures. Mathematics has played a role in improving X-ray imaging using techniques like Fourier transforms combined with filtering algorithms, which ultimately reduce noise and improve contrast [20]. The rise of cross-sectional imaging CT was introduced by Sir Godfrey Hounsfield and Allan Cormack (1970s) as a significant advancement in medical imaging that produced cross-sectional images of the human body [21]. 3D anatomical structures in CT imaging are reconstructed from multiple X-ray projections using mathematical methods, mainly the Radon transform. Modern CT imaging that incorporates AI utilizes DL denoising algorithms, thereby improving image quality and significantly increasing diagnostic accuracy [22].

1.2.2 Magnetic Resonance Imaging (MRI)

MRI is based on the principles of nuclear magnetic resonance (NMR), wherein hydrogen nuclei within tissues absorb and then re-emit radiofrequency signals while subjected to a strong magnetic field. The math that drives MRI can be complex using Fourier transforms to transpose signals from a time domain to a spatial domain [23]. Models such as k-space encoding, wavelet transforms, and compressed sensing techniques have improved MRI imaging through the decreased duration of scans while maintaining image quality. Moreover, some AI-based MRI systems are now using DL reconstruction strategies to improve image resolution, reduce artifacts, and support radiologists in examining complicated anatomical details [24].

1.2.3 Ultrasound Imaging

Modern ultrasound systems use AI-based, real-time feedback systems to help assist with fetal biometry, cardiac assessment, and liver fibrosis assessment using predictive models based on huge annotated patterns of echogenicity and have become a widely utilized, affordable method of imaging in a clinical setting. The mathematics underpinning ultrasound imaging uses wave equations, Doppler shift analysis, and various signal processing techniques to deliver real-time images of internal organs [25].

1.2.4 Positron Emission Tomography (PET) and Functional Imaging

PET is a form of functional imaging representing metabolic activities occurring in the body with radioactive tracers emitting positrons – a particularly important application in oncology and in neurology [26]. Different types of functional imaging techniques, including PET and single-photon emission computed tomography (SPECT), have made it possible to visualize metabolic and physiological processes at the level of the cell [27]. Functional imaging modalities depend on the detection of gamma rays emitted from radiotracers that are mathematically reconstructed from the median gamma ray count spatially detected, using different mathematical techniques including, for example, the maximum likelihood expectation maximization (MLEM) method [28].

1.2.5 Artificial Intelligence and the Future of Medical Imaging

By automating picture reconstruction, segmentation, and detection of anomalies, recent advances in AI and DL have changed the landscape of medical imaging. AI-oriented algorithms such as CNNs and transformer models have significantly impacted diagnostic accuracy and are changing workflows in radiology [29]. Mathematics lies at the heart of AI applications in medical imaging, where probabilistic models, optimization, algorithms, and frameworks for statistical learning facilitate decision-making by modifying uncertainty. AI-based medical imaging is being included in healthcare intelligence technologies so that diseases can be identified early, treatment plans can be personalized, and predictive analytics can be compiled for better patient outcomes [30]. While the development of imaging modalities provides the base, AI and DL are setting the stage for a new era of diagnostic accuracy.

1.3 The Role of Artificial Intelligence and Deep Learning in Modern Diagnostics

AI and DL techniques are also transforming modern diagnostics by automating complex tasks, improving the accuracy of diagnoses, and ultimately, diminishing the influence of human errors. These advancements rely heavily upon applied mathematics, which provides the foundation of image processing, pattern recognition, and predictive analytics within the realm of medical healthcare [31].

1.3.1 Mathematical Foundations of Artificial Intelligence in Medical Imaging

Current clinical diagnostic applications of AI hinge on mathematical theories such as linear algebra, probability, and optimization. One class of AI – deep learning – relies on artificial neural networks to recognize patterns in medical images and classify disorders accurately [30]. Mathematical foundations – convolution operations, gradient descent optimization, and loss functions – assist AI models in learning from labeled data and improving diagnostic performance with time [32].

1.3.2 Image Processing and Enhancement Using Artificial Intelligence

Medical images frequently experience noise and motion artifacts and poor contrast, requiring mathematical methods for enhancement and restoration. AI image processing algorithms are based on mathematical forms, including wavelet transforms, Fourier analysis, and DL super-resolution methods for improving the accuracy of the image [33]. DL denoising methods use autoencoders and generative adversarial networks (GANs) to remove artifacts from MRI and CT images while preserving their diagnostic values. GANs have two competing neural networks that develop realistic synthetic images. They are used, often, to improve images or to augment data [34].

1.3.3 Artificial Intelligence in Disease Detection and Classification

AI-based diagnostic systems have demonstrated remarkable success in detecting conditions such as cancer, diabetic retinopathy, and neurological disorders with high levels of sensitivity and specificity [35]. A convolutional neural network (CNN) is a DL model that corresponds with the way we humans see by identifying patterns and features in images on its own. They are commonly used in the diagnosis of disease from images in medical imaging. CNNs use backpropagation and stochastic gradient descent (SGD) to update weights and minimize classification errors [36]. AI-based breast cancer detection also utilizes DL models to accurately classify benign and malignant cancers on mammography images [37]. Similarly, AI retinal analysis uses mathematical models used for vascular segmentation and lesion detection of fundus images for diabetic retinopathy and glaucoma.

1.3.4 Predictive Analytics and Personalized Medicine with Artificial Intelligence

The power of AI to analyze and identify trends from large amounts of patient data and predict disease trajectories is an innovation in personalized medicine [38]. AI models targeting genomics to improve risk prediction use DL techniques for DNA sequence classification, mutation calling, and biomarker development, which has improved adjuvant therapy precision and ultimately precision medicine. AI predictive models for cardiovascular disease prevention use statistical learning algorithms to assess patient characteristics such as demographics, lifestyle, and clinical biomarkers in order to improve early intervention strategies in high-risk populations [39].

1.4 Importance of Precision and Automation in Medical Imaging

In medical imaging, precision and automation are critical to dependably drive diagnosis, workflow management, and improved patient outcomes. As images become more complicated and require faster and more accurate interpretations, medical imaging has begun to integrate mathematical models, AI, and DL to improve diagnostic accuracy and automate interpretation [40].

1.4.1 The Mathematical Basis of Accuracy in Medical Imaging

Accuracy in medical imaging refers to the ability to obtain very detailed, artifact-free, and repeatable images for accurate diagnosis [41]. High accuracy is achieved through mathematical models that optimize image acquisition, reconstruction, and enhancement.

1.4.1.1 Mathematical Models in Image Acquisition

Mathematical formulas are used in medical imaging modalities such as X-ray, CT, MRI, and ultrasound to improve images and reduce noise and distortion [42]. This can be seen in X-ray and CT imaging. A number of mathematical optimization methods, such as iterative reconstruction algorithms and model-based denoising using ML have been researched and investigated to improve image quality and minimize radiation exposure [43].

1.4.1.2 Image Processing and Enhancement Through Mathematics

Medical images can often contain noise, motion artifacts, or be poorly contrasted. With the help of mathematics, the following techniques can be employed to enhance image clarity:

i. **Wavelet Transforms:** Used to enhance visual contrast and detail resolution.
ii. **Histogram Equalization:** A mathematical technique to adjust image contrast to improve the visual representation of anatomical structures [44].
iii. **Deep Learning-Based Image Super-Resolution:** This includes CNNs and GANs that change low-resolution images to high-resolution images through the use of math optimization methods. These mathematical techniques, supported by science, enable medical imaging systems to derive clearer, high-contrast images that allow for a better level of diagnostic accuracy [45].

1.5 Medical Imaging Modalities: Principles and Applications

Medical imaging is essential to current diagnostic practice, providing noninvasive access to physiological and pathological states. The use of mathematical models and computational algorithms across a wide range of imaging modalities supports improvement in picture acquisition, reconstruction, and analysis. AI-based techniques use mathematical models to improve image processing, automating disease identification and enhancing clinical decision-making [46].

1.5.1 Optical Coherence Tomography (OCT)

Tomography employs low-coherence interferometry to obtain high-resolution, cross-sectional images, which can be advantageous, in particular, for eye health [47]. It is often used to evaluate glaucoma, diabetic retinopathy (DR), and age-related macular degeneration (AMD) [48]. Segmentation algorithms based on AI, such as CNNs, can improve the detection of retinal layers and disease features while also increasing accuracy [49]. Fourier-domain OCT uses Fourier transforms to transform spectral information into structural images for improved resolution and speed of acquisition [50].

1.5.1.1 Principle of OCT Imaging

OCT is an imaging technology that leverages low-coherence interferometry in order to produce high-resolution cross-sectional images of biological tissue, for example, the retina. The measurements are made based on how much intensity of the reflected light returns to the interferometer, while the resonant time delays are analyzed in three dimensions with Fourier-domain analysis and wavelet transforms [51].

1.5.1.2 Application in Ophthalmology

OCT is effectively used during clinical research and practice in ophthalmology particularly for detecting and monitoring retinal diseases, which may include glaucoma, (DR), AMD, etc. Mathematical models can assess retinal nerve fiber layer thickness while also forecasting the progression of these diseases to preclinical levels. With decision support using large labeled datasets of OCT images, AI can improve the accuracy and efficiency of automated diagnosis to facilitate earlier recognition and treatment [52].

1.5.1.3 The Role of AI in OCT

Evaluation: AI evaluation of OCT images utilizes DL methods such as CNNs for improving image segmentation, feature extraction, and disease classification. ML methods improve the ability to recognize patterns, which allows ophthalmologists to distinguish between normal and diseased conditions with better specificity and sensitivity. DL models rely on mathematical algorithms that contain elements such as gradient-based optimization and probability theory that allow for continued learning and improved accuracy for diagnoses [53].

1.5.2 Fundus Photography (FP) and Fluorescein Angiography (FA)

Fundus photography (FP) takes a high-resolution picture of the retina, while fluorescein angiography (FA) examines retinal vasculature with fluorescein dye [54]. AI/DL screening tools use DL to recognize DR and other retinal abnormalities with higher sensitivity and specificity. Image processing uses edge detection algorithms, including Sobel and Canny filters, to enhance the appearance of blood vessels [55].

1.5.3 Magnetic Resonance Imaging (MRI)

Contemporary MRI techniques utilize DL for reconstructions, achieving fourfold acceleration in imaging time, without diagnostic quality loss, via approaches like AUTOMAP and k-space-based unrolled networks. MRI is fundamental to neurology, oncology, and cardiology because it provides the ability to visualize brain tumors, strokes, and dysfunctions of the heart in fine detail [56]. Electromagnetic reconstructions of MRI images, driven by AI, leverage a mathematical construct called compressed sensing that enables high-resolution imaging from limited samples, thus reducing scan times. Neural networks also yield improved noise suppression and correction of artifacts to usability as a diagnostic tool [57].

1.5.4 Computed Tomography (CT)

CT utilizes X-ray attenuation to create cross-sectional images, commonly used in trauma, oncological imaging, and neuroimaging. Image quality is improved with AI by reducing noise and increasing contrast, mostly using DL models [58]. CT reconstruction uses Radon transformations and filtered back projection for image generation, in addition to iterative reconstruction techniques to reduce artifacts.

Figure 1.1 illustrates a liver segmentation method with a U-Net design. The CT images and associated ground truth masks are used to train the model, which predicts binary masks of the liver. Results enhance interpretability of predictions in medical decision-making [59].

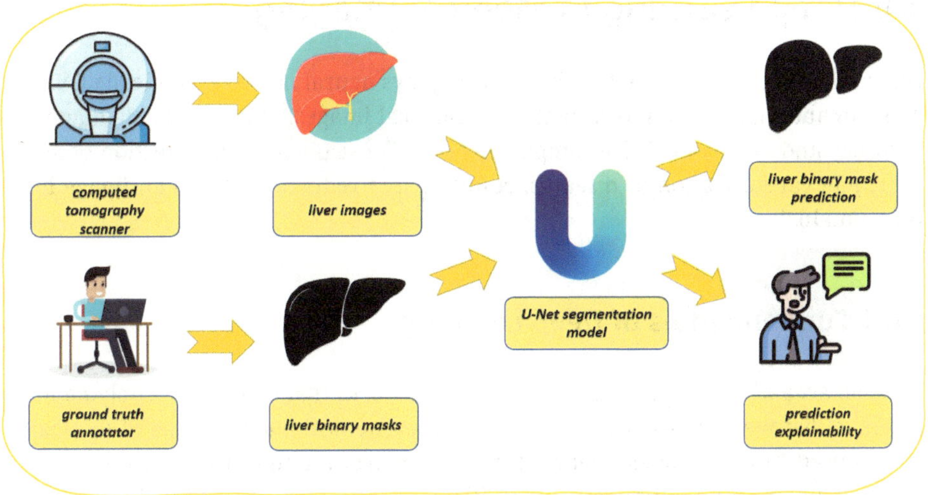

Figure 1.1: U-Net-based liver segmentation from CT images with explainable predictions.
Source: Reproduced from Brunese MC et al., 2025, Diagnostics (MDPI) under the Creative Commons CC BY license.

1.5.5 Ultrasound

Ultrasound imaging depends on the bounce of high-frequency sound waves to visualize inside organs. It is a vital component of fetal tracking, cardiology, and abdominal imaging. AI has automated fetal biometry and assessment of heart function, utilizing ML algorithms to track anatomical elements in real time. Signal processing methods, using Fourier and wavelet transformations, enhance resolution and reduce noise [60].

1.5.6 Positron Emission Tomography (PET)

PET utilizes radiotracers to discover metabolic activity and facilitates the identification of oncological and neurodegenerative disorders. AI is advancing tumor segmentation and metabolic analysis via probabilistic graphical models and DL [61]. Similarly, mathematical operations like MLEM are incorporated in PET image reconstruction for accuracy improvement and noise reduction. After discussing the basic principles of a few imaging techniques, we turn to how DL models can be used to extract, interpret, and enhance (clinical) meaning from these images.

1.6 Deep Learning in Medical Imaging

DL within the context of medical imaging employs neural networks, typically CNNs, to assist in automating the interpretation of medical images, thus enhancing diagnostic accuracy and efficiency. DL accomplishes tasks like segmentation, detection of anomalies, and categorization of diseases, resulting in a reduction of human image interpretation [62].

1.6.1 Fundamentals of Deep Learning

DL is an area of AI that uses artificial neural networks that operate by applying many layers to discover hierarchical information from medical images. Some of the mathematics used in DL includes linear algebra for matrix multiplications, probability for decision-making, and optimization algorithms, such as SGD, to learn [63]. For medical imaging, supervised learning is commonly used for disease classification, where labeled datasets are applied to teach models to identify abnormal features within the data. Unsupervised learning methods, including cluster analysis and dimensionality reduction methods such as principal component analysis (PCA), are useful for identifying hidden structure in the image. Lastly, reinforcement learning is applied in imaging settings for adaptive scanning and radiology reporting automation [64].

1.6.2 Deep Learning Models in Image Analysis

CNNs, which are used in medical imaging, rely on feature extraction methods (e.g., convolution operations) to identify patterns associated with diseases. These models have introduced mathematical functions with convolution integrals and activation functions, such as ReLU (rectified linear unit), to enhance classification accuracy in images. GANs improve data augmentation through the generation of synthetic medical images, thereby enhancing the variability of training datasets and reducing overfitting [65]. Long short-term memory (LSTM) networks, which are a type of recurrent neural network (RNN), are used within the context of sequential imaging identification, such as tracking tumor growth in longitudinal scans [66]. Together, mathematical principles help facilitate DL models in transforming raw medical images into meaningful diagnostic information, thus bridging healthcare and computing research. The overall DL workflow used in medical image analysis is illustrated in Figure 1.2.

1.6.2.1 Benchmark Architectures: U-Net and TransUNet

U-Net has become a key architecture for medical image segmentation, owing to an encoder-decoder architecture with skip connections as a way to enable spatial information preservation. This method is useful in biomedical images that have a limited amount of labeled data because it allows the network to use context and localize well [67]. The U-Net has been applied in liver segmentation, brain tumor segmentation (segmenting brain tumors), and obtaining retinal layers.

TransUNet extends U-Net by adding a transformer encoder at the bottleneck of the architecture. The adoption of a self-attention mechanism allows the model to determine global dependencies in the image, which is useful for complex anatomical systems [68]. Overall, these models are often evaluated by metrics such as Dice similarity coefficient (DSC), intersection over union (IoU), and area under the receiver operating characteristic curve (AUC). Recent work indicated that TransUNet achieved better Dice scores than U-Net based on overall performance, especially for multi-organ tasks [69].

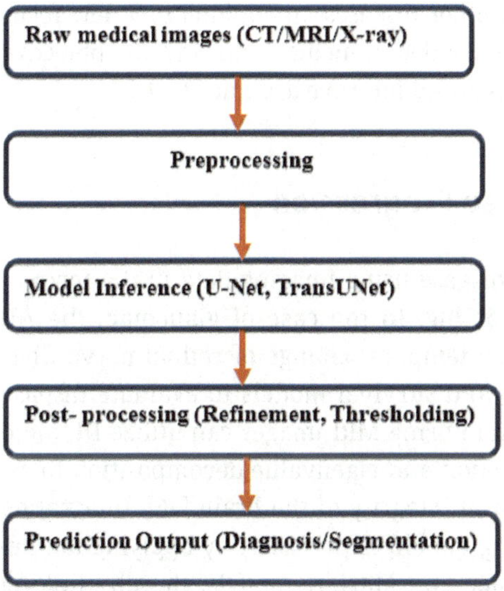

Figure 1.2: Workflow of deep learning in medical image analysis.
Source: Author-created image (no permission required)

1.6.2.2 Performance Indicators and Datasets in Deep Learning

Publicly available data has been essential in evaluating DL models in the medical imaging realm. For example, BraTS allows for segmentation of brain tumor images and U-Net-based models have produced Dice scores of >0.88 [70]. There has been similar application in chest CT scans with the LUNA16 dataset, where nodule detection is common and results are typically presented. A certain increased sensitivity was observed with low false-positive rates for both these examples [71]. Comparable results have been shown in DR classification by CNNs with the EyePac dataset, where AUC > 0.94. CheXpert is commonly being used with multi-label classification of chest X-rays also and has seen accuracies well above 85% in some instances [72].

1.6.3 Automated Image Processing and Feature Extraction

In medical imaging, image processing via automation and AI provides advancements in segmentation, object detection, and anomaly detection. Segmentation algorithms include edge detection methods (e.g., Sobel and Canny filters), which can be used in computational typography to shape contours around anatomical structures. Object detection models often employ a region-based CNN to discover and locate anomalies that occur in medical scans using further mathematical techniques such as non-maximum suppression and anchor box construction [73]. AI-initiated feature extraction approaches include quantitative mathematical transformations (e.g., Fourier or wavelet analysis) to locate disease markers found in images with promising results based on the potential for earlier diagnosis of cancer and other diseases. In addition to a new focus on diagnostic processes with mathematical formulas in medical imaging, the objective is to, where possible, simplify professional tasks and improve accuracy [74].

1.6.4 Predictive Analytics in Disease Progression

DL algorithms forecast the progression of disease using imaging data in the form of time-series datasets and probabilistic modeling. In the case of glaucoma, the AI-enhanced predictive models can analyze the temporal change of retinal nerve fiber layer thickness using regression modeling and survival models to evaluate disease progression [75]. Alzheimer's disease detection using MRI images can utilize DL models that leverage tensor-based feature extraction and eigenvalue decomposition to assess subtle signs of shrinkage of the structural integrity of the brain [76]. Integrating predictive analytics with treatment planning enhances precision by applying mathematical methods such as Bayesian inference and Markov models, thereby linking computer approaches to clinical decision processes. While picture interpretation is useful, AI has roles to play in the automation of image enhancement, the actual diagnosis, and working with entire systems of healthcare.

1.7 AI-Driven Enhancements in Medical Imaging

Advances in AI in the field of medical image analysis bring better accuracy in selecting the right diagnosis, improved quality of images, and improved affordability of imaging. Computer-aided diagnosis (CAD) algorithms are based on DL models that include CNNs. The algorithms can potentially separate patients with disease from healthy patients and avoid problems of misdiagnosis. AI-based real-time image processing with mathematical methods that include Fourier transforms and denoising improves MRIs and CT scans. Also, the integration of imaging data with electronic health records (EHRs) allows AI-based predictive analysis to offer tailored treatment. AI is being used in telemedicine and mobile health (mHealth) applications for remote diagnostics, improving healthcare access to underserved regions [77].

1.7.1 Computer-Aided Diagnosis (CAD) Systems

Radiologists and pathologists employ CAD systems to review medical images and detect abnormalities using ML algorithms under the successive refinement model of diagnostic accuracy. CAD systems are known to the field of radiology for common procedures in radiology like mammography and chest radiographs, which utilize mathematical models of CAD by incorporating CNN and statistical classifiers to increase their diagnostic accuracy [78]. While assessing pathology, CAD systems based on DL can evaluate histopathological images through cell structure analysis focusing on feature extraction methods such as wavelet transformations and PCA [79]. CAD systems based on DL are integrated as a means to gauge levels of disease and risk, such as cancer or cardiovascular disease, and used with probability distributions and regression models to enhance medical decision-making. These mathematical models serve as intermediaries of computational imaging and clinical decision-making, with the goal to reduce variability of diagnoses and improve patient care [80]. An integrated overview of AI-driven enhancements across imaging, CAD, telemedicine, and EHR workflows is shown in Figure 1.3.

1.7.2 Real-Time Image Processing and Augmentation

Using AI for real-time image processing to improve MRI and CT imaging involves improving contrast, decreasing noise, and removing artifacts. In MRI applications, super-resolution methods based on DL represent Fourier and wavelet transforms that serve as practical mathematical models to improve low-resolution images [81]. AI denoising methods rely on deep convolutional autoencoders that capture the probabilities of the images and local patterns, which are modeled to perform a degree of Bayesian inference to remove Gaussian noise from low-quality CT imaging while still capturing anatomy. These AI-enabled improvements yield improved quality to see anatomy, fewer repeated scans, and improved radiation dose in CT, all currently implemented mathematically to advance and modernize healthcare imaging [82].

1.7.3 Fusion of Imaging Data with Electronic Health Records (EHR)

Real-time image processing of MRIs and CTs using AI can enhance images; for instance, AI can improve contrast, decrease artifacts, and reduce noise in images. Digital Imaging and Communications in Medicine (DICOM) is a standardized format used for the storage and transfer of medical images throughout equipment and healthcare systems. Super-resolution approaches in MRIs utilize DL methods that apply Fourier and wavelet transforms as mathematical equations to improve low-resolution images [83]. Denoising methods using AI apply deep convolutional autoencoders and are based on probability modeling and Bayesian inference to reduce Gaussian noise while preserving anatomical structure found in CT images. The use of AI in imaging is a way to improve image quality, reduce repeat scans, and minimize radiation exposure in CT imaging, and it exemplifies the use of mathematics to improve imaging systems in healthcare [84].

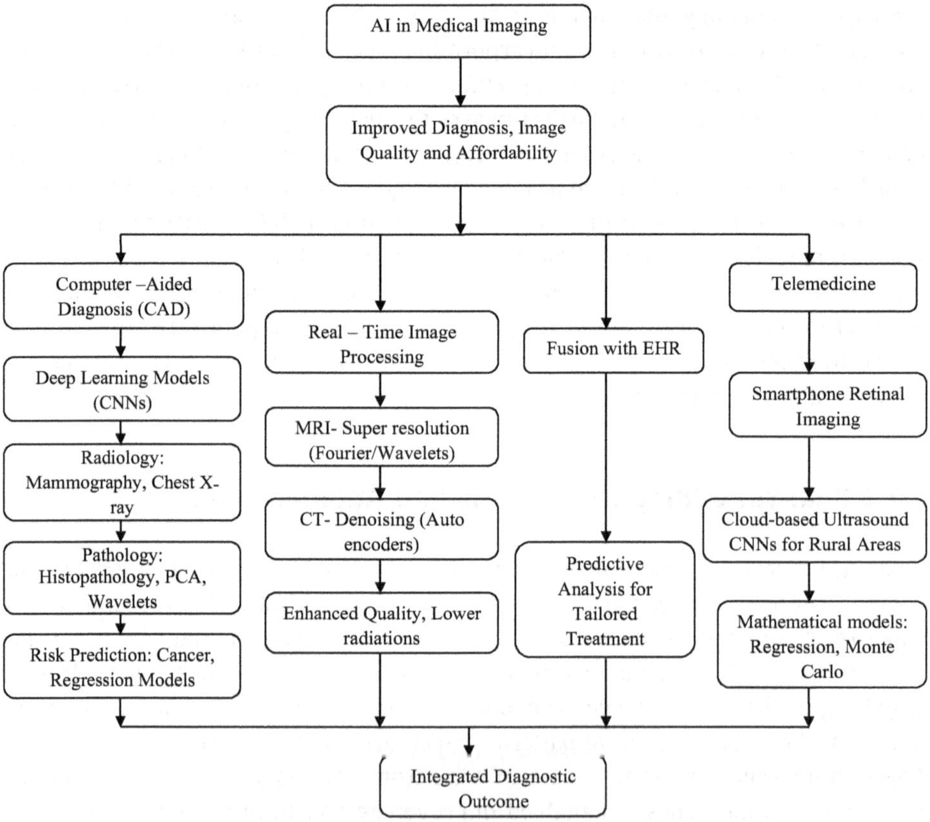

Figure 1.3: Integrated workflow of AI-driven enhancements in medical imaging.
Source: Author-created image (no permission required)

1.7.4 Telemedicine and Remote Diagnosis

AI-assisted telemedicine platforms enable remote healthcare by providing automated diagnosis through mobile health (mHealth) applications. ML approaches used in mHealth applications use decision trees and neural networks to classify diseases using smartphone-based retinal images, thereby improving access to ophthalmic care in resource-poor areas. In rural healthcare, AI-assisted ultrasound devices use cloud-based CNNs for immediate detection of abnormal findings and less reliance on onsite radiologists [85]. Remote diagnosis models use mathematical models in healthcare, including linear regression for predictive analysis and Monte Carlo simulations to estimate uncertainty. They also point to the essential role of math in improving care that is accessible through AI [86].

1.8 Challenges and Ethical Considerations

The integration of AI in medical imaging has many challenges, especially data security, bias, compliance with regulation, transparency, and combining care with people and AI. Addressing these challenges will require mathematical and computational models to improve AI performance, improve fairness, and assure clinical trustworthiness [87].

1.8.1 Data Privacy and Security

Medical imaging data requires encryption and federated learning to protect patient privacy and allow AI training on distributed datasets. Differential privacy techniques anonymize data to maintain privacy while allowing for analysis. Complying with laws like HIPAA and GDPR ensures lawful and ethical application of AI to healthcare [88].

1.8.1.1 Challenges in Handling Medical Imaging Data

Medical imaging data is sensitive and requires strict security processes to prevent unauthorized access and breaches. Mathematical processes like homomorphic encryption and federated learning promote privacy in data access by permitting AI models to learn from decentered datasets while retaining any raw medical images. Methodologies grounded in differential privacy modify data by adding stochastic noise that complicates reidentification and preserves statistical utility. Importantly, the protective measures in data management through AI-based practices in medical imaging serve to protect patient confidentiality necessary for HIPAA and GDPR [89].

1.8.1.2 Bias in AI Algorithms

AI systems constructed with unbalanced datasets often show bias with respect to race and gender, which leads to disparity among subpopulations in health outcomes. Mathematical approaches, including reweighting algorithms and fairness-aware ML, have been developed to mitigate bias and support equity. At the same time, an important priority is speeding up diverse and curated datasets to improve the equity disparity of diagnostics [90].

1.8.1.3 Concerns About Racial and Gender Bias in AI Models

AI models built on unrepresentative datasets often exhibit racial and gender differences, resulting in disparities in healthcare outcomes. Statistical approaches such as reweighting and adversarial debiasing can reduce bias by equalizing distributions at the dataset level. Fairness-aware ML models apply mathematical constraints to guarantee equalized odds and demographic parity, thus reducing bias within AI-enabled diagnoses [91].

1.8.2 Regulatory Approvals and Clinical Validation

AI in medical imaging is subject to strict regulatory requirements, including clearances from the FDA and EMA to provide a formal assessment of safety and efficacy. Clinical validation of AI must demonstrate performance measures based in statistics, such as specificity, sensitivity, and receiver operating characteristic (ROC) curves to assess accuracy. In addition, adaptive Bayesian designs provide enhancements of the AI trial and a reduction in patient care risks [92].

1.8.2.1 Regulatory Agents for AI in Medical Imaging

Health authorities such as the FDA (Food and Drug Administration) and EMA (European Medicines Agency) have enforced regulatory criteria impacting the incorporation of AI models in clinical practice. Statistically measurable evaluations by characteristics such as ROC curves and metrics evaluating sensitivity and specificity will be completed during the AI-enabled medical device's statistical evaluation rather than by direct clinical evaluation. Clinical trials often use Bayesian adaptive designs to maximize AI model assessments while minimizing patient exposure to an ineffective therapy. Regulatory compliance enables an impression that the AI is efficacious and safe prior to patient deployment; however, there is no regulatory or practice concern regarding clinical evaluations of AI in healthcare [93].

1.8.3 Interpretability and Trust in AI

Using eXplainable AI (XAI) models brings additional transparency to AI's applications in healthcare by using methods such as SHapley Additive exPlanations (SHAP) and local interpretable model-agnostic explanations (LIME) to explain the process by which AI arrived at its decisions. Trust in AI can also be achieved with interpretable mathematical frameworks (game theory models, attention models, etc.). Trust in physicians/clinicians is also needed for trust in AI to remain trustworthy and stable and the physician/clinician will take on the responsibility in the clinical context [94].

1.8.3.1 The Need for Explainable AI (XAI) in Clinical Settings

Clinical settings are looking for transparency when utilizing AI (XAI) because AI systems are often viewed as "black boxes," and it may not be easy for a clinician to see how the AI determined its output or deconstructed the rationale for the output interventions to be calculated. XAI methods, such as SHAP and LIME, utilize methods from mathematical game theory or perturbation-based methods to articulate how the AI processes rationale to explain how its reasoning model works [95].

1.8.4 Physician-AI Collaboration

AI is a tool to supplement human ability, not supplant it, through amelioration of the diagnostic decision-making process for clinicians. Hybrid AI-human systems improve accuracy and trustworthiness of outcomes using ensemble learning and uncertainty quantification. Collaborative models ensure AI recommendations align with clinical decisions, thus improving patient outcomes [96].

1.8.4.1 Ensuring AI Complements Rather than Replaces Human Expertise

AI should augment rather than replace human intelligence by supplying mechanisms to support clinical decision-making. Hybrid AI-human systems employ ensemble learning, which combines machine-based predictions with clinician input to improve the accuracy of diagnostic outputs. Game-theoretic models and uncertainty quantification models reinforce AI-human cooperation, establishing that the suggestions produced by the machine align with clinical reasoning [97]. The act of generating a collaborative environment allows for a satisfactory diagnostic and prescriptive experience for the physician while maintaining accountability over the AI.

1.9 Future Prospects and Innovations

The field of medical imaging is undergoing profound changes, driven by advancements in AI, quantum computing, and intelligent devices, thereby enhancing diagnostic accuracy and accessibility [98]. Mathematics is essential when accelerating the improvement of AI models, quantum algorithms, and the use of real-time health data (or a combination of both) to monitor health status, as it is also part of statistical inference and probability theory, to the structures of computational models.

1.9.1 Advancements in AI Algorithms

The development of AI algorithms aims to enhance clinical use through increased diagnostic accuracy, efficiency, and automation. The role of mathematics is to enhance AI learning through the mechanisms of optimization theory and probability theory (with the study of chance) as well as computational models [99].

1.9.1.1 Self-Learning AI Models in Medical Imaging

Self-learning AI models, such as reinforcement learning (RL) and self-supervised learning, improve their functions by identifying similar characteristics or patterns across a set of medical imaging without needing significant sample sizes of labeled data [100]. Self-learning models include mathematically based structures such as a Markov decision process (or distribution), Bayesian optimization/sampling, and deep variational inference to improve diagnosis in radiology and pathology. By employing differential equations and probability distributions, the AI algorithm can help improve tumor segmentation, detect anomalies, and enhance real-time (or live) imaging analysis [101].

1.9.2 Quantum Computing and Medical Imaging

Quantum computing has the potential to revolutionize medical imaging by solving complex computational problems and doing so at incredible speed. Mathematical concepts such as linear algebra, tensor operations, and quantum probability distributions are important for quantum-enhanced medical imaging types.

Figure 1.4 illustrates the uses of quantum computing in the field of healthcare. It enhances the drug discovery process, medical diagnoses, analytics with AI, genomics, and hardware processing, such as radiation. Enhancements in efficiency, accuracy, and individualized treatments have been achieved in each of these areas [102].

Figure 1.4: Quantum computing applications in healthcare.
Source: Reproduced from Chow JCL, 2025, Algorithms (MDPI) under the Creative Commons CC BY license.

1.9.2.1 Potential Impact on Real-Time Imaging and Deep Learning Efficiency

DL quantum methods, including quantum support vector machines (QSVM) and quantum Boltzmann machines, have improved the detection of anomalies in medical images by being able to process high-dimensional data better than classical computational techniques. Quantum-enhanced DL models use ideas from quantum physics, like quantum entanglement and superposition, to work faster and better when creating MRI and CT images. The processes enhance the use of large medical datasets through the use of a quantum Fourier transform and improve the speed of processing for real-time diagnosis [103].

Figure 1.5 illustrates the comparison of classical and quantum computing in terms of computation units, performance, error rates, and applicability. Classical com-

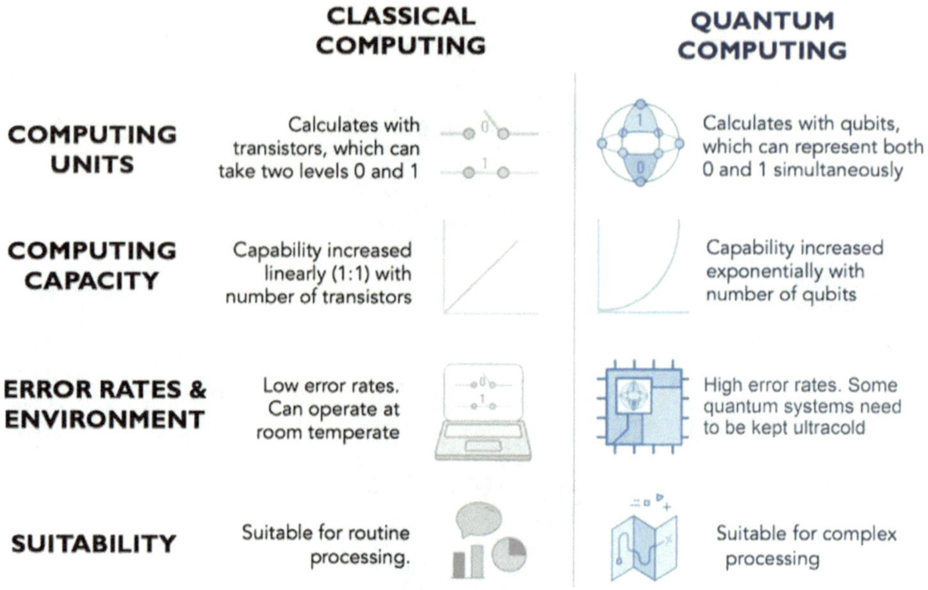

Figure 1.5: Classical computing vs. quantum computing.
Source: Reproduced from Ur Rasool R et al., 2023, Future Internet (MDPI) under the Creative Commons CC BY license.

puting is in the form of transistors and provides linear scaling, while quantum computing uses qubits and offers exponential scaling. Quantum systems can handle complex problems while having high error rates and requiring ultracold conditions [104].

1.9.3 Integration with Wearable and Smart Devices

The convergence of AI with wearable and smart devices has the potential to change healthcare by enabling real-time monitoring of patients. Mathematical methods, including signal processing, time-series analytic tools, and ML algorithms, make it possible to monitor physiological data in real time [105].

1.9.3.1 AI-Based Real-Time Health Monitoring

Wearable devices utilize AI analyses to monitor cardiovascular health, glucose, and neurological issues. Devices employ mathematical methods, like Kalman filtering, Fourier analysis, and RNNs, to process real-time sensor data. AI-based prediction modeling enhances early disease detection and personalized treatment options, involving differential equations and statistical regression [106].

1.9.4 Global Impact and Accessibility

AI-based medical imaging can close healthcare gaps internationally by enabling the possibility for diagnostics in marginalized communities. Mathematics is a critical enabler of federated learning, cloud-based AI systems, and cost-effective medical imaging solutions [107].

1.9.4.1 Expanding AI-Based Diagnostics in Low-Resource Settings

Improving AI-Based Diagnostics in Resource-Limited Settings: AI models trained through federated learning would allow healthcare institutions to have access to high-quality diagnostics in resource-limited settings without requiring significant computing. Optimization algorithms and probabilistic models enable AI-enabled mobile health applications to remotely diagnose diseases while maintaining data privacy. AI-based cloud systems could partner with edge computing and distributed learning to process imaging data, which allows for affordable and scalable delivery of healthcare to rural communities [108, 109]. These advances point to a changing mode of imaging research. Lastly, we summarize important findings and the role that AI stands to play in diagnostic imaging.

1.10 Conclusion

Medical imaging has changed due to innovations in AI and mathematical models, including improved diagnostic accuracy, real-time processing, and increased accessibility. These developments, enabled by DL, quantum computing, and wearable technologies, can facilitate automated diagnosis and treatment and greater worldwide accessibility to healthcare. The evolution of AI-based medical imaging depends on continual innovation, research on ethical AI, and interdisciplinary collaboration between healthcare and mathematics.

DL has not just improved medical imaging – it has changed the game. AI applications have become primary methods of imaging operations, such as automated detection, segmentation, classification, and outcome prediction. This change in the medical imaging field is advancing the way healthcare is moved from reactive diagnosis to proactive real-time interventions and personalized to patient needs. AI models are decreasing the time to diagnosis, increasing accuracy, and enhancing access to care no matter where the patient is located, from rural health initiatives to tertiary health systems where radiology departments are sometimes seeing the majority of patients in clinics.

This chapter provided an overview of several imaging modalities, identified their limitations, and stressed the fact that DL can improve their shortcomings. Architectures like U-Net and TransUNet represent that trajectory of technological advancement for segmentation tasks, while milestone datasets like BraTS, LUNA, and EyePACS provide rigorously tested benchmarking to implement in practice. The ongoing integration of AI with new technologies like wearables, quantum computing, and cloud-based platforms will continue to change medical imaging; however, successful integration will require ethical oversight and regulatory approval as well as transparency and explainability of AI models.

In summary, DL is not just personally advancing image processing; it is revolutionizing how healthcare can be delivered.

1.10.1 Summary of Key Insights

Medical imaging is evolving and improving in capabilities, thanks to the introduction of new technologies potentially capable of AI, for example, self-learning or self-learning models, quantum computing, and real-time wearable monitors. If some mathematic construct can be developed – like optimization algorithms, Bayesian inference, or quantum mechanics principles – this would improve image processing and diagnostics. AI-based models fortify risk stratification, enable real-time image quality enhancements, and integrate outputs into EHRs.

1.10.2 The Transformative Impact of AI in Medical Imaging

Technological advances in AI have transformed the discipline of medical imaging by automating disease identification, improving diagnosis efficiency, and enhancing care delivery in resource-limited environments. For example, the CNNs and GANs of DL frameworks have led to accurate segmentation and detection of abnormalities in MRI, CT, and fundus imaging. Innovations in quantum computing have also quickened sophisticated image reconstruction, and wearable AI systems provide real-time health monitoring to fill gaps from the absence of preventive healthcare. These innovations have streamlined workflows in radiology and pathology while lightening the load for physicians and improving patient care.

1.10.3 Future Roadmap for AI-Driven Healthcare Diagnostics

The future of AI in medical imaging will depend on self-learning AI, quantum-enhanced DL technologies, and their ethical use. Further research in federated learning will improve privacy-preserving AI applications, allowing medical institutions in decentralized organizations to collaborate on training and validating their AI models. The fusion of AI with cloud computing and smart hospital-designated devices will result in real-time, low-cost diagnostic interventions in advanced and resource-limited situations. Advances in XAI will help build trust and interpretability in the AI work, ensuring AI-derived healthcare depictions are transparent and aligned with clinical decision-making. AI and its application of mathematics will continue to shape the future of medical imaging by stimulating creative practices and improving diagnostic accuracy.

1.10.4 Future Outlook in Medical Imaging

1. **Federated Learning:**
   Federated learning will lead to the decentralization of AI model training by training models locally on patient data, which minimizes privacy concerns and maximizes collaborative usability [110].
2. **Synthetic Data from Generative Adversarial Networks:**
   Generative Adversarial Networks (GANs) can create synthetic medical data which will support the training of robust AI models when real datasets are scarce [111].
3. **Wearable AI Imaging Technologies:**
   Wearable and portable imaging technologies, such as handheld OCT and mobile ultrasound will enable point-of-care diagnostics [112].

References

[1] Grenander U, Miller MI. Pattern theory: From representation to inference. OUP Oxford; 2006 Dec 14.

[2] Kourou K, Exarchos TP, Exarchos KP, Karamouzis MV, Fotiadis DI. Machine learning applications in cancer prognosis and prediction. Computational and Structural Biotechnology Journal. 2015 Jan 1;13:8–17.

[3] Zhou SK, Greenspan H, Davatzikos C, Duncan JS, Van Ginneken B, Madabhushi A, Prince JL, Rueckert D, Summers RM. A review of deep learning in medical imaging: imaging traits, technology trends, case studies with progress highlights, and future promises. Proceedings of the IEEE. 2021 Feb 26;109(5):820–838.

[4] Zhao H, Liu Z, Tang J, Gao B, Qin Q, Li J, Zhou Y, Yao P, Xi Y, Lin Y, Qian H. Energy-efficient high-fidelity image reconstruction with memristor arrays for medical diagnosis. Nature Communications. 2023 Apr 20;14(1):2276.

[5] Litjens G, Kooi T, Bejnordi BE, Setio AA, Ciompi F, Ghafoorian M, Van Der Laak JA, Van Ginneken B, Sánchez CI. A survey on deep learning in medical image analysis. Medical Image Analysis. 2017 Dec 1;42:60–88.

[6] Topol E. Deep medicine: How artificial intelligence can make healthcare human again. Hachette UK; 2019 Mar 12.

[7] Shin Y, Lee M, Lee Y, Kim K, Kim T. Artificial Intelligence-Powered Quality Assurance: transforming diagnostics, surgery, and patient care – innovations, limitations, and future directions. Life. 2025 Apr 16;15(4):654.

[8] Goodfellow IJ, Pouget-Abadie J, Mirza M, Xu B, Warde-Farley D, Ozair S, Courville A, Bengio Y. Generative adversarial nets. Advances in Neural Information Processing Systems. 2014;27.

[9] Zhang H, Qie Y. Applying deep learning to medical imaging: a review. Applied Sciences. 2023 Sep 21;13(18):10521.

[10] Shen D, Wu G, Suk HI. Deep learning in medical image analysis. Annual Review of Biomedical Engineering. 2017 Jun 21;19(1):221–248.

[11] Braga J. Non-invasive imaging techniques. A Companion to Dental Anthropology. 2015 Nov;30:514–527.

[12] Jerrold T, Bushberg AS, Leidholdt EM, Boone JM The essential physics of medical imaging. North American Edition, ISBN-13. 2012:978-0781780575.

[13] Mumuni AN, Hasford F, Udeme NI, Dada MO, Awojoyogbe BO. A SWOT analysis of artificial intelligence in diagnostic imaging in the developing world: making a case for a paradigm shift. Physical Sciences Reviews. 2024 Jan 1;9(1):443–476.

[14] Kasban H, El-Bendary MA, Salama DH. A comparative study of medical imaging techniques. International Journal of Information Science and Intelligent System. 2015 Apr;4(2):37–58.

[15] Brant WE, Helms C. Diagnostic imaging methods. In U: Brant WE, Helms CA. Fundamentals of diagnostic radiology-4th ed. LWW 2012 Mar (vol. 20, pp. 2–5).

[16] Mazurowski MA, Buda M, Saha A, Bashir MR. Deep learning in radiology: an overview of the concepts and a survey of the state of the art with focus on MRI. Journal of Magnetic Resonance Imaging. 2019 Apr;49(4):939–954.

[17] Mahesh M. The essential physics of medical imaging. Medical Physics. 2013 Jul;40(7):077301.

[18] Esteva A, Robicquet A, Ramsundar B, Kuleshov V, DePristo M, Chou K, Cui C, Corrado G, Thrun S, Dean J. A guide to deep learning in healthcare. Nature Medicine. 2019 Jan;25(1):24–29.

[19] Nüsslin F. Wilhelm Conrad Röntgen: the scientist and his discovery. Physica Medica. 2020 Nov 1;79:65–68.

[20] Samei E, Peck DJ. Hendee's physics of medical imaging. John Wiley & Sons; 2019 Apr 23.

[21] Jan J. Medical image processing, reconstruction and analysis: Concepts and methods. CRC Press; 2019 Aug 30.

[22] Panchbhai AS. Wilhelm Conrad Röntgen and the discovery of X-rays: revisited after centennial. Journal of Indian Academy of Oral Medicine and Radiology. 2015 Jan 1;27(1):90–95.

[23] Hore PJ. Nuclear magnetic resonance. Oxford University Press; 2015.

[24] Ye JC. Compressed sensing MRI: a review from signal processing perspective. BMC Biomedical Engineering. 2019 Mar 29;1(1):8.

[25] Bercoff J. Ultrafast ultrasound imaging. Ultrasound imaging-Medical Applications. 2011 Aug;23:3–24.

[26] Jiang W, Chalich Y, Deen MJ. Sensors for positron emission tomography applications. Sensors. 2019 Nov 17;19(22):5019.

[27] Dolcetti G. *Remote monitoring of shallow turbulent flows based on the Doppler spectra of airborne ultrasound* (Doctoral dissertation, University of Sheffield).

[28] Thorwarth D. Functional imaging for radiotherapy treatment planning: current status and future directions – a review. The British Journal of Radiology. 2015 Jul 1;88(1051):20150056.

[29] Haselmann M, Gruber DP, Tabatabai P. Anomaly detection using deep learning based image completion. In 2018 17th IEEE international conference on machine learning and applications (ICMLA) 2018 Dec 17 (pp. 1237–1242). IEEE.

[30] Sharma D, Kaushik P. Applications of AI in Neurological Disease Detection – a review of specific ways in which AI is being used to detect and diagnose neurological disorders, such as Alzheimer's and Parkinson's. AI in Disease Detection: Advancements and Applications. 2025 Jan;8:167–189.

[31] Nichols JA, Herbert Chan HW, Baker MA. Machine learning: applications of artificial intelligence to imaging and diagnosis. Biophysical Reviews. 2019 Feb 7;11:111–118.

[32] Goodfellow I, Bengio Y, Courville A, Bengio Y. Deep learning. Cambridge: MIT press; 2016 Nov 18.

[33] Boopathi S, Pandey BK, Pandey D. Advances in artificial intelligence for image processing: Techniques, applications, and optimization. In Handbook of research on thrust technologies' effect on image processing 2023 (pp. 73–95). IGI Global.

[34] Shih FY. Image processing and mathematical morphology: Fundamentals and applications. CRC press; 2017 Jul 12.

[35] Kamnitsas K, Bai W, Ferrante E, McDonagh S, Sinclair M, Pawlowski N, Rajchl M, Lee M, Kainz B, Rueckert D, Glocker B. Ensembles of multiple models and architectures for robust brain tumour segmentation. In Brainlesion: Glioma, multiple sclerosis, stroke and traumatic brain injuries: Third international workshop, BrainLes 2017, held in conjunction with MICCAI 2017, Quebec City, QC, Canada September 14, 2017 Revised Selected Papers 3 2018 (pp. 450–462). Springer International Publishing.

[36] Gulshan V, Peng L, Coram M, Stumpe MC, Wu D, Narayanaswamy A, Venugopalan S, Widner K, Madams T, Cuadros J, Kim R. Development and validation of a deep learning algorithm for detection of diabetic retinopathy in retinal fundus photographs. jama. 2016 Dec 13;316 (22):2402–2410.

[37] LeCun Y, Bengio Y, Hinton G. Deep learning. nature. 2015 May 28;521(7553):436–444.

[38] Gargeya R, Leng T. Automated identification of diabetic retinopathy using deep learning. Ophthalmology. 2017 Jul 1;124(7):962–969.

[39] Angermueller C, Pärnamaa T, Parts L, Stegle O. Deep learning for computational biology. Molecular Systems Biology. 2016 Jul;12(7):878.

[40] Oyeniyi J, Oluwaseyi P. Emerging trends in AI-powered medical imaging: enhancing diagnostic accuracy and treatment decisions. International Journal of Enhanced Research in Science Technology & Engineering. 2024;13:2319–7463.

[41] Saleem F, Al-Ghamdi AS, Alassafi MO, AlGhamdi SA. Machine learning, deep learning, and mathematical models to analyze forecasting and epidemiology of COVID-19: a systematic literature review. International Journal of Environmental Research and Public Health. 2022 Apr 22;19(9):5099.

[42] Bushberg JT, Seiberg JA, Leidholdt EM Jr, Boone JM. The essential physics of medical imaging. Lippincott Williams & Wilkins; 2020 Nov 24.

[43] Burger W, Burge MJ. Digital image processing: An algorithmic introduction. Springer Nature; 2022 Jul 23.

[44] Ammari H, Garnier J, Kang H, Nguyen LH, Seppecher L. Multi-wave medical imaging. Mathematical Modelling & Imaging Reconstruction. 2017 May.

[45] Toennies KD. Guide to medical image analysis. Springer-Verlag London Ltd.; 2017.

[46] Soori M, Arezoo B, Dastres R. Artificial intelligence, machine learning and deep learning in advanced robotics, a review. Cognitive Robotics. 2023 Jan 1;3:54–70.

[47] Drexler W, Fujimoto JG. Retinal optical coherence tomography imaging. Optical Coherence Tomography. 2015;1685.

[48] Schlegl T, Waldstein SM, Bogunovic H, Endstraßer F, Sadeghipour A, Philip AM, Podkowinski D, Gerendas BS, Langs G, Schmidt-Erfurth U. Fully automated detection and quantification of macular fluid in OCT using deep learning. Ophthalmology. 2018 Apr 1;125(4):549–558.

[49] De Fauw J, Ledsam JR, Romera-Paredes B, Nikolov S, Tomasev N, Blackwell S, Askham H, Glorot X, O'Donoghue B, Visentin D, Van Den Driessche G. Clinically applicable deep learning for diagnosis and referral in retinal disease. Nature Medicine. 2018 Sep;24(9):1342–1350.

[50] Kalkman J. Fourier-Domain Optical Coherence Tomography Signal Analysis and Numerical Modeling. International Journal of Optics. 2017;2017(1):9586067.

[51] Pircher M, Zawadzki RJ. Review of adaptive optics OCT (AO-OCT): principles and applications for retinal imaging. Biomedical Optics Express. 2017 Apr 19;8(5):2536–2562.
[52] De Fauw J, Ledsam JR, Romera-Paredes B, Nikolov S, Tomasev N, Blackwell S, Askham H, Glorot X, O'Donoghue B, Visentin D, Van Den Driessche G. Clinically applicable deep learning for diagnosis and referral in retinal disease. Nature Medicine. 2018 Sep;24(9):1342–1350.
[53] Ting DS, Cheung CY, Lim G, Tan GS, Quang ND, Gan A, Hamzah H, Garcia-Franco R, San Yeo IY, Lee SY, Wong EY. Development and validation of a deep learning system for diabetic retinopathy and related eye diseases using retinal images from multiethnic populations with diabetes. Jama. 2017 Dec 12;318(22):2211–2223.
[54] Spaide RF, Fujimoto JG, Waheed NK. Image artifacts in optical coherence tomography angiography. Retina. 2015 Nov 1;35(11):2163–2180.
[55] Qummar S, Khan FG, Shah S, Khan A, Shamshirband S, Rehman ZU, Khan IA, Jadoon W. A deep learning ensemble approach for diabetic retinopathy detection. IEEE Access. 2019 Oct 15;7:150530–150539.
[56] Moser E, Laistler E, Schmitt F, Kontaxis G. Ultra-high field NMR and MRI – the role of magnet technology to increase sensitivity and specificity. Frontiers in Physics. 2017 Aug 17;5:33.
[57] Katti G, Ara SA, Shireen A. Magnetic resonance imaging (MRI)–a review. International Journal of Dental Clinics. 2011 Jan 1;3(1):65–70.
[58] McCollough CH, Leng S. Use of artificial intelligence in computed tomography dose optimisation. Annals of the ICRP. 2020 Dec;49(1_suppl):113–125.
[59] Brunese MC, Rocca A, Santone A, Cesarelli M, Brunese L, Mercaldo F. Explainable and robust deep learning for liver segmentation through U-Net network. Diagnostics (Basel). 2025;15(7):878. doi: 10.3390/diagnostics15070878.
[60] Tanter M, Fink M. Ultrafast imaging in biomedical ultrasound. IEEE Transactions on Ultrasonics, Ferroelectrics, and Frequency Control. 2014 Jan 2;61(1):102–119.
[61] Lu RC, She B, Gao WT, Ji YH, Xu DD, Wang QS, Wang SB. Positron-emission tomography for hepatocellular carcinoma: current status and future prospects. World Journal of Gastroenterology. 2019 Aug 28;25(32):4682.
[62] Ravì D, Wong C, Deligianni F, Berthelot M, Andreu-Perez J, Lo B, Yang GZ. Deep learning for health informatics. IEEE Journal of Biomedical and Health Informatics. 2016 Dec 29;21(1):4–21.
[63] Razzak MI, Naz S, Zaib A. Deep learning for medical image processing: overview, challenges and the future. Classification in BioApps: Automation of Decision Making. 2017 Nov;14:323–350.
[64] Ardila D, Kiraly AP, Bharadwaj S, Choi B, Reicher JJ, Peng L, Tse D, Etemadi M, Ye W, Corrado G, Naidich DP. End-to-end lung cancer screening with three-dimensional deep learning on low-dose chest computed tomography. Nature Medicine. 2019 Jun;25(6):954–961.
[65] Kermany DS, Goldbaum M, Cai W, Valentim CC, Liang H, Baxter SL, McKeown A, Yang G, Wu X, Yan F, Dong J. Identifying medical diagnoses and treatable diseases by image-based deep learning. cell. 2018 Feb 22;172(5):1122–1131.
[66] Wang J, Han L, Ran D. Architectures and applications of U-net in medical image segmentation: A review. In 2023 9th international symposium on system security, safety, and reliability (ISSSR) 2023 Jun 10 (pp. 84–94). IEEE.
[67] Chen J, Lu Y, Yu Q, Luo X, Adeli E, Wang Y, Lu L, Yuille AL, Zhou Y. TransUNet: transformers make strong encoders for medical image segmentation. arXiv preprint arXiv:2102.04306. 2021 Feb 8.
[68] Bhati D, Neha F, Amiruzzaman M. A survey on explainable artificial intelligence (XAI) techniques for visualizing deep learning models in medical imaging. Journal of Imaging. 2024 Sep 25;10(10):239.
[69] Bakas S, Reyes M, Jakab A, Bauer S, Rempfler M, Crimi A, Shinohara RT, Berger C, Ha SM, Rozycki M, Prastawa M Identifying the best machine learning algorithms for brain tumor segmentation, progression assessment, and overall survival prediction in the BRATS challenge. arXiv preprint arXiv:1811.02629. 2018 Nov 5.

[70] Setio AA, Traverso A, De Bel T, Berens MS, Van Den Bogaard C, Cerello P, Chen H, Dou Q, Fantacci ME, Geurts B, Van der Gugten R. Validation, comparison, and combination of algorithms for automatic detection of pulmonary nodules in computed tomography images: the LUNA16 challenge. Medical Image Analysis. 2017 Dec 1;42:1–3.

[71] Irvin J, Rajpurkar P, Ko M, Yu Y, Ciurea-Ilcus S, Chute C, Marklund H, Haghgoo B, Ball R, Shpanskaya K, Seekins J. Chexpert: a large chest radiograph dataset with uncertainty labels and expert comparison. In Proceedings of the AAAI Conference on Artificial Intelligence 2019 Jul 17 (Vol. 33, No. 01, pp. 590–597).

[72] Schlemper J. *Deep learning for fast and robust medical image reconstruction and analysis* (Doctoral dissertation, Imperial College London).

[73] Ronneberger O, Fischer P, Brox T. U-net: Convolutional networks for biomedical image segmentation. In Medical image computing and computer-assisted intervention–MICCAI 2015: 18th international conference October 5–9, 2015 proceedings, part III 18 2015 (pp. 234–241). Munich, Germany: Springer international publishing.

[74] Doneva M. Mathematical models for magnetic resonance imaging reconstruction: an overview of the approaches, problems, and future research areas. IEEE Signal Processing Magazine. 2020 Jan 17;37(1):24–32.

[75] Obuchowicz R, Strzelecki M, Piórkowski A. Clinical applications of artificial intelligence in medical imaging and image processing – a review. Cancers. 2024 May 14;16(10):1870.

[76] Ding Y, Sohn JH, Kawczynski MG, Trivedi H, Harnish R, Jenkins NW, Lituiev D, Copeland TP, Aboian MS, Mari Aparici C, Behr SC. A deep learning model to predict a diagnosis of Alzheimer disease by using 18F-FDG PET of the brain. Radiology. 2019 Feb;290(2):456–464.

[77] Santos MK, Ferreira JR, Wada DT, Tenório AP, Nogueira-Barbosa MH, Marques PM. Artificial intelligence, machine learning, computer-aided diagnosis, and radiomics: advances in imaging towards to precision medicine. Radiologiabrasileira. 2019 Sep 23;52(06):387–396.

[78] Armato SG III, Huisman H, Drukker K, Hadjiiski L, Kirby JS, Petrick N, Redmond G, Giger ML, Cha K, Mamonov A, Kalpathy-Cramer J. PROSTATEx Challenges for computerized classification of prostate lesions from multiparametric magnetic resonance images. Journal of Medical Imaging. 2018 Oct 1;5 (4):044501–.

[79] Rodríguez-Ruiz A, Krupinski E, Mordang JJ, Schilling K, Heywang-Köbrunner SH, Sechopoulos I, Mann RM. Detection of breast cancer with mammography: effect of an artificial intelligence support system. Radiology. 2019 Feb;290(2):305–314.

[80] Suganyadevi S, Seethalakshmi V, Balasamy K. A review on deep learning in medical image analysis. International Journal of Multimedia Information Retrieval. 2022 Mar;11(1):19–38.

[81] Zbontar J, Knoll F, Sriram A, Murrell T, Huang Z, Muckley MJ, Defazio A, Stern R, Johnson P, Bruno M, Parente M. fastMRI: an open dataset and benchmarks for accelerated MRI. arXiv preprint arXiv:1811.08839. 2018 Nov 21.

[82] McCollough CH, Bartley AC, Carter RE, Chen B, Drees TA, Edwards P, Holmes DR III, Huang AE, Khan F, Leng S, McMillan KL. Low-dose CT for the detection and classification of metastatic liver lesions: results of the 2016 low dose CT grand challenge. Medical Physics. 2017 Oct;44(10):e339–52.

[83] Chen X, Xie H, Tao X, Wang FL, Leng M, Lei B. Artificial intelligence and multimodal data fusion for smart healthcare: topic modeling and bibliometrics. Artificial Intelligence Review. 2024 Mar 15;57(4):91.

[84] Lee S, Jung JY, Mahatthanatrakul A, Kim JS. Artificial intelligence in spinal imaging and patient care: a review of recent advances. Neurospine. 2024 Jun 30;21(2):474.

[85] Hwang S, Song Y, Kim J. Evaluation of AI-assisted telemedicine service using a mobile pet application. Applied Sciences. 2021 Mar 17;11(6):2707.

[86] Hadziahmetovic M, Nicholas P, Jindal S, Mettu PS, Cousins SW. Evaluation of a remote diagnosis imaging model vs dilated eye examination in referable macular degeneration. JAMA Ophthalmology. 2019 Jul 1;137(7):802–808.

[87] Najjar R. Redefining radiology: a review of artificial intelligence integration in medical imaging. Diagnostics. 2023 Aug 25;13(17):2760.

[88] Kaissis GA, Makowski MR, Rückert D, Braren RF. Secure, privacy-preserving and federated machine learning in medical imaging. Nature Machine Intelligence. 2020 Jun;2(6):305–311.

[89] Choudhri AF, Chatterjee AR, Javan R, Radvany MG, Shih G. Security issues for mobile medical imaging: a primer. Radiographics. 2015 Oct;35(6):1814–1824.

[90] Mehrabi N, Morstatter F, Saxena N, Lerman K, Galstyan A. A survey on bias and fairness in machine learning. ACM Computing Surveys. 2021 Jul 13;54(6):1–35.

[91] Petrović A, Nikolić M, Radovanović S, Delibašić B, Jovanović M. FAIR: fair adversarial instance re-weighting. Neurocomputing. 2022 Mar 1;476:14–37.

[92] Benjamens S, Dhunnoo P, Meskó B. The state of artificial intelligence-based FDA-approved medical devices and algorithms: an online database. NPJ Digital Medicine. 2020 Sep 11;3(1):118.

[93] Harvey HB, Gowda V. How the FDA regulates AI. Academic Radiology. 2020 Jan 1;27(1):58–61.

[94] Begoli E, Bhattacharya T, Kusnezov D. The need for uncertainty quantification in machine-assisted medical decision making. Nature Machine Intelligence. 2019 Jan;1(1):20–23.

[95] Samek W, Wiegand T, Müller KR Explainable artificial intelligence: Understanding, visualizing and interpreting deep learning models. arXiv preprint arXiv:1708.08296. 2017 Aug 28.

[96] Dellermann D, Calma A, Lipusch N, Weber T, Weigel S, Ebel P The future of human-AI collaboration: a taxonomy of design knowledge for hybrid intelligence systems. arXiv preprint arXiv:2105.03354. 2021 May 7.

[97] Mäkelä E, Stephany F Complement or substitute? How AI increases the demand for human skills. arXiv preprint arXiv:2412.19754. 2024 Dec 27.

[98] Suresh P, Keerthika P, Devi MR, Kamalam GK, Logeswaran K, Kumar SC, Iyappan P, Baiju BV. Revolutionizing healthcare industry with quantum artificial intelligence (AI) and machine learning (ML) techniques. In The quantum evolution 2024 (pp. 159–183). CRC Press.

[99] Solanki A. Advancements in Artificial Intelligence: a comprehensive review and future prospects. International Journal of Artificial Intelligence Research and Development (IJAIRD). 2024 Oct 2;2 (1):53–64.

[100] Pohle R, Toennies KD. Self-learning model-based segmentation of medical images. Image Processing & Communications. 2001;7(3–4):97–113.

[101] Asker J, Fershtman C, Pakes A. Artificial intelligence, algorithm design, and pricing. In AEA papers and proceedings 2022 May 1 (vol. 112, pp. 452–456). 2014 Broadway, Suite 305 Nashville, TN 37203: American Economic Association.

[103] Chow JCL. Quantum computing and machine learning in medical decision-making: a comprehensive review. Algorithms. 2025;18(3):156. doi: 10.3390/a18030156.

[102] Schuld M, Petruccione F. Supervised learning with quantum computers. Quantum science and technology (Springer, 2018); 2018.

[105] Ur Rasool R, Ahmad HF, Rafique W, Qayyum A, Qadir J, Anwar Z. Quantum computing for healthcare: a review. Future Internet. 2023 Feb 27;15(3):94.

[104] Godin T, Sader L, Khodadad Kashi A, Hanzard PH, Hideur A, Moss DJ, Morandotti R, Genty G, Dudley JM, Pasquazi A, Kues M. Recent advances on time stretch dispersive Fourier transform and its applications. Advances in Physics: X. 2022 Dec 31;7(1):2067487.

[105] Castillejo P, Martinez JF, Rodriguez-Molina J, Cuerva A. Integration of wearable devices in a wireless sensor network for an E-health application. IEEE Wireless Communications. 2013 Aug;20(4):38–49.

[106] Dangi RR, Sharma A, Vageriya V. Transforming healthcare in low-resource settings with artificial intelligence: recent developments and outcomes. Public Health Nursing. 2025 Mar;42(2):1017–1030.

[107] Tjoa E, Guan C. A survey on explainable artificial intelligence (XAI): toward medical XAI. IEEE Transactions on Neural Networks and Learning Systems. 2020 Oct 20;32(11):4793–4813.
[108] Rieke N, Hancox J, Li W, Milletari F, Roth HR, Albarqouni S, Bakas S, Galtier MN, Landman BA, Maier-Hein K, Ourselin S. The future of digital health with federated learning. NPJ Digital Medicine. 2020 Sep 14;3(1):119.
[109] Singh S, Thakur R, Kumar K, Dogra A. New Perspective of Multi-dimensional Approach for the Management of Attention-deficit Hyperactivity Disorder: A Review. The Open Neuroimaging Journal. 2024 Oct 10;17(1).
[110] Frid-Adar M, Klang E, Amitai M, Goldberger J, Greenspan H. Synthetic data augmentation using GAN for improved liver lesion classification. In 2018 IEEE 15th international symposium on biomedical imaging (ISBI 2018) 2018 Apr 4 (pp. 289–293). IEEE.
[111] Singh B, Kaunert C, Vig K, Gautam BK. Wearable sensors assimilated with internet of things (IoT) for advancing medical imaging and digital healthcare: Real-time scenario. In Inclusivity and accessibility in digital health 2024 (pp. 275–297). IGI Global.
[112] Khan AO, Islam SM, Sarkar A, Islam T, Paul R, Bari MS. Real-time predictive health monitoring using AI-driven wearable sensors: enhancing early detection and personalized interventions in chronic disease management. International Journal for Multidisciplinary Research. 2024.

Rashi Rastogi

2 A Deep Learning Architecture for Detecting Pneumonia in Chest X-rays: An Approach for Processing Medical Images

Abstract: Diagnostic medical imaging relies heavily on deep learning models that help make disease diagnosis more efficient. This chapter examines how deep learning can contribute to early pneumonia detection, which greatly contributes to global mortality. It appears that this technique could solve problems from inconsistent interpretations among radiologists, reducing the risk of wrong diagnoses. This work wants to create essential diagnostic tools that ensure accuracy and help avoid mistakes. For this study, CNN technology is used on chest X-rays to determine which images in the dataset are from pneumonia cases. The chest X-ray dataset includes recognized images that facilitate the classification of pneumonia into three categories: normal, bacterial, and viral pathological patterns. The dataset exhibits an unbalanced class distribution, necessitating three processing techniques during preparation: normalization, image resizing, and augmentation. The deep learning system relies on specialized pretrained models, such as EfficientNet and InceptionV3, to enhance classification outcomes. The findings indicate that the developed classification system yields outcomes that surpass those generated by machine learning systems. Improvements in early-stage pneumonia detection are attainable due to the model's specific and sensitive characteristics. The analytical capabilities of deep learning reduce the dependence on manual post-analysis by pathologists, streamlining the diagnostic workflow. Future work could focus on extending the model to support multi-class classification of different pneumonia types. The integration of genomic data within multimodal analysis is central to the study aimed at developing a diagnostic tool. This research provides a basis for the assertion that advanced deep learning algorithms can significantly enhance pneumonia diagnosis, thereby improving patient treatment and planning.

Keywords: deep learning, medical image processing, CNNs, GANs, transformers, medical image segmentation, healthcare

2.1 Introduction

Medical image processing uses deep learning to automatically analyze complex medical data to diagnose disorders and improve research. Convolutional neural networks (CNNs) using ResNet and InceptionV3 pretrained models improve pneumonia and other medical imaging accuracy [1]. Scientific materials improve diagnosis accuracy with deep learning, but their ultimate goal is to produce mammogram-based screen-

 | https://doi.org/10.1515/9783112226223-002

ing systems that humans can make mistakes with. Image normalization, scaling, and augmentation allow deep learning framework to make correct diagnosis classifications without false positives or negatives. AI-based medical image analysis enhances disease detection accuracy while reducing the burden of manual evaluations by healthcare professionals, thereby improving patient care.

Medical image processing has advanced, forcing personnel to migrate from radiologist interpretation to AI-based deep learning analysis. Previous diagnostic imaging by medical professionals caused expert disagreements and misdiagnosis. Medical imaging advanced with machine learning and deep learning because they can detect characteristics and identify patterns and disorders [2]. Because CNNs excel at medical picture classification and segmentation, their use in medical imaging is essential. Deep learning algorithms quickly give accurate medical findings, making pneumonia diagnosis possible. Researchers struggle with uneven data samples and AI model evaluation; however recent AI progress improves applicability. Researchers use complex preprocessing methods, transfer learning, and multimodal evaluation to improve clinical healthcare and diagnostics.

Deep learning technology automatically retrieves medical imaging data from diagnostic equipment including X-rays, CT scans, MRIs, ultrasounds, and PET scans. Medical picture classification and segmentation rely largely on CNNs since they analyze enormous datasets. CNNs use their three primary layers to identify picture patterns locally before lowering spatial information for a diagnostic [3]. Forward propagation is needed to evaluate deep learning models on large datasets with backpropagation optimizations [4]. Transfer learning allows doctors to employ model collections to speed up development and perform well when medical data is constrained. Deep learning algorithms are opaque, thus they cannot generate data that healthcare experts can understand. This makes medical imaging diagnosis difficult. Researchers employ attention processes and saliency maps to create gradient-based visualizations for model transparency because healthcare professionals need model integrity assurances. Deep learning is vital for modern medical diagnosis and therapeutic planning.

Deep learning improves disease diagnostic precision, allowing patients to be diagnosed early, so doctors may intervene to improve outcomes [5]. Generative adversarial networks (GANs) provide fake datasets to protect patient privacy during training. Deep learning systems use multimodal analysis to combine imaging and clinical or genetic data to enhance illness detection and treatment planning in modern medicine.

Despite these gains, medical specialists struggle with solution model interpretability and biased training content. The research combines saliency maps and attention processes to explain AI approaches to medical professionals to build trust [6]. Martin stated that processes show that deep learning is improving medical imaging optimization, making better health care more accessible internationally.

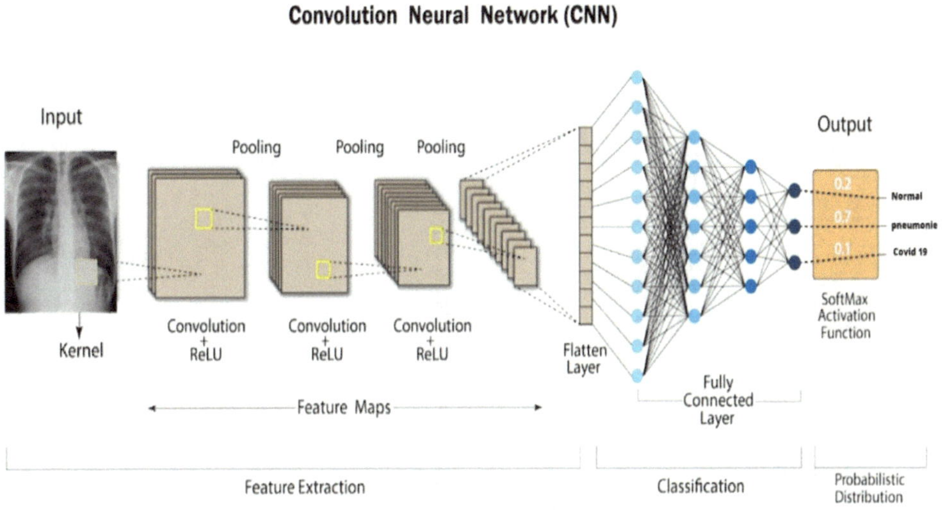

Figure 2.1: Schematic design of a convolutional neural network (CNN) architecture. Adapted from Rguibi et al. [7].

Figure 2.1 depicts the CNNs architecture specifically designed to process and analyze the image based on mimicking that of the human's visual system. CNN is usually composed of many layers such as convolutional layers, pooling layers, and fully connected layers, and some activation function that collaboratively extracts meaningful features from input images. The convolutional layers extract patterns similar to edges, textures, and shapes and pass them through several layers for more complex representation. Max pooling and other pooling layers reduce the amount of spatial dimensions of the feature maps to pay a computational cost, helping to reduce overfitting. The fully connected layers learn the patterns of classification using the extracted features, performing high level reasoning and classification. Rectified linear unit (ReLU) activation functions like the previous one introduce nonlinearity in the network, thus allowing it to learn very complex relationships in the data. For classification task, the final layer does a softmax or sigmoid function to produce probability scores. As they are able to automatically learn hierarchical features, CNNs are widely used in medical image analysis for accurate disease diagnosis, segmentation, and anomaly detection in medical imaged modalities such as MRI, CT, and X-rays [7].

Figure 2.2 shows the framework of ResNet-50, which stands for residual network with 50 layers as a specially designed deep convolutional neural network built to overcome the vanishing gradient problem in deep network architectures. Interestingly, the key innovation in ResNet-50 lies in the introduction of residual learning through short cuts/ skip connections. A direct connection between layers skips multiple stages, so each stage provides input to the following stage in sequence. Identity mappings provide the model with an easier gradient flow because they closely resemble inputs. Therefore, the

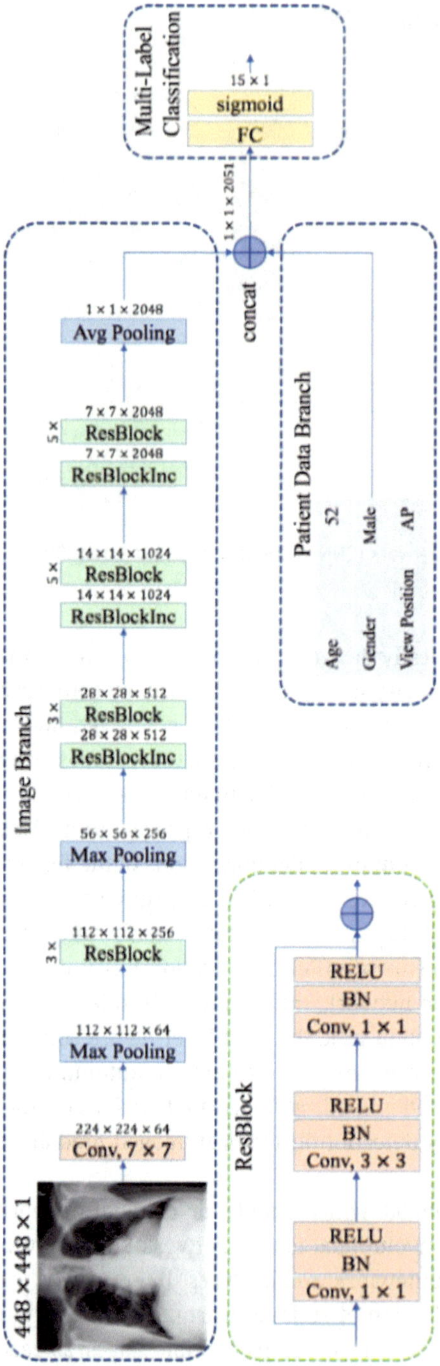

Figure 2.2: Simplified schematic of the ResNet-50 architecture, highlighting residual connections. Adapted from Baltruschat et al. [8].

gradient illumination becomes possible. The network obtains an opportunity to bypass training errors that grow together with increased layer count [8].

A bottleneck design separates the three layers within each residual block of the ResNet-50 representation. Beneath the bottleneck structure, the model operates by using a 1 × 1 convolution for dimensionality reduction to 64, while 3 × 3 performs spatial features and a final application of 1 × 1 convolution to bring back dimensionality. Nonlinear ReLU activation functions are added before the implementation of batch normalization to stabilize training operations. The network contains one max pooling layer and one average pooling layer, together with 48 convolutional layers, before proceeding to a fully connected layer for classification. The deep architecture and efficient performance of ResNet-50 makes it perform optimally in image classification, together with object detection and medical imaging applications.

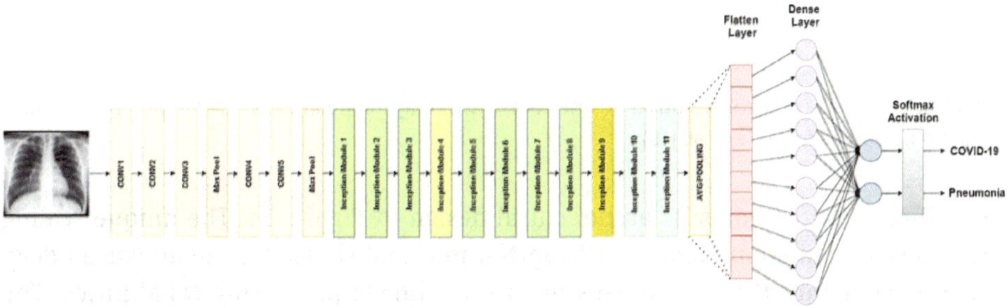

Figure 2.3: Conceptual representation of the Inception architecture with parallel convolutional paths. Adapted from Shazia et al. [9].

As depicted in Figure 2.3, the GoogleNet model features an Inception architecture that serves as a deep convolutional neural network for feature extraction as well as computational efficiency [9]. Inception modules are the special feature of this design because they perform parallel convolution operations across paths of various kernel sizes including 1 × 1, 3 × 3, and 5 × 5, and add a 3 × 3 max pooling operation branch. The network simultaneously executes multiple operations, which enable the capture of details at various scales through single layers that process fine elements and extensive patterns. The paths run in parallel to create one feature map that contains information at various spatial and channel levels.

The crucial innovation introduced by Inception involves performing 1 × 1 convolution operations that precede the utilization of bigger filters. These 1 × 1 filters provide bottleneck layers that shrink input channels from substantial amounts while reducing the computational load while maintaining performance levels. The performance capability is enhanced through implementation of auxiliary classifiers at various intermediate stages. Smaller classifiers serve as methods to direct the learning process, so training becomes quicker and maintains superior stability. Factorized convolutions divide big convolution filters into smaller ones and batch normalization techniques in Inception V3, allowing for simpler training procedures.

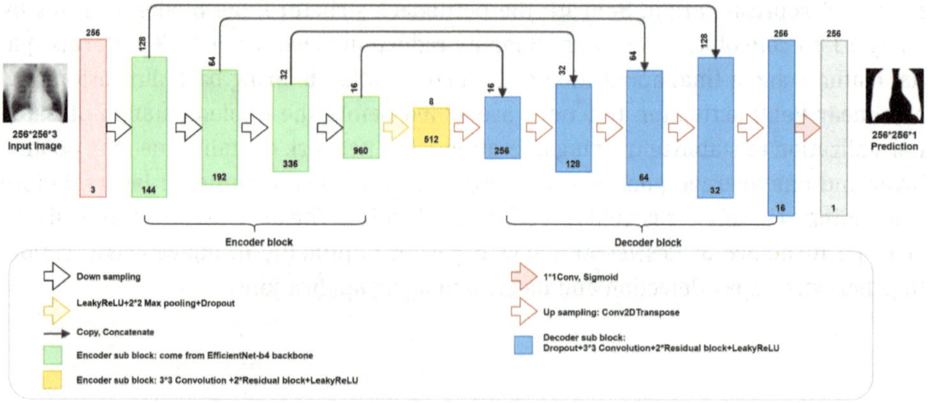

Figure 2.4: Overview of the EfficientNet architecture, illustrating compound scaling of depth, width, and resolution. Adapted from Liu et al. [10].

The EfficientNet contains its architectural design in Figure 2.4. The unique scaling procedure of compound allows EfficientNet to establish itself as an advanced deep learning structure that transforms healthcare image processing capabilities. The key dimensions of network depth, together with width and input resolution, receive simultaneous optimization in this system. The computational modeling achieves balanced scaling through sequential grid searches that enable better spectral performance and computing speed [10].

The fundamental structure of EfficientNet is based on mobile inverted bottleneck convolution (MBConv) blocks. The blocks first apply 1 × 1 convolution to increase input channels and follow up with depthwise 3 × 3 convolution before running a 1 × 1 convolution to decrease channels back to their original number. Squeeze and excitation optimization is done in every block for dynamic channel response adjustment according to inter channel dependence. To stabilize training and increase the performance outcome, the model uses batch normalization in all its layers and swish activation functions.

As the blocks mature, they continue to evolve in terms of resource consumption but in a systematic manner to the network architecture through the addition of MBConv blocks. The structure afterwards is an initial convolutional layer, and then we have some other number of MBConv blocks each with a varying expansion ratio and varying channel size. The final components for classification are fully connected layers and global average pooling network. From its experimental results, this architecture proves superior to other architectures in terms of consistent and high accuracy in medical image analysis tasks.

The use of deep learning methods in medical image analysis faces various constraints and problems that medical practitioners need to overcome before its clinical implementation becomes possible. Healthcare professionals maintain that knowledge of deep learning model functioning remains essential because they need to understand the bases for its medical decisions. The priority is to develop a system that ensures privacy protection during data processing and medical control, along with patient authorization, for safely deploying this technology. Explainable AI remains essential to gain trust and administrative approval since it reveals the prediction process to both users and administrators. Medical imaging applications of deep learning technology require teamwork between researchers and healthcare professionals, together with policymakers, to establish ethical and effective procedures.

This chapter has five parts. This section starts with basic medical image processing fundamentals followed by an explanation of intelligent deep learning's needed role in pneumonia diagnostic improvement. Section 2.2 investigates medical image analysis techniques, alongside their suggested material issues and practical enhancements. This section presents an improved method for data preprocessing, along with CNN model development, through transfer learning, to solve class imbalance and bad data quality issues. Section 2.4 details how the proposed models were evaluated, together with their assessment metrics, and discusses methods through which the proposed models can improve classification accuracy and sensitivity. The main findings of the research are presented before the research examines medical diagnostic applications and proposes some more possible directions for multimodal integration and multi class categorization on illness management.

2.2 Related Works

This section provides a critical and detailed analysis of the 2020–2022 work due to its relevance. The studies employ modern segmentation and classification plugins like ResNet, InceptionV3, EfficientNet, and others to increase diagnosis accuracy. Results from a detailed review of the evolving medical imaging landscape focus on current technological concerns.

In reviewing 120 medical imaging research publications, Suganyadevi et al. [11] provided a thorough assessment of current improvements that might benefit radiologists worldwide. It analyses current obstacles, significant issues, and possible pathways, concluding that ResNet performs better. Chan et al. [12] explored these difficulties and how to strengthen deep-learning-based CAD tools. It was found that intelligent patient care solutions must be easily incorporated into the clinical process. Wang et al. [13] reviewed 100 sample works from IEEE, Elsevier, Google Scholar, Web of Science, and other platforms, from 2000 to 2020. This document provides a comprehensive overview of CNN architecture, transfer learning foundations, transfer learning strategies, transfer learning applications in medical image analysis, and future directions.

Puttagunta et al. [14] developed artificial neural networks and analyzed DLA, which has medical imaging applications. X-ray, CT, mammography, and digital histopathology pictures are the most common DLA deployments. A structured review of DLA-based medical image classification, detection, and segmentation publications. The DLA evaluation suggests medical image analysis changes. Li et al. [15] explore several architectures and upsampling strategies to develop deep learning methods for super resolution issues. Next, we will examine deep learning approaches and their use in medical imaging super resolution concerns, along with the challenges.

Liu et al. [16] investigate medical picture segmentation using deep learning. The first portion covers deep learning-based medical picture segmentation basics. The present research state and the three main medical picture segmentation methods and their limitations are summarized to clarify future progress. The two diseased tissues and organs demonstrate the differences between them and the typical segmentation techniques used for comparison. Wang et al. [17] analyze medical picture segmentation using deep learning, which is the first of two distinct contributions. Second, surveys frequently divide deep learning in medical image segmentation into many groups and give extensive literature for each group sequentially. In contrast, the material found here is initially ordered from broad to specialized. This study reviews supervised and weakly supervised learning methods, ignoring the unsupervised approach, which has been thoroughly covered in earlier surveys and is currently less popular. This study systematically reviews supervised and weakly supervised learning paradigms, while deliberately excluding unsupervised approaches, as they have been comprehensively addressed in prior surveys and presently exhibit comparatively lower research momentum. The review critically examines existing literature with particular emphasis on three core dimensions of supervised learning: backbone network selection, architectural block design strategies, and loss function optimization techniques. These components are analyzed in terms of their contribution to representation quality, convergence behavior, and overall model generalization performance.

TL architectures for automated medical image processing are examined by Kora et al. [18]. Medical imaging tasks, including segmentation, object recognition, disease categorization, and severity assessment, have been done with TL. Compared to standard deep learning, TL may improve decision support with less learning data. AlexNet, ResNet, VGGNet, and GoogleNet are popular medical image analysis models. These models can learn from medical images, and the modified version enhances transfer learning models' medical image analysis. Li et al. [19] developed multimodal medical image fusion. It generally improves medical diagnosis accuracy. These texts have long stressed the need of multimodal imagery for information clarity. Medical photos are needed for image fusion, using supervised deep learning, and for early research into autonomous diagnosis technologies.

Chen et al [20] reviewed medical image analysis-specific unsupervised/self-supervised and semi-supervised learning methods. Deep learning experiments on four medical image processing tasks are extensively reviewed in this study. Transformer-based segmentation models were used in task-specific architectures. It studied aspects that potentially affect clinical deployment of large-scale deep learning applications. Castiglioni et al. [21] then discuss practical challenges in developing AI clinical decision support systems. ML/radiomics feature selection, training, validation, and testing are also covered. Deep learning models using multilayer artificial and convolutional neural networks for direct image processing are introduced. The data curation part includes picture labelling, annotation (segmentation is important in radiomics), data harmonization (addressing noise from different imaging procedures), and federated learning. Based on Singh et al. [22], this work reviews explainable deep learning in medical imaging applications. This paper discusses a realistic process for designing a clinical end-user system, including issue characterization, tactics, clinical deployment problems, and future research.

Cai et al. [23] combined intelligent imaging and deep learning with medical image classification and segmentation advances to analyze huge data and diagnose early diseases. Singh et al. [24] gave a brief mathematical review of 3D CNNs and medical picture preprocessing. This study reviews 3D medical imaging analysis research using 3D CNNs for classification, segmentation, detection, and localization. The review by Van der Velden et al. [25] covered almost 200 works on deep learning for medical image analysis using explainable AI (XAI). The papers are XAI-organized. The document discusses medical image analysis, XAI developments, and future directions. Karimi et al. [26] reviewed deep learning label noise research and medical image analysis labeling noise research. This review shows that medical image analysis has neglected deep learning for label noise. Three medical imaging datasets, with different label noise, are used to assess several existing approaches. Additionally, Xie et al. [27] present a fresh set of simple approaches to mitigate label noise's negative effects, improving understanding of its scope and potential solutions, in addition to providing an overview of the integration of medical domain knowledge with deep learning across multiple tasks, such as disease diagnosis, lesion and organ detection, and segmentation.

Zhou et al. [28] evaluate medical imaging features, identify clinical demands and technological hurdles, and examine how deep learning trends address these concerns. This includes network architecture, sparse and noisy labelling, federated learning, and interpretability, showcasing many clinically relevant instances, including digital pathology and chest, brain, abdomen, and cardiovascular imaging. Exploration and articulation of future prospects have ended. Modern medical image segmentation methods use data augmentation on CT and MRI images to train deep learning models, according to Chlap et al. [29]. The papers were categorized using basic methodologies, adaptive frameworks, sophisticated learning algorithms, and data improvement procedures. This comprehensive analysis seeks to provide a thorough grasp of these strategies and inspire trust in augmented data models for clinical AI training as they develop. Fu et al. [30] examined deep learning-based medical picture registration algorithms. Medicine's recent deep learning-based registration advances and applications are summarized. The categorization has seven categories based on methods, functions, and popularity. The document examined each issue and its important contributions and challenges. After a detailed review, each area was evaluated briefly, highlighting its achievements and future potential. Each deep learning-based lung and brain registration approach was tested on benchmark datasets.

Nalluri and Sasikala [31] introduced a deep learning model based on NASNet (neural architecture search network), which was pretrained on the ImageNet dataset to enable early detection of pneumonia from chest X-ray images. Featuring 2.6 million trainable parameters, the model is lightweight enough to be deployed on mobile devices, while still delivering high performance in terms of precision, recall, and F_1 score. In a similar approach, Asnake et al. [32] developed a deep learning pipeline that begins with preprocessing X-ray images to extract relevant features, followed by segmentation using a thresholding method. The YOLOv3 algorithm is then employed to detect normal and pneumonia-affected cases, which are subsequently classified using support vector machine (SVM) and softmax classifiers. The model was trained and tested on a chest X-ray image dataset and achieved outstanding results, with 99% accuracy, precision, recall, and F_1 score. Table 2.1 shows the comparison among existing techniques from the various researchers' works.

Inferences drawn from the literature:

- According to the reviewed literature, deep learning has a substantial impact on medical imaging, which is rightfully translated in a better accuracy for the medical imaging diagnostic tasks, segmentation, and classification.
- Architectures like ResNet, InceptionV3, and EfficientNet always lead in research such that they show the effectiveness in achieving excellent results for a multitude of medical imaging challenges.
- Many reviews have demonstrated that transfer learning has been an effective approach of reducing reliance on extensive annotated dataset, while maintaining high accuracy.
- At the same time, there are still a lot of obstacles, like noises in label, class imbalance and model interpretability, and the hard work of fighting through these obstacles has been done using sophisticated methods like saliency maps, attention mechanisms, and adding domain knowledge.
- Beyond this, multimodal approaches and supervision learning frameworks are important strategies to improve generalizability as well as clinical applications.
- While there are certainly notable advancements in this area, it is clear that there remains much work to be done, in that all these medical diagnostics can be infused with ethical considerations, more diverse data, and increased transparency or transparency in AI-driven models.

The summary of various deep learning models and their classification is illustrated in Table 2.2.

Table 2.1: Comparison of the existing techniques.

References	Focus area	Contributions	Techniques used	Dataset used	Results	Research gaps
[11]	Survey on medical imaging improvements	Assessed 120 papers; highlighted ResNet as superior	Review/analysis	120 imaging papers	Identified current barriers and pathways	Lack of unified framework and deployment issues
[12]	CAD tools in deep learning	Explored challenges and clinical integration	Deep learning-based CAD	Various CAD tools	Clinical process incorporation emphasized	Clinical workflow adaptation remains a gap
[13]	Transfer learning in CNN	Detailed review on TL applications from 2000 to 2020	CNN and transfer learning	IEEE, Elsevier, Google Scholar, and Web of Science	Comprehensive TL strategies overview	Limited empirical testing on newer datasets
[14]	Artificial neural networks in DLA	Reviewed DLA on various imaging modalities	ANN and DLA	X-ray, CT, mammography, and histopathology	Suggested DLA performance variations	Segmentation/ classification integration lacking
[15]	Medical image super-resolution	Reviewed architectures and upsampling methods	DL for super-resolution	Not specified	Challenges in SR-based imaging outlined	Lacks standardized evaluation datasets
[16]	DL-based image segmentation	Reviewed segmentation types and issues	Deep learning segmentation	Two disease-specific datasets	Summarized segmentation methods	Limited review on hybrid segmentation methods
[17]	Taxonomy of segmentation methods	Grouped DL segmentation methods from general to specialized	Supervised/weakly supervised DL	Multiple sources	Provided group-wise detailed review	Ignored unsupervised methods
[18]	Transfer learning for diagnosis	Reviewed TL for classification, segmentation, and severity scoring	AlexNet, ResNet, VGGNet, and GoogleNet	Various medical datasets	Enhanced performance with TL	Need for TL customization for small datasets

(continued)

Table 2.1 (continued)

References	Focus area	Contributions	Techniques used	Dataset used	Results	Research gaps
[19]	Multimodal image fusion	Emphasized supervised DL in fusion	Supervised DL fusion	Multimodal medical images	Improved diagnosis via fusion	Early fusion lacks automation research
[20]	Semi-self-supervised learning	Applied SSL to four key medical imaging tasks	Transformer-based SSL	Multiple imaging tasks	Studied scalability and deployment	Few clinical trial validations
[21]	Clinical AI decision support	Addressed practical issues in AI deployment	CNNs, radiomics, and FL	Radiomics datasets	Frameworks for real deployment	Scalability and annotation quality issues
[22]	Explainable AI in medical imaging	Reviewed XAI and clinical deployment strategies	XAI + deep learning	XAI-focused imaging datasets	Proposed design process for XAI systems	Real-time explainability models underexplored
[23]	DL + intelligent imaging	Early disease detection using large data	DL for classification/ segmentation	Large medical image collections	Integrated data and imaging for diagnosis	Interpretability and real-time constraints
[24]	3D CNNs in medical imaging	Reviewed 3D CNN applications and preprocessing	3D CNN	3D imaging datasets	Detailed 3D task review	Optimization strategies for 3D data lacking
[25]	XAI with DL	Organized 200 papers by XAI role in imaging	XAI + DL	XAI-related datasets	Summarized role of explainability in DL	Model robustness under label noise not covered
[26]	Label noise in medical imaging	Reviewed DL methods for noisy labels	DL + label noise	Three medical datasets with label noise	Tested multiple noise mitigation methods	Needs more robust label handling strategies

[27]	Noise mitigation techniques	Introduced simple methods to reduce label noise effects	Lightweight DL models	Not specified	Improved understanding of label noise	Needs large-scale validation
[28]	DL trends and clinical needs	Evaluated deep learning for medical demands and architecture design	Sparse labels, FL, and interpretability	Multiple imaging modalities	Mapped trends to clinical needs	Deployment case studies are limited
[29]	Augmentation in segmentation	Categorized strategies for learning and data improvement	DL + augmentation techniques	CT and MRI images	Inspired trust in augmented clinical models	Clinical acceptance of augmentation techniques is low
[30]	DL-based medical image registration	Summarized registration methods and applications	DL registration models	Lung and brain datasets	Grouped seven method types for registration	Inconsistent benchmarking practices
[31]	Early pneumonia detection	Proposed NASNet (pretrained on ImageNet) for early pneumonia prediction via chest X-rays	NASNet and transfer learning	Chest X-ray images	Lightweight model with good precision, recall, and F_1 score	Limited scalability testing across diverse devices/ datasets
[32]	Pneumonia detection and classification	Preprocessing, segmentation detection with YOLOv3, and classification using SVM and softmax	YOLOv3, Threshold segmentation, SVM, Softmax	Chest X-ray dataset	Accuracy, precision, recall, F_1 score: all 99%	Lack of comparison with other hybrid or transformer-based models

Table 2.2: Summary of deep learning models and their categorization.

Model	Primary task	Special use case
CNN	Classification and segmentation	General feature extraction and anomaly detection
ResNet-50	Classification	Deep architecture for pneumonia detection
InceptionV3	Feature extraction	Multiscale feature detection in chest X-rays
EfficientNet	Classification and transfer learning	Optimized scaling and high-performance detection
GANs	Data augmentation	Synthetic image generation for privacy protection
U-Net	Segmentation	Pixel-level labeling in medical imaging
Transformers	Segmentation and classification	Context-aware modeling and volumetric image analysis

2.3 Proposed Methodology

Deep learning has brought in a great leap forward in the medical field in terms of disease identification, even if it's automated, involving accurate segmentation and disease classification. Deep learning models such as CNNs are efficient at analyzing medical image patterns that are beyond human interpretative ability and enable all of these massive amounts of medical imaging data to be made comprehensible. Deep learning also helps improve accuracy in medical diagnostics, detecting early disease and aiding in designing of a better treatment plan. It outlines the process of preparing medical images data for use with deep learning models in medical images analysis, proposes models' architecture, training strategy, and evaluation methods that aim to enhance the reliability of medical diagnosis.

The flowchart in Figure 2.5 and the algorithm of the implementation of the deep learning model is as described in algorithm 2.1. Then, it sets up the model in different phases, starting with data loading till the getting of the desired result. Firstly, research objectives are set and the terms of the problem statement as well as the desired results are identified at the first step. The first phase outlines the objectives through which the workflow is executed. Then, it gets the set from a credible medical imaging collection, such as a dataset. The data provided should be of sufficient volume and of high enough quality in order to support the effective model development and testing processes. The analysis requires three preprocessing steps: the first one is for normalizing values and scaling pixel intensities, the second one is about reducing images to get an image that fits the model input dimensions, and the third one is creating augmented samples by any method like rotation, flipping, zooming, etc. The phase of data preparation handles noise, data inconsistency, and class imbalances issues in the selected dataset. The model exhib-

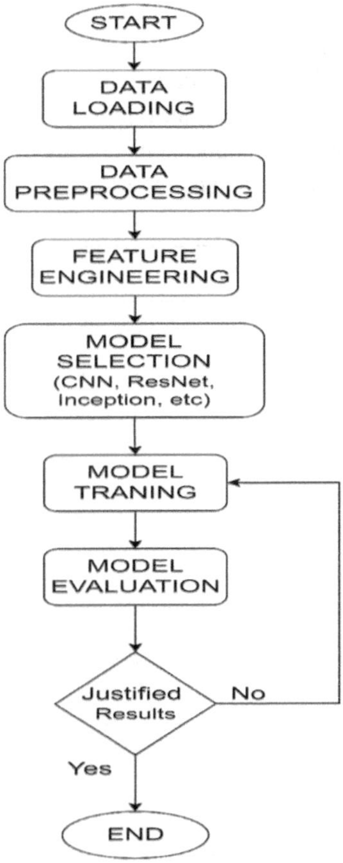

Figure 2.5: Flowchart of the deep learning model implementation.

its enhanced class separation abilities because improvements are implemented to some critical data patterns. A vital extraction process is identification of pertinent medical diagnosis pattern like color, texture, and shapes. The data characteristics and problem need result in the selection of deep learning models. Basically, CNNs are the most popular deep learning model for building the spatial hierarchy. ResNet helps avoid gradient vanishing and Inception is for feature extraction across multiple scales. Backpropagation and gradient descent are used to iteratively optimize this data and obtain results during the model training process, which is done on the preprocessed data. In this phase, these rates are fine tuned in conjunction with batch size and the number of epochs to achieve optimal model performance. The performance is evaluated using a distinct dataset trained model to assess its performance metrics. Accuracy refers to evaluating if a particular system is correct or not, which is combined with precision, i.e., how many true positives or recall, which is sensitivity. In addition, it employs the F_1 score for precision and recall, loss, as to what constitutes effectiveness. They lead to the evaluation of their validity. When performance assessments outcomes are a match to both, what they will result in and what has been set as the objective, then process improvements can be made. When objectives are not met, the procedure performs their essential modifications on the preprocessing, model selection, and training. This workflow culminates with detailed documentation of the methods used and a summary of results, with an outline of how research deployment can be strategically used toward practical applications. During this workflow phase, the findings must be structured in order to implement and replicate.

Algorithm 2.1: Algorithm for the deep learning model implementation.

```
START
// Step 1: Preprocess Images
IF medical_images_available THEN
   FOR each image IN dataset DO
      Resized = Resize(image, W, H)
      Normalized = Resized / 255
      Augmented = Augment(Normalized)
      Add to preprocessed_images
   END FOR
ELSE
   RETURN "Error: No images available"

// Step 2: Feature Extraction
IF preprocessed_images NOT NULL THEN
   FOR each image IN preprocessed_images DO
      Features = CNN_ExtractFeatures(image)
      Activated = ReLU(Features)
      Add to features_list
   END FOR
ELSE
   RETURN "Error: No preprocessed images"

// Step 3: Train Model
IF training_data NOT NULL THEN
   Train model (e.g., EfficientNet, ResNet)
   WHILE not converged DO
      Loss = CrossEntropy(predicted, actual)
      Weights = Update(Weights, Loss, Adam)
   END WHILE
ELSE
   RETURN "Error: No training data"

// Step 4: Validate and Evaluate
IF validation_data NOT NULL THEN
   Accuracy = (TP + TN) / Total
   Precision = TP / (TP + FP)
   Recall = TP / (TP + FN)
   F1 = 2 * (Precision * Recall) / (Precision + Recall)
```

```
  Confusion_Matrix = Generate(validation_data)
ELSE
  RETURN "Error: No validation data"

// Step 5: Test and Analyze
IF test_images NOT NULL THEN
  FOR each image IN test_images DO
    Class, Confidence = Classify(image, model)
    Mark image as "Disease Detected" or "Healthy"
  END FOR
  Display Results
ELSE
  RETURN "Error: No test images"

// Step 6: Result Justification
IF results_justified THEN
  Save Model and Report
ELSE
  Perform Retraining or Hyperparameter Tuning
END
```

2.3.1 Proposed Mathematical Model

The structure of the EfficientNet model is described to show how it adjusts itself for various scales. Anyone looking to form a clearer picture of the workings of this model ought to take the equations into account, with an eye on convolutional techniques, squeeze-and-excitation features, and compound scaling. The complete expression allows us to easily repeat the model and comprehend EfficientNet's superior performance in organizing medical pictures.

EfficientNet uses compound scaling to make its model architecture efficient across width, depth, and image resolution.

It starts by choosing the image parameters and scaling the whole image in the same proportions with the compound method.

- Input Representation

Let $x \in R^{H \times W \times C}$ represent a medical image:

- H is the image height
- W is the image width
- C is the number of channels (e.g., Grayscale $C = 1$ *and RGB* $C = 3$)

EfficientNet creates a fixed size by scaling the image with the compound scaling rule:

$$W' = \alpha^k H, \beta^k W \quad \text{for } k \geq 0 \tag{2.1}$$

where
- α, β are scaling factors for resolution
- k is the scaling depth

- Compound Scaling
  EfficientNet uses a compound scaling formula to balance three dimensions: depth (d), width (w), and resolution (r):

$$d = \alpha^\phi, \quad w = \beta^\phi, \quad r = \gamma^\phi \tag{2.2}$$

where
- ϕ is the global scaling factor.
- α, β, γ are constants satisfying: $\alpha \cdot \beta^2 \cdot \gamma^2 \approx 2$

- Mobile Inverted Bottleneck Convolution (MBConv)
  The core of EfficientNet is built on MBConv. How the system is built and how it identifies and shrinks significant details are described in these equations.
  - Expansion Phase:
    The input feature map $z_{l-1} \in R^{H_l \times W_l \times C_l}$ is expanded to a higher-dimensional space:

$$z_l^{\text{expand}} = \sigma(W_e {}^* z_{l-1}) \tag{2.3}$$

    where
    - $W_e \in R^{1 \times 1 \times C_l \times k_e \cdot C_l}$ are expansion weights.
    - k_e is the expansion ratio
    - σ is the nonlinear activation (e.g., Swish: $\sigma(x) = x/(1 + e^{-x})$).
  - Depthwise Convolution Phase:
    A depthwise convolution is applied to the expanded feature map:

$$z_l^{\text{depthwise}} = \sigma\left(W_d {}^* z_l^{\text{expand}}\right) \tag{2.4}$$

    where
    - $W_d \in R^{k \times k \times 1 \times C_l \cdot k_e}$ is the depthwise convolution kernel
    - k is the kernel size
  - Squeeze-and-Excitation (SE) Optimization:
    Channel-wise attention is applied via SE blocks. The feature map is squeezed to a single channel via global average pooling:

$$s_c = \frac{1}{H_l \cdot W_l} \sum_{i=1}^{H_l} \sum_{j=1}^{W_l} z_l^{\text{depthwise}}[i, j, c] \tag{2.5}$$

Then, excitation is performed:

$$e_c = \sigma(W_2 \cdot \text{ReLU}(W_1 \cdot s_c)) \tag{2.6}$$

where

  - $W_1 \in R^{C_l/r \times C_l}$, $W_2 \in R^{C_l \times C_l/r}$ are fully connected layers.
  - r is the reduction ratio.
- The SE-scaled feature map is:

$$z_l^{\text{SE}}[i, j, c] = e_c \cdot z_l^{\text{depthwise}}[i, j, c] \tag{2.7}$$

- Projection Phase:
  The high-dimensional feature map is projected back to its original dimension:

$$z_l = W_p {}^* z_l^{\text{SE}} \tag{2.8}$$

where

  - $W_p \in R^{1 \times 1 \times (k_e \cdot C_l) \times C_l}$ are projection weights
- Residual Connection (if applicable):
  If the input and output dimensions match, a skip connection is applied:

$$z_l = z_{l-1} + z_l \tag{2.9}$$

- Layer Stacking
  EfficientNet consists of multiple MBConv blocks stacked across different stages. Let L_s denote the number of layers in stage s. The input to stage s is:

$$z_s = \text{MBConv}_{s, L_s}(\ldots(\text{MBConv}_{s,1}(z_{s-1}))) \tag{2.10}$$

Each stage reduces the spatial resolution by a factor of 2 using strided convolutions:

$$H_{s+1} = \frac{H_s}{2}, \quad W_{s+1} = \frac{W_s}{2} \tag{2.11}$$

The number of filters at stage sss is scaled according to the width multiplier w:

$$C_{s+1} = w \cdot C_s \tag{2.12}$$

- Classification Head
  The final feature map $z_L \in R^{H_L \times W_L \times C_L}$ is processed by the classification head:
  - Global Average Pooling (GAP):
    Reduces the spatial dimensions to a single vector:

$$z^{\mathrm{GAP}} = \frac{1}{H_L \cdot W_L} \sum_{i=1}^{H_L} \sum_{j=1}^{W_L} z_L[i,j,:] \tag{2.13}$$

- Fully connected layer:
  Projections of the GAP vector to the class probabilities:

$$\hat{y} = \mathrm{Softmax}\left(W_c \cdot z^{\mathrm{GAP}} + b_c\right) \tag{2.14}$$

where
- $W_c \in R^{K \times C_L}$ are classification weights.
- is the bias vector.

The softmax function computes probabilities:

$$\widehat{y_k} = \frac{e^{z_k}}{\sum_{j=1}^{K} e^{z_j}}, \quad k = 1, 2, \ldots, K \tag{2.15}$$

- Loss Function
  To measure the classification error, categorical cross-entropy is used, as shown below.
  For multi-class classification, the categorical cross-entropy loss is used:

$$\mathcal{L} = -\frac{1}{N} \sum_{i=1}^{N} \sum_{k=1}^{K} y_{i,k} \log(\widehat{y}_{i,k}) \tag{2.16}$$

where $y_{i,k}$ is the true label and $\widehat{y_i}, {}_k$ is the predicted probability.

- Scaling for Medical Image Processing
  These scaling parameters are especially critical in medical imaging, where high resolution is needed to capture subtle diagnostic patterns
  For medical image processing, EfficientNet's compound scaling is optimized for:
  - Higher resolution (r): To capture fine details in medical images, like lesions or abnormalities.
  - Deeper networks (d): To handle complex spatial patterns.
  - Wider networks (w): To process larger feature maps.

By balancing r, d, w according to the scaling formula, EfficientNet achieves state-of-the-art performance while being computationally efficient.

When deep learning models are applied to medical images, broadly adopted optimization concepts are vital for making learning both accurate and efficient. Extraction of features depends on convolution, helping the network notice various edges, textures, and parts of an image in medical sciences. Then, activation functions such as the ReLU are put in to make the model learn more difficult patterns. The Swish activation, which is especially useful in deeper networks, is the function used in efficient architectures. When max pooling is applied, the size of the feature map decreases, which helps generalize the features and prevents overfitting. The Adam optimizer is frequently applied to maximize the results of the model. Based on the first and second moments of gradients, it decides on appropriate learning rates and embeds the helpful features of momentum and RMSProp ideas in one way. To improve accuracy during training, the categorical cross-entropy loss function calculates and measures mistakes made on multi-class classification issues. By using these techniques, deep learning models become firmer, usable in new cases, and able to predict diagnoses well from complex scans.

2.4 Results

This experimental study aims to investigate the applicability of contemporary deep learning techniques for processing medical images, specifically to identify pneumonia in chest X-rays, utilizing an open-source database accessible on Kaggle. Further details of the dataset are explained in Section 2.4.1, i.e., dataset description. To address class imbalance, the proposed method utilized CNNs, in conjunction with pretrained models such as fine-tuned CNN, ResNet-50, EfficientNet, and InceptionV3. The preprocessing involved normalization, resizing, and data augmentation, all of which were executed to accomplish this task. The evaluated models demonstrated a notable enhancement in accuracy, precision, recall, and F_1 score, as seen in Section 2.4.2. The study's findings in Section 2.4.3 indicate that deep learning can enhance diagnostic outcomes while minimizing reliance on human assessment.

2.4.1 Dataset Description

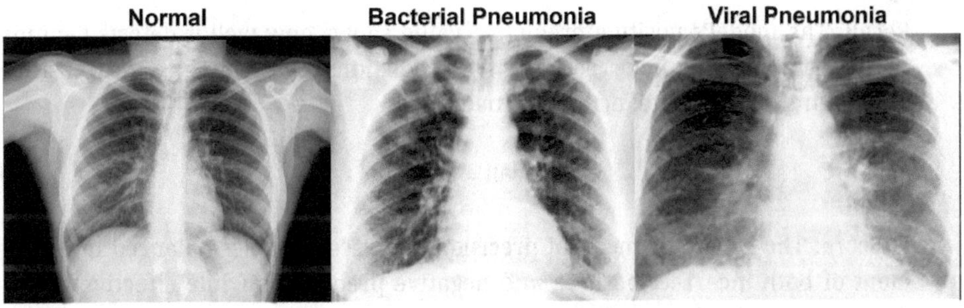

Figure 2.6: Sample chest X-ray images (normal, bacterial pneumonia, and viral pneumonia) obtained from a publicly available Kaggle dataset (Guangzhou Women and Children's Medical Center), used for research illustration purposes only.

Figure 2.6 shows the sample images from the dataset. The normal chest X-ray (left panel) depicts clear lungs without any areas of abnormal opacification in the image.

Bacterial pneumonia (middle) typically exhibits a focal lobar consolidation, in this case in the right upper lobe (white arrows), whereas viral pneumonia (right) manifests with a more diffuse "interstitial" pattern in both lungs. There are 5,863 X-ray images (JPEG) and 2 categories (pneumonia/normal). Chest X-ray images (anterior-posterior) were selected from retrospective cohorts of pediatric patients of one to five years old from Guangzhou Women and Children's Medical Center, Guangzhou. All chest X-ray imaging were performed as part of patients' routine clinical care.

This study made use of the "Chest X-Ray Images (Pneumonia)" dataset from Kaggle, which was initially gathered by Guangzhou Women and Children's Medical Centre. This collection, which includes 5,863 children's anterior-posterior chest X-rays, classified as either normal or pneumonia, has provided support for numerous research on pneumonia.

2.4.2 Metrics

The model is evaluated using metrics such as:

– Accuracy: A measure of model correctness, which calculates accurate positive and negative predictions in proportion to the total number of instances:

$$\text{Accuracy} = \frac{\text{TP} + \text{TN}}{\text{TP} + \text{TN} + \text{FP} + \text{FN}} \tag{2.17}$$

– Precision: The positive prediction rate measures actual correct positives among all predicted positive instances to minimize errors:

$$\text{Precision} = \frac{\text{TP}}{\text{TP} + \text{FP}} \tag{2.18}$$

– Recall: The model's positive predictive value shows how well it detects genuine cases while avoiding incorrect negative predictions, by dividing correct positive predictions over the total actual positives:

$$\text{Recall} = \frac{\text{TP}}{\text{TP} + \text{FN}} \tag{2.19}$$

– F_1 Score: The harmonic mean of precision and recall enables balanced measurement of both incorrect positive and negative predictions while effectively handling datasets that have imbalanced classes:

$$F_1 = 2 \cdot \frac{\text{Precision} \cdot \text{Recall}}{\text{Precision} + \text{Recall}} \tag{2.20}$$

When there are significantly more instances of one class in the data than another, benchmark evaluations for classifiers frequently use AUC-ROC in addition to Accuracy, Precision, Recall, and F_1 score to assess performance. This statistic may be utilized in the future to better understand the model.

2.4.3 Discussion

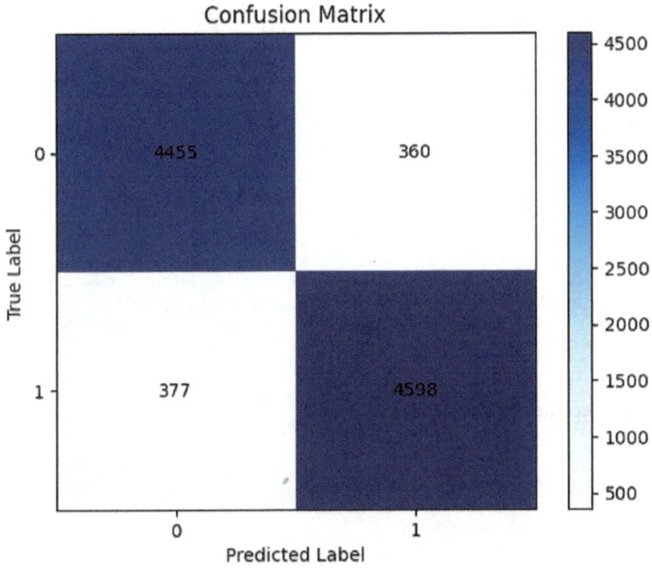

Figure 2.7: Confusion matrix of the proposed framework.

Figure 2.7 depicts the confusion matrix of proposed framework. It shows the number of false negatives, true positives, false positives, and true negatives. The confusion matrix helps to verify the number of actual disasters correctly predicted as a disease (true positive), as compared to those that were not identified as diseases (false negative).

As shown in Figure 2.8, it compares how accurate the different models were. The accuracy of the CNN model is 87.60% and the accuracy of ResNet-50 model is 74.60%. Inception ends up with an accuracy of 89.90% and EfficientNet, with an accuracy of 98.40%. The accuracy reported by Manickam et al. [33] is 93.06%, while Hasan et al. [34] achieved an accuracy of 96.20%. Zein et al. [35] achieved an accuracy of 96.70%.

Figure 2.9 shows a graphical depiction of the various models' levels of precision, with the comparison, and it allows one to compare the level of precision at which all the models operate. With respect to precision, the ResNet-50 model obtains a precision of 72.70%, while the CNN model reaches a precision of 83.80%. They achieve a performance of 90.50% in terms of their precision, compared to 99.10% that Inception is able to achieve. The precision reported by Manickam et al. [33] is 88.97%, while Hasan et al. [34] demonstrate a precision of 97.70%. Zein et al. [35] demonstrate a precision of 99.90%.

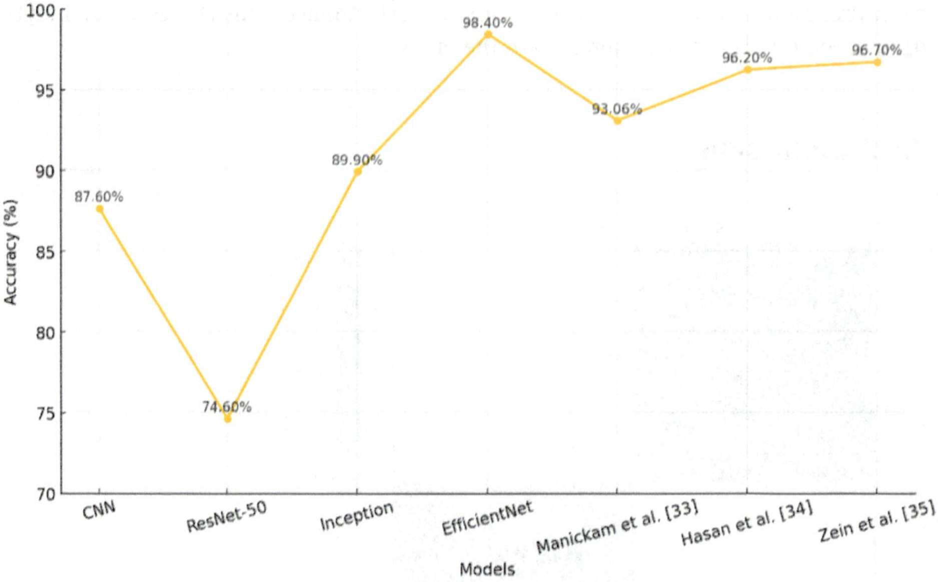

Figure 2.8: Comparison graph of the accuracies of the model.

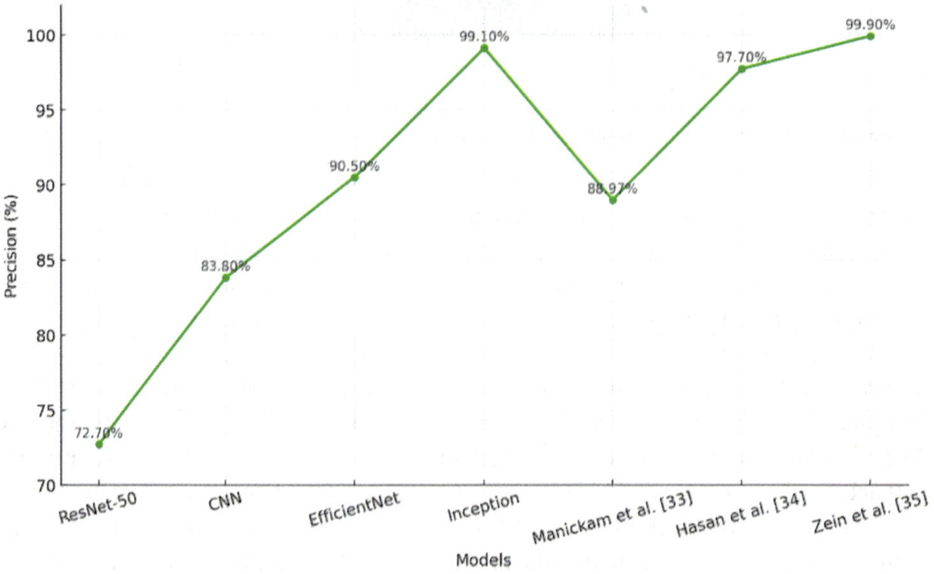

Figure 2.9: Precision comparison graph among models.

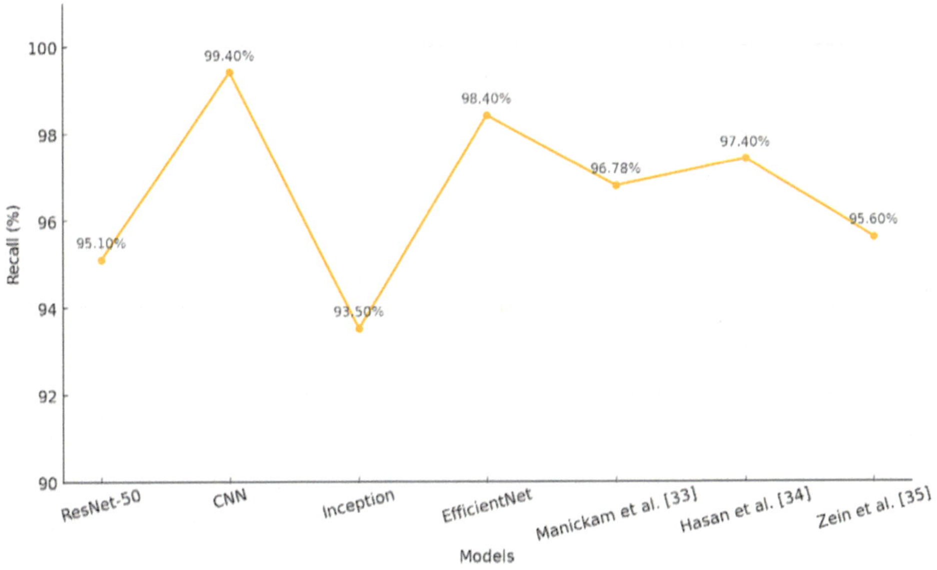

Figure 2.10: Comparison graph of the recall value.

In Figure 2.10, the graphical representation of the comparison of different degrees of recall is presented with respect to the different models. Recall of the ResNet-50 model is 95.10%, while that of the CNN model is 99.40%. The differences are also compared between the two models. Recall wise, when comparing Inception with EfficientNet, Inception comes up with a recall of 93.50% and EfficientNet achieves a recall of 98.40%. The Recall linked to the Manickam et al. [33] is 96.78%, while Hasan et al. [34] report a recall of 97.40%. Zein et al. [35] achieved a recall of 95.60%.

The different degrees of F_1 score on the different models are represented graphically and shown in Figure 2.11. The F_1 score for ResNet 50 model was 76.90% and for CNN model was 90.70%. F_1 scores of 91.10% and 98.70% were achieved by Inception and EfficientNet, respectively, as illustrated in Table 2.3. The F_1 score for Manickam et al. [33] is 92.71%, while Hasan et al. [34] achieved an F_1 score of 97.30%. The study by Zein et al. [35] reports an F_1 score of 97.70%.

The results of CNN, ResNet50, Inception, and EfficientNet models, which are the models used to obtain the results with the description of the metrics (Accuracy, Precision, Recall, and F_1 score) are shown in the loaded data (Table 2.4).

The results present the comparative results of the previously existing models. The metric values (accuracy, precision, recall, and F_1 score) organized in rows, with each column indicating the corresponding value of the model for that specific metric (Table 2.4).

The performance of the suggested models on various datasets, in comparison to other industry-leading methods, is shown by the figures in Table 2.4. Chest X-rays were analyzed by researchers in the field, such as Manickam et al. and Hasan et al.,

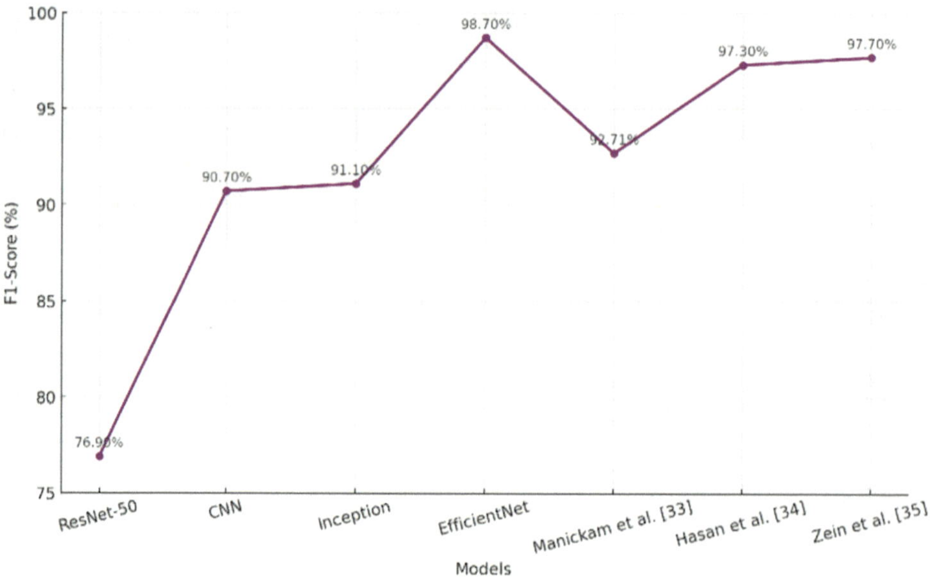

Figure 2.11: Comparison graph of the F_1 score.

Table 2.3: The metric values of the models.

Metric	CNN (%)	ResNet-50 (%)	Inception (%)	EfficientNet (%)
Accuracy	87.60	74.60	89.90	98.40
Precision	83.80	72.70	90.50	99.10
Recall	99.40	95.10	93.50	98.40
F_1 score	90.70	76.90	91.10	98.70

Table 2.4: Comparison of existing models.

Metric	Manickam et al. [33] (%)	Hasan et al. [34] (%)	Zein et al. [35] (%)
Accuracy	93.06	96.20	96.70
Precision	88.97	97.70	99.90
Recall	96.78	97.40	95.60
F_1 score	92.71	97.30	97.70

utilizing a variety of file formats that included both adult and differentially acquired patients. However, when compared to alternative approaches, our research with EfficientNet on pediatric data shows superior or comparable outcomes in accuracy, precision, recall, and F_1 score. It is evident that EfficientNet and the other models included in this study have strong generalization abilities and are adept at analyzing various sets of X-ray pictures because the models perform well across all datasets.

Grad-CAM and saliency maps are used to illustrate the problems with deep learning models to healthcare professionals. They demonstrate which aspects of the original X-ray image have the greatest impact on the model's conclusion. Figure 2.12 illustrates how the model accurately identifies lung opacity regions with the use of Grad-CAM. Radiologists now use these technologies to identify biases or defects in the AI model and to accept AI suggestions. By doing this, support workers can employ AI in healthcare processes and models become more accountable.

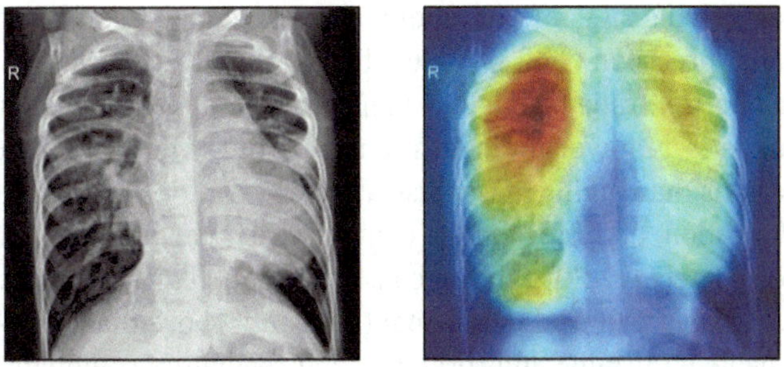

(a) Original Image **(b)** Grad-CAM 'Chest'

Figure 2.12: Grad-CAM visualization for pneumonia detection showing (a) an original chest X-ray image from a publicly available open-access Kaggle dataset (Guangzhou Women and Children's Medical Center) and (b) the corresponding Grad-CAM heatmap highlighting regions contributing to the model's prediction. The visualization was generated by the authors using the trained model for explanatory analysis.

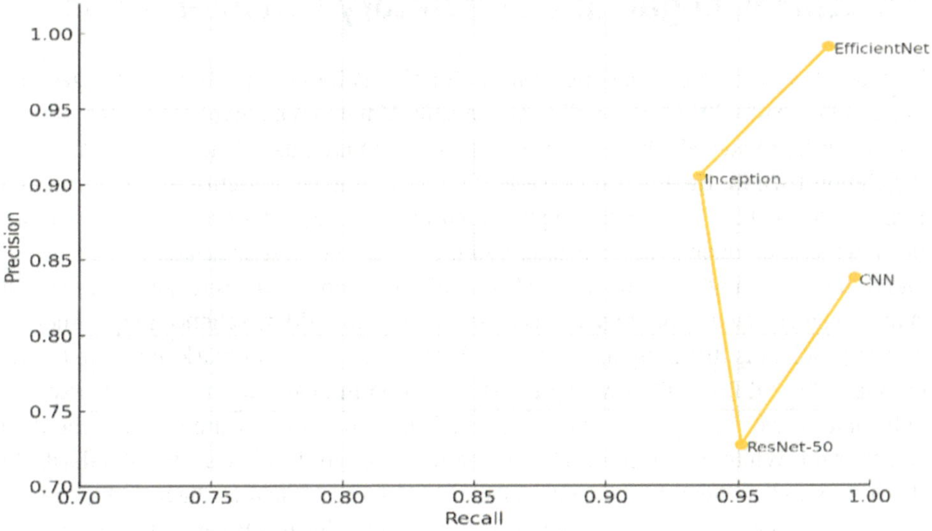

Figure 2.13: Precision versus recall trade-off among models.

The choice between higher precision and recall is an important factor in assessing deep learning models for interpreting medical images. Figure 2.13 makes clear that precision means the model recognizes only relevant cases (i.e., reduces false positives) and recall means it recognizes all actual cases (i.e., avoids mistakes by omitting any actual cases). Because missing an actual case of pneumonia can be very serious, clinical settings put a high priority on recall. If an excessive amount of attention is given to recall, it can result in more false positives, drop the precision, and possibly cause extra testing or treatment. Another way to look at it is that models that give the best precision might eliminate some true cases of pneumonia, creating dangers of their own. Both the recall (98.4%) and precision (99.1%) values showed that EfficientNet has a strong balance. Still, balancing the trade-off plays a major role in picking a model and choosing a threshold, since in real-world diagnostic use, wrongly not finding a critical issue is often more serious than giving more false alarms.

While CNN, ResNet-50, Inception, and EfficientNet exhibit varied performance on chest X-ray data for pneumonia detection, EfficientNet consistently outperforms others due to its optimized scaling. However, clinical deployment demands more than accuracy – it requires models to be generalizable, interpretable, and regulatory-compliant. Recent trends such as Vision Transformers, federated learning, multimodal fusion, and XAI are reshaping medical imaging by improving transparency and cross-institutional usability. Yet, challenges persist, including dataset bias, class imbalance, system integration barriers, and unclear accountability in AI decisions.

2.5 Ethical, Legal, and Regulatory Considerations

Bringing AI into medical imaging creates ethical, legal and regulatory challenges that need to be dealt with to make clinical use safe. Major issues involve keeping patient information private, which means strict adherence to laws like General Data Protection Regulation (GDPR), Health Insurance Portability and Accountability Act (HIPAA), and India's DPDPA 2023. Such risks are prevented in this study by processing data that has been made anonymous and is available to the public. Because misclassifications in AI-aided diagnosis might occur, medical practitioners should not rely solely on AI and should instead review the model's decisions with Grad-CAM or salience maps. Another ethical problem is that systems learned on limited data may not work the same for everyone. FDA and EU MDR now require AI tools used in healthcare to be easily explainable, able to work properly under most conditions and able to show where the data comes from. While this framework is being investigated, it follows such standards. In the future, people need to trust physicians for AI to be trusted in diagnostics, so honesty, proper consent, and following medical, legal and ethical rules are vital.

2.6 Conclusion

This chapter presents a deep learning-based approach for pneumonia detection in chest X-rays using multiple architectures, including CNN, ResNet-50, Inception, and EfficientNet. As a result of an optimized architecture, EfficientNet achieved the best performance with 98.4% accuracy and 98.1% precision. CNN was simpler (88.9%, 90.5% F_1 score) and did well (89.9% accuracy, 90.5% precision), though Inception was also useful in that regards. High F_1 score (98.4%) was maintained as well by EfficientNet although it has higher overall metrics. The findings indicate that medical image analysis with newer architectures like EfficientNet is a good idea, and it is much dependent on clinical needs. Their high accuracy implies that the top performers may be good clinical diagnostic support tools. This chapter also describes several avenues for future work. It includes improving model interpretability via attention mechanisms or explainable AI, and widening the models to allow multi class classification of pneumonia types, beyond viral pneumonia and bacterial pneumonia. It is also important to increase such diversity in the dataset to include additional patient populations and imaging conditions. Last, these models should be combined with clinical workflows and extended to multimodal model analysis, including other patient data, to enable a more comprehensive diagnostic pipeline. Overall, this work proves the feasibility of deep learning, namely, the use of EfficientNet, for the detection of pneumonia and provides the groundwork for future development of AI-based diagnostic tools to assist healthcare.

References

[1] Abdou MA. Literature review: Efficient deep neural networks techniques for medical image analysis. Neural Computing and Applications. 2022 Apr;34(8):5791–5812.

[2] Sharma S, Guleria K. A systematic literature review on deep learning approaches for pneumonia detection using chest X-ray images. Multimedia Tools and Applications. 2024 Mar;83(8):24101–24151.

[3] Singhal A, Phogat M, Kumar D, Kumar A, Dahiya M, Shrivastava VK. Study of deep learning techniques for medical image analysis: A review. In Materials Today: Proceedings 2022 Jan 1 (vol. 56, pp. 209–214).

[4] Siddiqi R, Javaid S. Deep learning for pneumonia detection in chest x-ray images: A comprehensive survey. Journal of Imaging. 2024 Jul 23;10(8):176.

[5] Khan SS, Rupak AU, Faieaz WW, Jannat S, Prova NN, Gupta AK. Advances in medical imaging: Deep learning strategies for pneumonia identification in chest x-rays. In 2024 15th International Conference on Computing Communication and Networking Technologies (ICCCNT) 2024 Jun 24 (pp. 1–7). IEEE.

[6] Shilpa N, Banu WA, Metre PB. Revolutionizing Pneumonia Diagnosis: AI-Driven Deep Learning Framework for Automated Detection from Chest X-Rays. IEEE Access. 2024 Nov 15.

[7] Rguibi Z, Hajami A, Zitouni D, Elqaraoui A, Bedraoui A. Cxai: Explaining convolutional neural networks for medical imaging diagnostic. Electronics. 2022 Jun 2;11(11):1775.

[8] Usman C, Rehman SU, Ali A, Khan AM, Ahmad B. Pneumonia Disease Detection Using Chest X-Rays and Machine Learning. Algorithms. 2025 Feb 3;18(2):82.
[9] Shazia A, Xuan TZ, Chuah JH, Usman J, Qian P, Lai KW. A comparative study of multiple neural network for detection of COVID-19 on chest X-ray. EURASIP Journal on Advances in Signal Processing. 2021 Jul 27;2021(1):50.
[10] Liu W, Luo J, Yang Y, Wang W, Deng J, Yu L. Automatic lung segmentation in chest X-ray images using improved U-Net. Scientific Reports. 2022 May 23;12(1):8649.
[11] Suganyadevi S, Seethalakshmi V, Balasamy K. A review on deep learning in medical image analysis. International Journal of Multimedia Information Retrieval. 2022 Mar;11(1):19–38.
[12] Chan HP, Samala RK, Hadjiiski LM, Zhou C. Deep learning in medical image analysis. Deep Learning in Medical Image Analysis: Challenges and Applications. 2020 Feb;7:3–21.
[13] Wang J, Liu W, Luo J, Yang Y, et al. Automatic lung segmentation in chest X-ray images using improved U-Net. Scientific Reports. 2022;12:8649. https://doi.org/10.1038/s41598-022-12743-y.
[14] Zhu H, Wang SH, Zhang YD. A review of deep learning on medical image analysis. Mobile Networks and Applications. 2021 Feb;26(1):351–380.
[15] Puttagunta M, Ravi S. Medical image analysis based on deep learning approach. Multimedia Tools and Applications. 2021 Jul;80(16):24365–24398.
[16] Li Y, Sixou B, Peyrin F. A review of the deep learning methods for medical images super resolution problems. Irbm. 2021 Apr 1;42(2):120–133.
[17] Liu X, Song L, Liu S, Zhang Y. A review of deep-learning-based medical image segmentation methods. Sustainability. 2021 Jan 25;13(3):1224.
[18] Wang R, Lei T, Cui R, Zhang B, Meng H, Nandi AK. Medical image segmentation using deep learning: A survey. IET Image Processing. 2022 Apr;16(5):1243–1267.
[19] Kora P, Ooi CP, Faust O, Raghavendra U, Gudigar A, Chan WY, Meenakshi K, Swaraja K, Plawiak P, Acharya UR. Transfer learning techniques for medical image analysis: A review. Biocybernetics and Biomedical Engineering. 2022 Jan 1;42(1):79–107.
[20] Li Y, Zhao J, Lv Z, Li J. Medical image fusion method by deep learning. International Journal of Cognitive Computing in Engineering. 2021 Jun 1;2:21–29.
[21] Chen X, Wang X, Zhang K, Fung KM, Thai TC, Moore K, Mannel RS, Liu H, Zheng B, Qiu Y. Recent advances and clinical applications of deep learning in medical image analysis. Medical Image Analysis. 2022 Jul 1;79:102444.
[22] Castiglioni I, Rundo L, Codari M, Di Leo G, Salvatore C, Interlenghi M, Gallivanone F, Cozzi A, D'Amico NC, Sardanelli F. AI applications to medical images: From machine learning to deep learning. Physica Medica. 2021 Mar 1;83:9–24.
[23] Singh A, Sengupta S, Lakshminarayanan V. Explainable deep learning models in medical image analysis. Journal of Imaging. 2020 Jun 20;6(6):52.
[24] Cai L, Gao J, Zhao D. A review of the application of deep learning in medical image classification and segmentation. Annals of Translational Medicine. 2020 Jun;8(11):713.
[25] Singh SP, Wang L, Gupta S, Goli H, Padmanabhan P, Gulyás B. 3D deep learning on medical images: A review. Sensors. 2020 Sep 7;20(18):5097.
[26] Van der Velden BH, Kuijf HJ, Gilhuijs KG, Viergever MA. Explainable artificial intelligence (XAI) in deep learning-based medical image analysis. Medical Image Analysis. 2022 Jul 1;79:102470.
[27] Karimi D, Dou H, Warfield SK, Gholipour A. Deep learning with noisy labels: Exploring techniques and remedies in medical image analysis. Medical Image Analysis. 2020 Oct 1;65:101759. Xie X, Niu J, Liu X, Chen Z, Tang S, Yu S. A survey on incorporating domain knowledge into deep learning for medical image analysis. Medical Image Analysis. 2021 Apr 1;69:101985.

[28] Zhou SK, Greenspan H, Davatzikos C, Duncan JS, Van Ginneken B, Madabhushi A, Prince JL, Rueckert D, Summers RM. A review of deep learning in medical imaging: Imaging traits, technology trends, case studies with progress highlights, and future promises. Proceedings of the IEEE. 2021;109 (5):820–838.
[29] Chlap P, Min H, Vandenberg N, Dowling J, Holloway L, Haworth A. A review of medical image data augmentation techniques for deep learning applications. Journal of Medical Imaging and Radiation Oncology. 2021 Aug;65(5):545–563.
[30] Fu Y, Lei Y, Wang T, Curran WJ, Liu T, Yang X. Deep learning in medical image registration: A review. Physics in Medicine & Biology. 2020 Oct 7;65(20):20TR01.
[31] Nalluri S, Sasikala R. A deep neural architecture for SOTA pneumonia detection from chest X-rays. International Journal of System Assurance Engineering and Management. 2024 Jan;15(1):489–502.
[32] Tajbakhsh N, Jeyaseelan L, Li Q, Chiang JN, Wu Z, Ding X. Embracing imperfect datasets: A review of deep learning solutions for medical image segmentation. Medical Image Analysis. 2020 Jul 1;63:101693.
[33] Manickam A, Jiang J, Zhou Y, Sagar A, Soundrapandiyan R, Samuel RD. Automated pneumonia detection on chest X-ray images: A deep learning approach with different optimizers and transfer learning architectures. Measurement. 2021 Nov 1;184:109953.
[34] Hasan MM, Kabir MM, Haque MR, Ahmed M. A combined approach using image processing and deep learning to detect pneumonia from chest X-ray image. In 2019 3rd international conference on electrical Computer & Telecommunication Engineering (ICECTE) 2019 Dec 26 (pp. 89–92). IEEE.
[35] El Zein OM, Soliman MM, Elkholy AK, Ghali NI. Transfer learning based model for pneumonia detection in chest X-ray images. International Journal of Intelligent Engineering and Systems. 2021 Sep 1;14(5):56–66.

Chander Prabha, Shalini Kumari

3 Data Handling, Data Representation, and Preprocessing for Medical Image Processing

Abstract: Medical images in any form viz., X-ray, CT-scan, or MRI scans play an important role in assessing any medical condition. These form part of the data. This chapter deals with the data handling, its representation, and further preprocessing to make it suitable for any medical diagnosis. Several techniques are available for processing medical images. This chapter explores such techniques and their transition from traditional machine learning to deep learning (DL) techniques. The DL methods including convolutional neural networks, recurrent neural networks, generative adversarial networks, and long short-term memory networks are covered in this chapter. The chapter also addresses DL concerns for medical image analysis, particular data annotation issues, and transfer learning and data augmentation methods.

Keywords: machine learning, deep learning, artificial intelligence, computer-aided design

3.1 Introduction

Medical imaging is crucial for the noninvasive detection, diagnosis, monitoring, and therapy assessment of various medical disorders. Various ways are used to visualize biological processes. Imaging techniques are categorized into anatomical or structural modalities and functional modalities. Across medical imaging, these two specialties are computed tomography (CT) and positron emission tomography (PET). Most image interpretations in the medical field rely on competent professionals. Image interpretation can be tedious, sometimes even expensive, time-consuming, and involve human error, necessitating a move toward computer-assisted therapies for such specialists. ML is a set of algorithms that, when given input data, automatically detects patterns based on that data, using the patterns to project future data or make decisions under uncertainty [1]. In a broad sense, this must be taken as a subset of artificial intelligence (AI). They can also include some approaches based on genetic algorithms. There are three main AI approaches: symbolic (rule-based, like IBM Watson), connectionism (network-based, like deep learning (DL) or artificial neural net), and Bayesian (based on the Bayesian theorem). ML techniques rely on data and how little input is needed to enable the machine to make decisions. In simple terms, a machine learns on the job, i.e., it uses the training dataset to make predictions concerning new inputs. ML and image processing techniques have become widespread in medical imaging di-

 | https://doi.org/10.1515/9783112226223-003

agnosis and are integrated to assist in image interpretation. Experienced professionals require significant time to analyze medical imaging data. Human interpretation remains complex because professional radiologists with multiple years of practice still make errors with image assessments. Medical professionals use computer-aided design (CAD) systems that help them understand medical images better. Complex ML techniques underpin these systems [2]. DL falls under ML and resembles the artificial neural network (ANN), modeled after a multilayered human cognition system. The growing interest in DL stems from its application on very large datasets of healthcare information. ANN emerged in the 1950s and then could not solve real-world problems due to problems of vanishing gradient and overfitting, limited computer processing power, and a dearth of data on which to train a deep architecture. The recently developed capabilities of big data, processing power enabled by GPUs, and knowledge of training deep neural networks have lifted many restraints. DL approaches have become highly successful in mimicking human behavior in different fields, including medical imaging.

Common duties performed by radiologists include recognizing structural abnormalities to ascertain whether they are a manifestation of a disease. From the 1980s onwards, ML methods have been widely used to perform classification tasks, offering an approach, methodology, and theoretical underpinning. Beginning in the early 2000s, many CAD technologies were put into clinical use. However, clinical trials have reported negative effects of these systems [3]. CAD systems generated more false positives than human counterparts, leading to extended evaluation times and biopsies. Thus, the net advantage afforded by CAD remained uncertain. DL technology is likely to improve CAD systems, increase accuracy in detection, and allow humans to offload repetitive radiological work to AI. DL is one of the best ways to get helpful knowledge from large amounts of medical data. This would allow detection of the lesions, diagnostic suggestions, and initial report writing. The global company IBM is developing radiological applications. It is also used for automatic lesion recognition and quantitative feature analysis of medical imaging, among other things.

The research field of medical image processing creates interpretation systems through algorithm development for medical images [4]. Medical image processing creates diagnostic tools from images needed in medicine for healthcare assessment purposes and for planning therapy and delivering treatments [5]. The medical imaging process includes segmentation, registration feature extraction, classification, and visualization stages. Healthcare imaging processing methods function to extract essential data from medical images for diagnostic help, therapeutic planning, and medical interventions. The revolutionary application of medical image processing enables doctors to view body structures and their functional analysis without invasive procedures. The technique enables physicians to identify diseases early, create precise therapeutic plans, and monitor treatment responses. Medical image processing technologies deliver better health results at lower expenses while enhancing patient healthcare delivery. The medical image analysis through DL algorithms with convolutional neural networks (CNNs) follows a layered design, which starts by detecting basic features from image edges and textures before moving to more sophisticated patterns to enable autonomous identification of important medical image information for detection segmentation and classification purposes. Recurrent neural networks (RNNs) visual models built for medical image processing through DL algorithms contain network components that understand image temporal connections and sequential patterns while analyzing videos and medical image time series data. Medical image analysis improvements through DL algorithms can be observed in medical examples. Healthcare specialists predict this technological breakthrough will significantly improve medical diagnosis, reduce personnel workloads, and enable prompt patient treatments. This section presents multiple investigational examples together with real-life implementations. Table 3.1 presents the various case studies with medical images.

Table 3.1: Various case studies with medical images.

References	Case study	Description
[6]	Skin cancer detection case study	The development of DL analysis for skin cancer detection based on skin lesion images was done. The system proved equally precise to that of specialist doctors in diagnosis, thus indicating its potential value as a primary tool for identifying early skin cancer.
[7]	Diabetic retinopathy screening case study	Moorfields Eye Hospital developed a DL algorithm for retinal picture assessment of diabetic retinopathy. The algorithm obtained training from an extensive database exceeding 128,000 photographs to match the detection performance of expert ophthalmologists.
[8]	Tumor segmentation in MRI case study	Segmentation of brain tumors from MRI scans by Massachusetts General Hospital researchers was automated through DL techniques. The algorithm shortened the process of tumor delineation, thus speeding up treatment planning duration for patients.
[9]	Automated bone fracture detection case study	Stanford University researchers Meena and Roy developed a DL system to identify bone fractures in X-ray images. The system achieved excellent accuracy rates for detecting bone fractures, which is better than regular rule-based systems. Medical image analysis that uses DL algorithms heavily depends on ML algorithms for accurate and efficient segmentation tasks. Medical images benefit from DL approaches, which implement convolutional neural networks (CNNs) and recurrent neural networks (RNNs) to successfully analyze spatial dependencies and symmetrical properties that naturally exist in medical images. These algorithms analyze medical images, showing possible symmetrical structures by identifying their built-in symmetrical patterns and structures. Medical image analysis with DL mechanisms employs three main practical approaches: GANs, hybrid models, and CNNs utilized jointly with RNNs in this field.

3.1.1 Data Preparation in Medical Images

Several safety measures must be implemented before applying medical images for AI algorithm creation. Legal approval from a local ethical committee normally must be obtained separately when researchers or developers want to use medical information in their algorithms for research and commercial purposes. The assessment of potential patient risks and benefits constitutes one of the responsibilities of institutional review boards during study evaluation. Retrospective investigations should be performed to analyze existing data for developmental purposes. The patients who participate in this research do not require additional procedures, so informed consent is typically waived from the procedure. The principal investigator must approve the distribution of participant data as a requirement for clinical trials. Prospective studies require informed consent. Data accessibility starts immediately after receiving ethical approval, followed by the de-identification process and proper security measures to protect the information. Digital Imaging and Communications in Medicine (DICOM) metadata and images require protection by removing all health information according to the DICOM standard [10]. Free-form annotations in some images need manual review since automated cleaning methods prove ineffective for their removal. The objectives of the current task and domain requirements determine the quantities and qualities of images within this process. The following step involves standardizing and making data computable. Conclude the process by linking digital images with their corresponding ground-truth data such as labels and segmentations or electronic phenotype (biopsy or laboratory results). Figure 3.1 shows the process of medical image data handling.

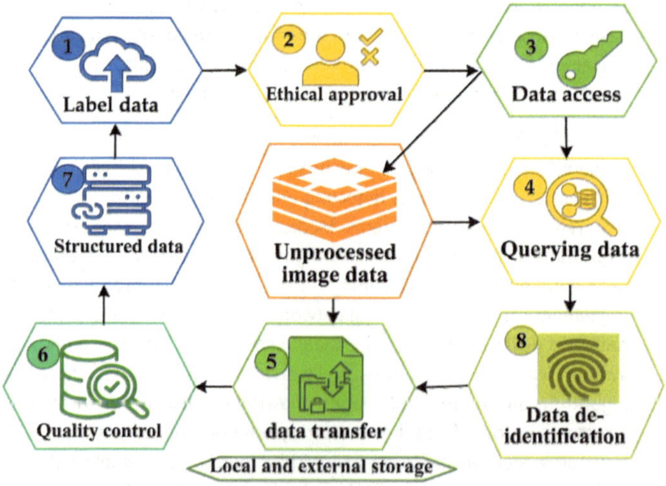

Figure 3.1: Process of medical image data handling.

Structure of This Chapter: This chapter comprises four sections, including the introductory section. This section examines the present security the data preparation and data handling in medical images and discusses the related case studies. Section 3.2 presents the transition from conventional ML Techniques to DL and their benefits in medical images. Section 3.3 presents fundamental principles and methodologies of medical image analysis that use DL algorithms. Section 3.4 presents challenges of the DL applications in medical image analysis. Finally, Section 3.5 concludes the chapter.

3.2 Transitioning from Conventional ML Techniques to DL

ML learning approaches denote the mode of algorithm training: supervised or unsupervised. In supervised learning, a function is characterized by inferring output from training data. In this method, training data are created using numerical or nominal vectors to describe input and output attributes: if the output data are continuous, the training procedure is called regression; if the output data are categorized, it is called classification. In contrast to output data consideration, unsupervised learning infers a function to describe hidden structures from unlabeled input data. There is no objective appraisal of correctness since examples are unlabeled. Digital image processing is incorporated into different computer systems. Each image processing method's main motive is implementing different operations impacting system performance. This was earlier a decision or an action on medical imaging methodologies. It provides basic and advanced tools for image analysis and visualization [11]. Figure 3.2 presents the AI that represents the uppermost domain of ML, and DL could be deemed a subsystem. Therefore, DL is an implementation technology for ML, while AI is a field in which exerting human-like intelligence in machines was officially stated. Medical images go through several steps, including image output detection. Medical images are fed into both ML and DL algorithms. Next, segmentation is performed on the selected image so that the area one wishes to examine can be analyzed. Information retrieval techniques will obtain the needed features from each of these segments. Selection of the required features occurs following the removal of noise.

Ultimately, the classifier classifies the retrieved data to make predictions based on them. The above procedures are used in all ML experiments. The categories of ML are the following: supervised, semi-supervised, unsupervised, reinforcement, and active algorithms. DL is the next stage of ML, and through the application of neural networks, it is intended to classify and forecast data more accurately.

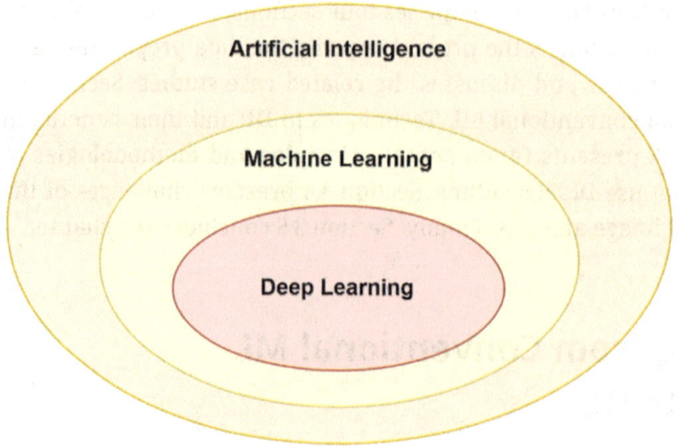

Figure 3.2: AI, ML, and DL domains.

i. **Supervised learning**

The training set consists of computer data that contains relevant objectives, allowing the system to acquire training. The system demonstrates precise response capabilities with the utilization of this training dataset. Classification and regression represent the two forms of supervised learning. A classification algorithm segments inputs into groups that can later be processed by trained systems using hidden inputs. The procedure is multilabeling. When classifying email messages as spam or non-spam, the technical process operates through classification. Such trained models exhibit a supervised behavior that generates continuous numeric results. Both regression predictions receive RMSE evaluation, but classification accuracy is the performance measure.

ii. **Unsupervised learning**

The system will make decisions using no preprogrammed datasets. The system receives predictions without receiving any designated labels from its operators. This method detects obscured patterns through the analysis of provided data features. As an unsupervised learning method, clustering enables the grouping of input data. Groupings occurred independently of any initial recognition by human foreknowledge. Groups shape up according to shared characteristics.

iii. **Semisupervised learning**

Training in semisupervised learning depends on incomplete datasets. The training technique utilizes trained data to find missing results. The untagged data training framework benefits from this approach. Semi-supervised learning combines supervised and unsupervised learning because its training incorporates tagged and untagged dataset information.

3.2.1 Benefits of Medical Image Processing

The noninvasive deep internal human anatomy examination benefits most from medical image processing methods. Medical image processing allows healthcare professionals to develop three-dimensional models, which facilitate improved patient care while designing better medical technology for drug distribution systems and disease recognition. Modern medical imaging technology and advanced software capabilities enable the creation of authentic digital constructions of diverse anatomical structures using different resolutions and varying tissues and bone features. Machine-based measurement and statistical analysis of genuine anatomical forms within simulation models enhance understanding medical device interaction with human bodies.

3.2.2 Medical Image Processing Work

CT and MRI scan data first require reconstruction to create programming software-compatible formats. Typically, doctors provide 3D grayscale intensity bitmap files with voxel grids for image processing. CT X-ray absorption rates generate image intensity, but signal strength from proton particles is the intensity determinant in MRI scans. Medical experts extract and alter bone tissue from image volumes that were reconstructed. Synopsys Simpleware allows users to conduct both 2D and 3D image processing functions:

- Noise reduction and artifact removal with image filters.
- The processing of images can be accelerated by performing cropping operations and subsequent data resampling steps.
- A system based on AI and ML methods automatically segments human anatomical regions
- The tool enables users to measure visual data elements, particularly centerlines, through statistics and measurement tools.
- The system can assess implant or medical equipment interaction with anatomical regions through imported CAD models.
- The processed models serve as an export to enable simulation through physics-based analysis or design or enable 3D printing of anatomical replicas.

3.3 Fundamental Concepts and Terminology

The fundamental principles and methodologies of medical image analysis that use DL algorithms rely on the terminology explored in the section. Table 3.2 presents the fundamental concepts and terminology of medical images.

Table 3.2: Fundamental concepts and terminology of medical images.

Terminology	Descriptions
Image analysis in medical healthcare	DL field algorithms referred to as CNN demonstrate superior effectiveness in performing medical image segmentation alongside feature extraction while conducting classification operations. DL algorithms allow medical condition and anomaly detection alongside localization and diagnosis procedures through analysis of annotated medical image databases to interpret hidden patterns [12]. Under these circumstances healthcare providers obtain faster and more precise interpretations of images, which drive better hospital patient care while advancing treatment creation and operational output. The programs assist healthcare professionals in making prompt medical decisions and detecting diseases early through their support of research efforts by examining large image databases. The integration of DL techniques into modern image analysis provides simultaneous benefits for healthcare treatment quality and the improvement of professional abilities and image reading processes [13].
Medical image analysis applications	DL approaches in medical image analysis enable various healthcare applications to happen. The DL approach through CNNs enables researchers to accomplish image segmentation tasks and object recognition together with disease classifying and reconstruction methods [14]. The algorithms are used to detect tumors together with identifying lesions as well as find pathological changes and anatomical abnormalities present in medical imaging data. DL models enable medical practitioners to evaluate both the advancement of disease conditions and therapeutic outcomes and future disease trajectories. Medical images serve as input for DL methods to extract crucial information automatically for precise and reliable reading purposes [15]. Medical applications of this technology would result in better healthcare resource deployment and improved clinical decisions together with enhanced patient outcomes.
Various aspects of medical image analysis	Medical image analysis solves numerous vital healthcare problems because it facilitates detailed diagnostic examination and evaluation [16]. Image preprocessing dictates the image quality and uniformity through its sequence of steps that include noise removal and augmentation as well as normalization. Image registration provides several exact image comparisons of the same patient or between different imaging modalities for fusion purposes [17]. The process of detecting and identifying abnormalities or specific anatomical patterns within images requires identifying essential traits and recognition patterns. The procedure of segmentation allows practitioners to identify and measure both anatomical features along with tumors and lesions [18]. Classification alongside recognition systems assists medical diagnosis because it allows the identification of normal against abnormal regions. Medical image analytics has reached impressive outcomes through DL algorithms especially CNNs by extracting complex patterns, which exist in large datasets. The success of medical image analysis in healthcare requires resolving data variability together with interpretability issues and across patient-group and imaging-modality generalization.

3.3.1 Machine Learning in Medical Imaging

ML algorithms supported by effective systems help medical researchers study specific disorders through imaging techniques. The precise representation of complex medical elements, including lesions and organs, remains impossible through basic mathematical methods in medical image processing. Pixel-based analysis in [5] evaluated medical photographs for diagnosing illnesses. Using medical imaging ML, numerical data is extracted directly from images rather than from sections of input data. This approach performs better than basic classifiers in dealing with particular problems. Analyzing the qualities of this image becomes challenging due to its low level of contrast. Pixel-based ML functions without requiring feature calculation or segmentation compared to traditional classifiers, which avoid faulty outcomes caused by imprecise segmentation and calculation. Adobe requires prolonged training periods because it works with images containing many pixels. The research utilized low-contrast medical images as study material [19]. Histogram Equalization (HE) is the most efficient method among all possible approaches to enhance contrast. Researchers presented an image contrast enhancement system named Modified Histogram-Based Contrast Enhancement using Homomorphic Filtering (MHFIL). Two sequential procedures constituted the approach: global contrast improvement through histogram modification.

Homomorphic filtering supports image sharpening tasks in the second operational phase. The research analyzes 10 medical chest X-ray images with low contrast. The MHFIL method needs minimum parameters throughout all 10 images to perform its functions. The responsibility for precise medical imaging analysis and interpretation falls on prime essential radiologists for photographs of equivalent diagnostic quality. ML expert systems transform patient data into premises. The rules inside expert systems derive from specialized information. Intelligent systems use ML to create clinical descriptions accurately representing patient conditions for clinical problem solutions. Through simple rules and decision trees, information becomes accessible. KARDIO is an example of how this technique translates ECG data [9]. Statistics represent the established standard for identifying medical image characteristics in computer-aided analysis. Hospital workers widely adopt the Channelized Hoteling Observer (CHO) method for nuclear medicine imaging procedures. The channel design starts from the perspective of simple perceptual targets available during human vision observation. This detection technique through CHO explores image quality performance while delivering improved medical imaging capabilities. CSVM process implements an algorithm directly after that. Two medical physicists conducted flaw discernibility tests on 100 noisy medical images. Then, the confidence score of a lesion actuality is measured on a six-point scale. Then, a training session will be conducted, and additional examinations will be conducted using 60 photographs. Humans performed the evaluation task under six flattening filter options through two rounds of OS-EM reconstruction methods.

3.3.2 Deep Learning in Medical Imaging

Computer systems need training to locate data attributes that pertain to specific issues. Several DL approaches draw their foundations from this concept. The models convert input images into disease-specific outputs through layers, embedding high-level structural features. Image analysis reaches its best results with CNNs. CNNs operate through multiple layers, which apply convolution filters during their input conversion process. Current medical research utilizes DL approaches to train network architectures to handle formats such as 3D data inputs. CNNs were originally developed to work with massive datasets, but DL methods had to disconnect the analyzed area to perform suboptimal 3D convolutional analysis. Figure 3.3 illustrates various ML and DL steps in medical imaging.

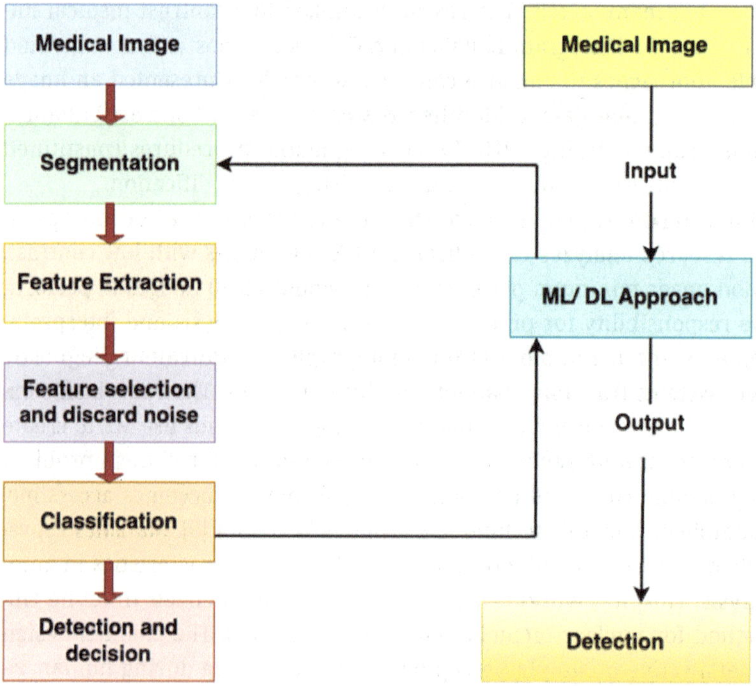

Figure 3.3: ML and DL workflow in medical images.

i. **Image classification**

The DL approach uses image classification to identify clinical difficulties before providing necessary therapy to patients. A combination of multiple images serves as the classification's input while it delivers a single diagnostic output that shows the presence or absence of disease. Diagnostic tests function as models, yet show smaller dataset sizes than those used in computer vision applications [20]. The authors of reported how the fine modification method delivered 57.6% testing accuracy for knee osteoarthritis multiclass scores that exceeded the feature extraction results at 53.4%. The retrieval method using CNN features improved cytopathology image diagnosis with 70.5% accuracy, superior to the traditional 69.1%.

ii. **Object classification**
Medical images will be analyzed through object categorization, concentrating on important sections. Medical image pieces can be categorized into multiple categories. The accuracy of medical image diagnosis depends on identifying important conceptual information at the local level and the global level. The study in [21] implemented three DL CNN approaches for repairing images with different object scales.

iii. **Detection: Organ and Region**
After classification come object localization and identification. Segmentation success demands identifying important objects because this enables target-focused analysis while filtering out insignificant data (noise). Our DL approaches process 3D data to accomplish this goal. Numerous independent groups divided between 2D and 3D MRI components formed the medical imaging dataset. Segmentation models find paths between interconnected elements seeking particular targets, including heart structures, aortic arches, and descending aorta.

iv. **Segmentation**
Segmentation arranges medical imaging data into organs along with their substructures. This methodology enables precise numerical analysis of medical data measurements. Example: heart or brain exam. It is also used in CAD operations. The method requires detecting single pixels throughout the focus object area. The U-Net architecture implements both upscaling operations and downscaling network layers. The system implemented a combination of layer convolutions and deconvolutions [22].

v. **Registration**
The registration process transforms multiple medical datasets into a unified coordinate system. Data integration between multiple perspectives and time sequences, depth measurements, and sensory data remains essential for medical imaging applications. Choose particular parameter types that function as our standard through an iterative approach. Researchers use DL techniques to determine similarity metrics between two photographic images. This method provides practical use within medical settings that process computed tomography (CT) and NMR data. Patient information acquisition holds significance when blended with a standard coordinate system for monitoring tumor progression while validating cures through parallel patient data assessments against anatomical atlases. The mutual information recorded in using the MR/CT registration methods developed by Powell and Brent contrasts from [23] about breast MR image analysis.

3.3.3 ML/DL Techniques for Medical Image Analysis

The five main techniques that make up the network models are CNNs, RNNs, GANs, LSTMs, and hybrid approaches that combine multiple methods. Figure 3.4 shows the ML/DL techniques for medical image analysis, and Table 3.3 describes the techniques.

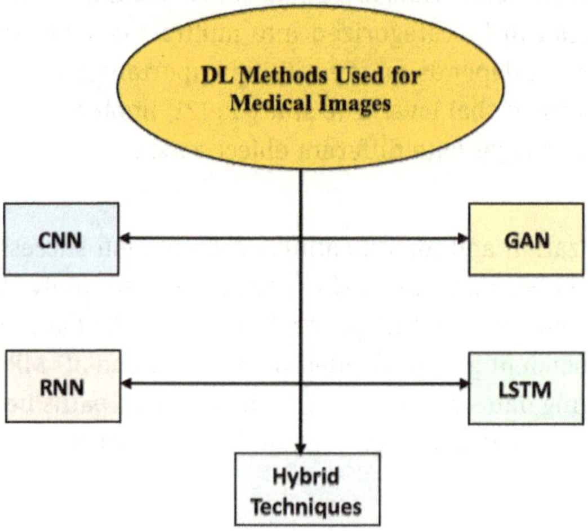

Figure 3.4: ML/ DL techniques.

Table 3.3: ML/DL techniques description.

ML/DL techniques	Description
Convolutional Neural Networks	CNN models excel at medical image tasks, which include localization of objects while segmenting and classifying areas because they have an inherent ability to extract vital image features. The detection and analysis of medical image anomalies together with tumor identification and organ segmentation happen with precision via CNNs. Multiple tier analysis is possible because CNNs have a hierarchical design that enables acquisition of significant features. Medical image processing has benefited from CNN application through more precise diagnostic processes, which lead to enhanced efficiency and greater automation and thereby delivers better patient results [11].

Table 3.3 (continued)

ML/DL techniques	Description
Generative adversarial network	The GAN technology offers crucial benefits to medical image analysis through DL because it generates authentic synthetic imagery, which improves dataset quantities while boosting diagnosis accuracy for multiple medical conditions. The research of Vaccari et al. [24] presented a GAN methodology for addressing the synthetic medical data requirements in Internet of Medical Things (IoMT) applications. The authors explained their proposed methodology to produce medical data samples with various time-based and non-time-based data types. GAN-based methodology offers several benefits for IoMT systems by producing realistic data that enhances their effectiveness. The scientists validated their proposed method by conducting tests on authentic medical information such as electrocardiogram (ECG) data and healthcare imaging data. Multiple statistical measures verified that their GAN-based method produced synthesized medical data that matched real medical data in appearance and statistical metrics. The proposed technique by the authors has proven to function as a powerful tool for creating synthetic medical data used in IoMT applications.
Recurrent neural network	RNNs represent vital components for DL-based medical image analysis because they effectively extract both temporal dependencies together with contextual information. The solution capabilities of RNNs excel when performing sequential or time-series data analysis on medical image sequences and dynamic imaging modalities. These networks enable both long-term sequence modeling and previous-time-step information application, which makes them suitable for pattern detection and disease evolution monitoring and tumor progression assessment. Modern RNN variants, specifically LSTM and GRU, help healthcare professionals extract complex temporal patterns from medical image sequences, which make these developments crucial for finding valuable insights from image sequences [25].
Long short-term memory	LSTM represents an essential method for medical image analysis through DL techniques since it identifies and replicates the visual sequential patterns in data. Medical imaging requires contextual information to properly understand complex spatial as well as temporal patterns. LSTM functions as an ideal RNN type for time series analysis as well as illness progression modeling and image sequence analysis by following long-range relationships and temporal dynamics. LSTMs incorporate memory and gating procedures, which enable the model to learn essential information thus improving the processing of medical images as well as diagnostic accuracy and personalized treatment delivery in healthcare settings [26].
Hybrid techniques	DL algorithms need new techniques or data types for performing useful medical image assessment. DL achieves its strength from processing images to identify category assignments. This approach becomes difficult to employ because training data is insufficient coupled with problems of interpretability. The implementation of hybrid approaches enables researchers to handle limitations, which create performance improvements. Examination results and readability improvement become achievable by merging traditional analytical methods with statistical patterns and domain-wisdom expertise. Clinical diagnosis requires image evaluations and text documentation analysis along with drug indicator monitoring for better patient health comprehension [27].

3.4 Challenges of the DL Applications in Medical Image Analysis

The main challenge in DL systems, which interprets medical images, stems from unsatisfactory annotation inputs. The operation and generalization capability of DL models heavily depend on detailed data labels. Various factors make high-quality medical photo annotation challenging since the process demands costly resources and significant time commitment, limiting available annotated image data [28]:

- Transfer learning provides an enhancement for smaller datasets through the application of pre-trained models learned from extensive big datasets. This technique reduces the amount of annotated data requirements for DL model training while simultaneously building generalizable models.
- Data augmentation works by modifying available annotated data, including the creation of synthetic data. This method elevates medical picture stability and boosts model annotations through increased quantity and variety of training data accessibility for DL models.
- Instead of annotating every data point, active learning selects the most informative samples among those uncertain to achieve better results. The annotation workload decreases while the training efficiency for DL models improves by implementing this approach.
- Medical experts, patients, and other involved stakeholders participate in collaborative annotation activities to ensure correct and consistent annotation data match clinical requirements and professional values.

3.5 Conclusion

DL has revolutionized medical image processing and analysis. More accurate and efficient medical image interpretation may reduce clinical error and processing time with the shift from standard ML to DL. The chapter showed how CNNs, RNNs, GANs, and LSTMs individually contribute to medical picture analysis. CNNs excel at feature extraction and classification, while RNNs analyze sequential medical data temporal relationships to great effect. GANs may generate synthetic medical images to supplement training datasets, and LSTMs can grasp long-term dependencies in medical image sequences. Data annotation is a major issue. Medical image annotation is time-consuming and needs expertise; therefore large, well-annotated datasets are a major obstacle. Transfer learning, data augmentation, and active learning are realistic solutions to these limits, but they are complicated. In the future, DL in medical image processing could greatly improve healthcare. These technologies enable faster, more accurate diagnosis and aid decision-making of medical professionals. The continuing development of hybrid methodologies that combine standard analytical methods with

DL predicts that AI-assisted medical image analysis will become essential in clinical practice. Future research should improve model interpretability, reduce the need for large annotated datasets, and develop more robust systems that can generalize across medical imaging modalities. These systems must also remain clinically relevant and practical for real-world implementation while maintaining high accuracy and reliability.

References

[1] Afshari S, BenTaieb A, Hamarneh G. Automatic localization of normal active organs in 3D PET scans. Computerized Medical Imaging and Graphics. 2018 Dec 1;70:111–118. Available from: http://dx.doi.org/10.1016/j.compmedimag.2018.09.008.

[2] Ait Skourt B, El Hassani A, Majda A. Lung CT image segmentation using deep neural networks. Procedia Computer Science. 2018 Jan 1;127:109–113. Available from: http://dx.doi.org/10.1016/j.procs.2018.01.104.

[3] Ali M, Son DH, Kang SH, Nam SR. An accurate CT saturation classification using a deep learning approach based on unsupervised feature extraction and supervised fine-tuning strategy. Energies. 2017 Nov 10;10(11):1830. 2017;10(11):1830. Available from: http://dx.doi.org/10.3390/en10111830.

[4] Kim J, Hong J, Park H. Prospects of deep learning for medical imaging. Precision and Future Medicine. 2018;2(2):37–52. Available from: http://dx.doi.org/10.23838/pfm.2018.00030.

[5] Shen D, Wu G, Zhang D, Suzuki K, Wang F, Yan P. Machine learning in medical imaging. Comput. Medical Imaging Graph. 2015 Apr 1;41:1–2. Available from: http://dx.doi.org/10.1016/j.compmedimag.2015.02.001.

[6] Vijayalakshmi M. Department of Information Science and Engineering GSSSIETW, Mysuru, Karnataka, India. Melanoma skin cancer detection using image processing and machine learning. International Journal of Trend in Scientific Research and Development [Internet]. 2019;3(4):780–784. Available from: http://dx.doi.org/10.31142/ijtsrd23936.

[7] De Fauw J, Keane P, Tomasev N, Visentin D, Van den Driessche G, Johnson M, Hughes CO, Chu C, Ledsam J, Back T, Peto T. Automated analysis of retinal imaging using machine learning techniques for computer vision. F1000Research. 2017 Jun 22;5:1573. Available from: http://dx.doi.org/10.12688/f1000research.8996.1.

[8] Guo Z, Li X, Huang H, Guo N, Li Q. Deep learning-based image segmentation on multimodal medical imaging. IEEE Transactions on Radiation and Plasma Medical Sciences. 2019 Jan 1;3(2):162–169. Available from: http://dx.doi.org/10.1109/trpms.2018.2890359.

[9] Meena T, Roy S. Bone fracture detection using deep supervised learning from radiological images: A paradigm shift. Diagnostics. 2022 Oct 7;12(10):2420. Available from: http://dx.doi.org/10.3390/diagnostics12102420.

[10] Hwang EJ, Park S, Jin KN, Im Kim J, Choi SY, Lee JH, Goo JM, Aum J, Yim JJ, Cohen JG, Ferretti GR. Development and validation of a deep learning–based automated detection algorithm for major thoracic diseases on chest radiographs. JAMA Network Open. 2019 Mar 1;2(3):e191095-. Available from: http://dx.doi.org/10.2139/ssrn.3289800.

[11] Singh AV, Rosenkranz D, Ansari MH, Singh R, Kanase A, Singh SP, Johnston B, Tentschert J, Laux P, Luch A. Artificial intelligence and machine learning empower advanced biomedical material design to toxicity prediction. Advanced Intelligent Systems. 2020 Dec;2(12):2000084. Available from: http://dx.doi.org/10.1002/aisy.202000084.

[12] Popescu AB, Taca IA, Vizitiu A, Nita CI, Suciu C, Itu LM, Scafa-Udriste A. Obfuscation algorithm for privacy-preserving deep learning-based medical image analysis. Applied Sciences. 2022 Apr 14;12 (8):3997. Available from: http://dx.doi.org/10.3390/app12083997.

[13] Singha A, Thakur RS, Patel T. Deep learning applications in medical image analysis. Biomedical data mining for information retrieval: Methodologies. Techniques and Applications. 2021 Aug;24:293–350. Available from: http://dx.doi.org/10.1002/9781119711278.ch11.

[14] Qureshi I, Yan J, Abbas Q, Shaheed K, Riaz AB, Wahid A, Khan MW, Szczuko P. Medical image segmentation using deep semantic-based methods: A review of techniques, applications and emerging trends. Information Fusion. 2023 Feb 1;90:316–352. Available from: http://dx.doi.org/10.1016/j.inffus.2022.09.031.

[15] Shin HC, Lu L, Summers RM. Natural language processing for large-scale medical image analysis using deep learning. Deep Learning for Medical Image Analysis. 2017 Jan;1:405–421.

[16] Uwimana A, Senanayake R Out of distribution detection and adversarial attacks on deep neural networks for robust medical image analysis. arXiv preprint arXiv:2107.04882. 2021 Jul 10. Available from: http://arxiv.org/abs/2107.04882

[17] Al-Galal SA, Alshaikhli IF, Abdulrazzaq MM. MRI brain tumor medical images analysis using deep learning techniques: A systematic review. Health and Technology. 2021 Mar;11(2):267–282. 2021;11 (2):267–82. Available from: http://dx.doi.org/10.1007/s12553-020-00514-6.

[18] Duncan JS, Insana MF, Ayache N. Biomedical imaging and analysis in the age of big data and deep learning [scanning the issue]. Proceedings of the IEEE. 2019 Dec 27;108(1):3–10. Available from: http://dx.doi.org/10.1109/jproc.2019.2956422.

[19] Dhiman G, Juneja S, Viriyasitavat W, Mohafez H, Hadizadeh M, Islam MA, El Bayoumy I, Gulati K. A novel machine-learning-based hybrid CNN model for tumor identification in medical image processing. Sustainability. 2022 Jan 27;14(3):1447. Available from: http://dx.doi.org/10.3390/su14031447.

[20] Zheng Y, Comaniciu D. Marginal space learning. In Marginal space learning for medical image analysis: Efficient detection and segmentation of anatomical structures. 2014 Mar 11 (pp. 25–65). New York, NY: Springer New York.

[21] Aubreville M, Krappmann M, Bertram C, Klopfleisch R, Maier A A guided spatial transformer network for histology cell differentiation. arXiv preprint arXiv:1707.08525. 2017 Jul 26. Available from: http://arxiv.org/abs/1707.08525

[22] Andermatt S, Pezold S, Cattin P. Multi-dimensional gated recurrent units for the segmentation of biomedical 3D-data. In International workshop on deep learning in medical image analysis 2016 Sep 27 (pp. 142–151). Cham: Springer International Publishing.

[23] Poudel RP, Lamata P, Montana G. Recurrent fully convolutional neural networks for multi-slice MRI cardiac segmentation. In International workshop on reconstruction and analysis of moving body organs 2016 Oct 17 (pp. 83–94). Cham: Springer International Publishing.

[24] Vaccari I, Orani V, Paglialonga A, Cambiaso E, Mongelli M. A generative adversarial network (GAN) technique for internet of medical things data. Sensors. 2021 May 27;21(11):3726. Available from: http://dx.doi.org/10.3390/s21113726.

[25] Sridhar C, Pareek PK, Kalidoss R, Jamal SS, Shukla PK, Nuagah SJ. Optimal medical image size reduction model creation using recurrent neural network and GenPSOWVQ. Journal of Healthcare Engineering. 2022;2022(1):2354866. Available from: http://dx.doi.org/10.1155/2022/2354866.

[26] Dutt UM, Letchmunan S, Ali M, Hassan FH, Baqir A, Sherazi HH. Machine learning based diabetes classification and prediction for healthcare applications. Journal of Healthcare Engineering. 2021;2021(1):9930985. Available from: http://dx.doi.org/10.1155/2021/9930985.

[27] Srikantamurthy MM, Rallabandi VS, Dudekula DB, Natarajan S, Park J. Classification of benign and malignant subtypes of breast cancer histopathology imaging using hybrid CNN-LSTM based transfer learning. BMC Medical Imaging. 2023 Jan 30;23(1):19. Available from: http://dx.doi.org/10.1186/s12880-023-00964-0.
[28] Shen G, Zeng W, Han C, Liu P, Zhang Y. Determination of the average maintenance time of CNC machine tools based on type II failure correlation. Eksploatacja I Niezawodność. 2017;19(4). Available from: http://dx.doi.org/10.17531/ein.2017.4.15.

Monika Sethi, Saravjeet Singh, Shivani Sood, Jatin Arora

4 Deep Learning in Medical Imaging Modalities: A Comprehensive Analysis for Enhanced Diagnosis

Abstract: Artificial intelligence (AI) in healthcare imaging is transforming Medicare through improved treatment for patients, efficacy, and accuracy in diagnosis. AI-powered devices are capable of quickly assessing enormous volumes of imaging data, minimizing human error and thus facilitating early disease identification. This chapter focuses on how AI is employed in medical image analysis, specifically focusing on deep learning methods like convolutional neural networks. Such techniques are faster and more accurate than conventional approaches for automatically identifying complex features in healthcare images. This chapter focuses on how these advances may be employed in a variety of imaging modalities, such as computed tomography, magnetic resonance imaging, and X-rays, to enhance segmenting of images, disease recognition, and clinical assessment. AI is also making it simpler to develop individualized treatment plans by finding trends in disease progression that doctors may not recognize at once. Research also addresses important issues in AI-driven medical imaging, including model interpretability, reliability of data, and ethical concerns. This chapter also provides a detailed review of the state and future prospects of AI in healthcare imaging through analysis of previous developments and real-world case studies, highlighting the technology's increasing influence on diagnostic and treatment methods. Additionally, this chapter discusses explainable AI (XAI) in medical diagnosis, emphasizing how it could promote transparency and confidence in AI-powered healthcare. The development of AI and its integration with other cutting-edge technologies, including as robotic-assisted surgery and augmented reality, should significantly improve patient care and personalized healthcare.

Keywords: deep learning, medical imaging, image processing, convolutional neural network, XAI

4.1 Introduction

Modern medical care has been revolutionized by the use of medical imaging, which provides noninvasive, high-resolution visions of organs and tissues for health monitoring, assessment, and subsequent treatment plans. Healthcare providers can now detect, analyze, and monitor diseases with unprecedented precision, owing to innovations in imaging methods across the last couple of decades, including computed tomography (CT), magnetic resonance imaging (MRI), positron emission tomography

 | https://doi.org/10.1515/9783112226223-004

(PET), single photon emission computed tomography (SPECT), ultrasound, X-rays, and mammography. Applied mathematics provides fundamental frameworks for image acquisition, restoration, and evaluation, establishing the basis for various imaging techniques' performance. Mathematical techniques solve complex inverse problems in CT and MRI, allowing precise restoration of images from indirect measurements. In this process, iterative approaches and Fourier transformations are essential. Partial differential equations (PDEs) can extract valuable details from data that is distorted by improving the clarity of images via noise reduction and deblurring [1]. Large deformation diffeomorphic metric mapping (LDDMM) is one technique that offers tools for identifying biological structures, whereas statistical shape analysis and differential geometry are two techniques utilized in computational anatomy to analyze anatomical variability [2]. In order to improve imaging techniques, especially in MRI, optimal control theory has been used to shorten acquisition times and improve picture contrast [4]. By incorporating previous experience from image processing, Bayesian probabilistic techniques boost the efficiency of medical image analysis [3].

The integration of deep learning (DL) with several medical imaging modalities is shown in Figure 4.1. By providing crucial details required for precise and timely decision-making during treatment, medical imaging is essential in a number of fields, including neurological disorders, cardiovascular disease, and in the field of orthopedics and cancer treatment [4]. However, there are now major issues with interpretation, processing speed, and diagnostic reliability due to the exponential expansion of medical imaging data. Conventional imaging analysis relies on experienced medical professionals, and their interpretations tend to be laborious, prone to inter-observer variation, and subject to psychological constraints. The requirement for computerized, smart approaches to improve medical imaging analysis's accuracy pace and reliability has grown as a consequence.

A subset of artificial intelligence (AI), DL has become an integral component in health care imaging in the past decade [5]. DL algorithms have demonstrated incredible abilities in pattern identification, classification of images, segmentation, and detecting anomalies, especially convolutional neural networks (CNNs), recurrent neural networks (RNNs), vision transformers (ViTs), and generative adversarial networks (GANs) [6]. In applications including recognizing tumors, lesion segmentation, and disease classification, such algorithms can derive high-level features from raw imaging data, often outperforming professionals. DL allows end-to-end learning where models automatically discover significant anomalous patterns in massive amounts of medical data, as opposed to conventional machine learning approaches, which employ explicitly extracted features [7]. DL has transformed into an efficient diagnostic tool that may decrease load and improve the health of patients and the availability of high-quality healthcare as big data, high-performance computing, and cloud-based AI platforms become more widely available.

Conventional practices are transforming, as a consequence of DL integration with medical imaging, strengthening diagnostic precision as well as efficiency. Physicians

Figure 4.1: Deep learning with medical image processing.

and radiologists are relying progressively on tools driven by AI to help with evaluation, prioritize emergencies, and provide automated recommendations. By examining imaging patterns that might be unnoticeable to the naked eye, AI-driven tools have shown exceptional proficiency in identifying early identification of diseases, including Alzheimer's disease, cancer, diabetic retinopathy, and Parkinson's [8]. In CT, MRI, and PET images, DL-based segmentation methods like U-Net [9] allow for the reliable segmentation of cancerous tumors, tissues, and lesions, substantially minimizing human labor and variability. GANs and other modern DL algorithms are used for resolution upscaling, intensity enhancement, and noise reduction, which enhances healthcare image quality and reducing distortions. DL makes it simpler to integrate medical and genetic data with multimodal imaging data (CT, MRI, and PET) to provide an extensive view of the course of disease, and offer tailored therapies [10].

This chapter explores how DL has revolutionized medical imaging by exploring its applications in identifying various diseases such as tumors, cancer, diabetic retinopathy, Alzheimer's disease, and Parkinson's disease. DL analyzes structural and functional changes to the brain using MRI, PET, and SPECT imaging, to offer early detection of neurodegenerative illnesses like Alzheimer's and Parkinson's. For instance, DL-powered detection methods in oncology improve cancer diagnosis using modalities such as histology, CT, and MRI, minimizing errors in diagnosis and supporting individualized medication for treatment.

4.2 Image Processing Techniques for Medical Imaging

With the introduction of radiation from X-rays in 1895, medical imaging has gone through significant evolution. As a result, modalities like CT, MRI, PET, and ultrasound have been developed, providing advanced diagnostic, treatment planning, and monitoring abilities. Processing of images, which includes methods like image enhancement, segmentation, registration, and quantification, is essential for improving medical images for better analysis. The significance of supplement drugs in differentiating between organs with identical densities is demonstrated by CT imaging, employing Hounsfield units. Advanced structural systems can be better recognized with the use of powerful imaging approaches like polygonal iso-surface modeling and volumetric representation. Structures like circulatory systems and lesions can be identified with the use of segmentation techniques like region growth and skeletal construction. Through the analysis of biological parameters like metabolic and circulatory movement, functional imaging – including dynamical scanning techniques – offers views beyond anatomical features [11]. By using pattern recognition algorithms that imitate human vision, object-based visual analysis improves diagnosis accuracy. In order to enhance healthcare decisions with automating disease recognition, this chapter emphasizes the use of AI and machine learning in medical imaging.

Numerous image processing techniques are frequently used in medical image analysis, each with its specific merits and functions. Numerous segmentation techniques have been investigated by researchers, especially for MRI images, such as knee bone analysis. Research goals, image type, and other features, all have a role in deciding the best segmentation technique. The four categories of these techniques are classifier-based, hybrid, region-based, and clustering [12]. Although they are easy to use, area-based techniques such as region growth and thresholding techniques (global, local, and Otsu) are extremely vulnerable to distortion [13]. Because there is less distortion in CT scans, these methods perform effectively there, but they might not work well in MRI examinations with no preprocessing, although for soft tissue analysis, MRI works well. Training approaches, such as grouping and categorization, rely on data for training, which can be complicated and laborious, and do not employ spatial or pattern data. Out of all the clustering methods, fuzzy *C*-means has shown excellent accuracy on a range of medical imaging data [14]. Hybrid approaches combine boundaries and region features to enhance segmentation. The graph-cut approach, which makes use of a cost function, works especially well for *N*-dimensional segmentation and provides high precision and optimum effects. In order to assess segmentation results, many validation methods are employed. The five most widely used validation metrics – DSI, TNF, FNF, TPF, and FPF – are used to evaluate the performance of various approaches and compare segmentation results [15].

Using the cardiotocography dataset, researchers investigated how principal component analysis (PCA) and linear discriminant analysis (LDA) affected predictive classifiers. The features were reduced from 36 to 26 by PCA and to 1 by LDA. The findings demonstrated that PCA outperformed LDA in classifier performance, with decision tree and random forest consistently exhibiting higher accuracy. Furthermore, PCA enhanced classifier performance for the IDS dataset, but the lower size of the diabetic retinopathy dataset had a negative impact on final results for both PCA and LDA [16].

4.3 Deep Learning in Medical Imaging

Radiologists now find it hard to complete readings on time and produce accurate findings due to the increasing complexity of diagnostic tests. By providing quantitative analysis of problematic lesions, DL technique is expected for better diagnosis reliability. It is also anticipated to improve workflow efficiency by lowering reading time through voice recognition and automatic report production. Because AI has of late outperformed humans in several computer vision and speech recognition tasks, digital assistants and autonomous driving technologies have advanced. Big tech firms like Baidu, Microsoft, Facebook, and Amazon are all rapidly exploring DL applications in healthcare diagnostics. Though its precise potential in radiography is yet unresolved, artificial intelligence is probably going to have a big impact over the next few decades [17]. The use of AI in radiology and diagnostic imaging has grown in popularity as it develops further. Its function in identifying anomalies, improving image analysis, and fostering patient care is being investigated more and more by academics and healthcare providers. Moreover, DL techniques demonstrated success in medical imaging by improving diagnostic accuracy and reducing doctors' workloads. Data dependability, overfitting, and ethical issues require to be resolved in spite of its potential uses. Clinical adoption was severely slowed by DL's black-box nature and legal liability concerns. Subsequently, it was anticipated that AI will help doctors rather than replace them, by enhancing quantitative analysis and optimizing efficiency. Clinicians and computer scientists had to work together to ensure successful integration, and legal and ethical considerations had to be taken into account [18].

DL has been used extensively in medical imaging to help interpret scans like mammograms, CT scans, MRIs, and X-rays. AI systems have shown encouraging accuracy in radiology, outperforming radiologists in the detection of malignant lung nodules and pneumonia. DNNs have enhanced diagnosis in cases of emergency, including the detection of spinal displacement injuries and fractures in the wrist. AI has also been applied to coronary calcium scoring, liver tumors, cerebral hemorrhages, and tumors in the lungs. Clinical validation is difficult, nevertheless, as seen by differences in study designs and performance indicators like AUC values. Further validation and regulatory approval are still needed for wider clinical usage, even though several AI applications, including diabetic retinopathy screening and breast cancer metastasis detection, have been proven in real-world scenarios [8].

Transfer learning (TL), which employs pretrained models from enormous databases like ImageNet, has become an effective strategy in medical imaging to overcome the problem of inadequate labeled data. The degree of similarity between the source and destination domains determines how successful TL is, even though it can boost efficiency. By utilizing pretrained models from many domains, researchers have applied TL to improve medical imaging problems. In order to create a deep convolutional neural network (DCNN), conventional and parallel convolutional layers were combined with global average pooling and residual connections. When the model was evaluated on animal classification tasks and diabetic foot ulcer (DFU) tests, it was shown that TL from the same domain greatly enhanced performance. The efficacy of domain-specific TL was demonstrated by the DFU dataset, which obtained an F_1 score of 86.6%, when trained from scratch, 89.4%, with TL from a different domain, and 97.6% with TL from the same domain [19]. There are many pretrained models available such as GoogleNet, LeNet, MobileNet, and AlexNet. Many researchers are currently using DL to get positive results in an extensive spectrum of medical imaging research for almost every disease, covering orthopedic conditions, Alzheimer's disease, Parkinson's disease, pathological diseases, and cancer of various types.

4.4 Explainable AI (XAI) for Medical Diagnosis

Automated image analysis has greatly advanced since the advent of DL techniques, which has replaced conventional techniques that focused on manually generated features. DL algorithms, as opposed to traditional statistical classifiers, automatically extract pertinent information from medical images to increase the precision of diagnosis. However, since these models operate as "black boxes," it is challenging to understand the decision-making process. This lack of accountability is alarming, especially for medical applications where inaccurate or biased forecasts could have severe implications. In order to improve the interpretability of DL models, explainable AI (XAI) techniques have been established. In order to ensure confidence, dependability, and moral application in medical diagnosis, XAI seeks to offer insights into model decisions [20].

AAOXAI-CD, or Adaptive Aquila Optimizer with Explainable Artificial Intelligence (XAI) Enabled Cancer Diagnosis, is a medical imaging technology developed by researchers at [21]. Using the Faster Squeeze Net model for feature extraction and the AAO method for hyperparameter tuning, the AAOXAI-CD technique attempts to boost the classification of colorectal and osteosarcoma cancers. Three DL classifiers – RNN, gated recurrent unit (GRU), and bidirectional long short-term memory (BiLSTM) – are integrated into a majority-weighted voting ensemble model for classification. Furthermore, to enhance interpretability and transparency in cancer detection, the AAOXAI-CD technology integrates the XAI approach LIME. Experiments using medical cancer imaging databases showed that the AAOXAI-CD methodology worked better than previous methods, assuring increased reliability and precision. Additionally, utilizing dynamic contrast-enhanced magnetic resonance imaging (DCE-MRI) of the breast, researchers in [22] used XAI visualization to examine the characteristics acquired by a DCNN for the classification of estrogen receptor status (ER+ vs. ER–). 1395 ER+ and 729 ER– regions of interest (ROIs) from 148 individuals, whose pre-contrast and post-contrast scans were included within their dataset. Based on the AlexNet model, they created a unique transfer-trained dual-domain DCNN architecture that combines dynamic and spatial imaging components. The AUC with leave-one-case-out cross-validation was applied to assess the model's performance. XAI methods like Smooth-Grad and Integrated Gradients were utilized to improve interpretability; these methods revealed pertinent spatial and dynamic information, in addition to unintentional learning from preprocessing aberrations. Borys et al. [23] discussed XAI in medical imaging, with an emphasis on XAI methods that are based on non-saliency. The approaches are categorized into case-based, textual, and auxiliary explanations, including a range of examples provided. The aim is to improve interdisciplinary comprehension between healthcare practitioners and AI developers by providing a nontechnical summary.

In order to classify brain tumors using MRI scans, research [24] suggests combining XAI with a collaborative federated learning model. To improve data privacy while classifying cases of glioma, meningioma, pituitary tumors, and no tumor, their method makes use of GoogLeNet in a federated learning architecture. In addition to achieving 94% accuracy, they use Grad-CAM and saliency maps to increase model transparency and give medical experts visually understandable explanations. Five essential requirements for maximizing XAI in clinical contexts are defined by the authors' proposed clinical XAI guidelines [25]. They introduce new evaluation criteria and conduct a systematic assessment of multimodal medical image interpretations. Their evaluation of 16 heatmap-based XAI systems shows flaws in informational coherence and veracity, underscoring the need for better approaches to guarantee clinical feasibility.

4.5 Challenges and Future Directions

Modern advances in DL have streamlined disease identification and the process of segmentation and categorization, revolutionizing the interpretation of medical images. There are still issues to be resolved, regardless of DL's efficacy, like improving early detection of diseases accuracy, incorporating multimodal imaging to enhance diagnostics, and designing models that are interpretable, which encourage trust around AI-driven treatment. Although few-shot learning and knowledge transfers might be helpful in situations with unusual diseases and inadequate data, data augmentation and synthesized image creation are essential to addressing dataset limitations. Medical imaging class imbalance is still an issue that demands inventive reproducing and augmenting methods. Hardware optimizations and transfer learning can help minimize the amount of time required for complicated model development. Another issue involves minimal annotations and actual labeling, which emphasizes the necessity of semisupervised and weakly supervised learning to maximize expert resources. By using pretrained models that have already discovered important features from enormous datasets, transfer learning serves an important part in dealing with data deficits. By fine-tuning such algorithms on certain medical imaging datasets, they may adjust to domain-specific patterns utilizing fewer labeled data, which lowers computing expenses and boosts efficiency. To overcome such challenges, multidisciplinary engagement and ongoing technological development are needed to improve DL models for medical imaging. AI can improve disease identification accuracy and dependability through better data augmentation, transfer learning, and interpretability. Reliability in AI-driven diagnoses will be further reinforced by enhanced anomaly detection, standardized diagnostic methods, and effective security measures. Real-world deployment can be accelerated by utilizing multimodal integration and training methodologies that are optimized. DL will continue to transform medical image analysis with these initiatives, improving the accuracy, effectiveness, and accessibility of diagnostics.

4.6 Conclusion

Medical imaging has been substantially enhanced by the combination of XAI, DL, and image processing techniques. This has allowed for better clinical decision-making and more precise diagnosis. While DL automates feature extraction and helps with diagnosis and segmentation, traditional image processing techniques improve image quality. But issues, including a lack of data, the interpretability of the model, and ethical decisions, persist as major challenges. By using methods like Grad-CAM and SHAP to make AI-driven decisions visible and reliable, XAI plays a critical role in addressing these problems. The necessity for federated learning and consistent evaluation metrics is highlighted by the persistence of concerns about data privacy, security, and generalizability, despite advancements. These methods will continue to be further refined by continuing multidisciplinary collaboration between AI researchers and medical practitioners, which will ultimately improve healthcare for patients in general, early disease recognition, and personalized therapies.

References

[1] Angenent S, Pichon E, Tannenbaum A. Mathematical methods in medical image processing. Bulletin of the American Mathematical Society. 2006;43(3):365–396.

[2] Ceritoglu C, Oishi K, Li X, Chou MC, Younes L, Albert M, Lyketsos C, Van Zijl PC, Miller MI, Mori S. Multi-contrast large deformation diffeomorphic metric mapping for diffusion tensor imaging. Neuroimage. 2009 Aug 15;47(2):618–627.

[3] Ma J, Miller MI, Trouvé A, Younes L. Bayesian template estimation in computational anatomy. NeuroImage. 2008 Aug 1;42(1):252–261.

[4] Litjens G, Kooi T, Bejnordi BE, Setio AA, Ciompi F, Ghafoorian M, Van Der Laak JA, Van Ginneken B, Sánchez CI. A survey on deep learning in medical image analysis. Medical Image Analysis. 2017 Dec 1;42:60–88.

[5] Ravì D, Wong C, Deligianni F, Berthelot M, Andreu-Perez J, Lo B, Yang GZ. Deep learning for health informatics. IEEE Journal of Biomedical and Health Informatics. 2016 Dec 29;21(1):4–21.

[6] Shen D, Wu G, Suk HI. Deep learning in medical image analysis. Annual Review of Biomedical Engineering. 2017 Jun 21;19(1):221–248.

[7] Esteva A, Chou K, Yeung S, Naik N, Madani A, Mottaghi A, Liu Y, Topol E, Dean J, Socher R. Deep learning-enabled medical computer vision. NPJ Digital Medicine. 2021 Jan 8;4(1):5.

[8] Topol EJ. High-performance medicine: The convergence of human and artificial intelligence. Nature Medicine. 2019 Jan;25(1):44–56.

[9] Ronneberger O, Fischer P, Brox T. U-net: Convolutional networks for biomedical image segmentation. In International conference on medical image computing and computer-assisted intervention 2015 Oct 5 (pp. 234–241). Cham: Springer international publishing.

[10] Lundervold AS, Lundervold A. An overview of deep learning in medical imaging focusing on MRI. Zeitschrift Fuer Medizinische Physik. 2019 May 1;29(2):102–127.

[11] Ritter F, Boskamp T, Homeyer A, Laue H, Schwier M, Link F, Peitgen HO. Medical image analysis. IEEE Pulse. 2011 Dec 1;2(6):60–70.

[12] Norouzi A, Rahim MS, Altameem A, Saba T, Rad AE, Rehman A, Uddin M. Medical image segmentation methods, algorithms, and applications. IETE Technical Review. 2014 May 4;31(3):199–213.

[13] Sahoo PK, Soltani SA, Wong AK. A survey of thresholding techniques. Computer vision, graphics, and image processing. 1988 Feb 1;41(2):233–260.

[14] Ji Z, Xia Y, Chen Q, Sun Q, Xia D, Feng DD. Fuzzy c-means clustering with weighted image patch for image segmentation. Applied Soft Computing. 2012 Jun 1;12(6):1659–1667.

[15] Dice LR. Measures of the amount of ecologic association between species. Ecology. 1945 Jul 1;26(3):297–302.

[16] Reddy GT, Reddy MP, Lakshmanna K, Kaluri R, Rajput DS, Srivastava G, Baker T. Analysis of dimensionality reduction techniques on big data. Ieee Access. 2020 Mar 16;8:54776–54788.

[17] Lee JG, Jun S, Cho YW, Lee H, Kim GB, Seo JB, Kim N. Deep learning in medical imaging: General overview. Korean Journal of Radiology. 2017 Aug 1;18(4):570–584.

[18] Kim M, Yun J, Cho Y, Shin K, Jang R, Bae HJ, Kim N. Deep learning in medical imaging. Neurospine. 2019 Dec 31;16(4):657.

[19] Alzubaidi L, Fadhel MA, Al-Shamma O, Zhang J, Santamaría J, Duan YR, Oleiwi S. Towards a better understanding of transfer learning for medical imaging: A case study. Applied Sciences. 2020 Jun 29;10(13):4523.

[20] Van der Velden BH, Kuijf HJ, Gilhuijs KG, Viergever MA. Explainable artificial intelligence (XAI) in deep learning-based medical image analysis. Medical Image Analysis. 2022 Jul 1;79:102470.

[21] Alkhalaf S, Alturise F, Bahaddad AA, Elnaim BM, Shabana S, Abdel-Khalek S, Mansour RF. Adaptive Aquila Optimizer with explainable artificial intelligence-enabled cancer diagnosis on medical imaging. Cancers. 2023 Feb 27;15(5):1492.

[22] Papanastasopoulos Z, Samala RK, Chan HP, Hadjiiski L, Paramagul C, Helvie MA, Neal CH. Explainable AI for medical imaging: Deep-learning CNN ensemble for classification of estrogen receptor status from breast MRI. In Medical imaging 2020: Computer-aided diagnosis 2020 Mar 16 (vol. 11314, pp. 228–235). SPIE.

[23] Borys K, Schmitt YA, Nauta M, Seifert C, Krämer N, Friedrich CM, Nensa F. Explainable AI in medical imaging: An overview for clinical practitioners–Beyond saliency-based XAI approaches. European Journal of Radiology. 2023 May 1;162:110786.

[24] Latif S, Brohi S, Ahmad J, Alqhatani A, Alshehri MS, Al Mazroa A, Ullah R. Explainable AI in medical imaging: An interpretable and collaborative federated learning model for brain tumor classification. Frontiers in Oncology. 2025 Feb 27;15:1535478.

[25] Jin W, Li X, Fatehi M, Hamarneh G. Guidelines and evaluation of clinical explainable AI in medical image analysis. Medical Image Analysis. 2023 Feb 1;84:102684.

Neha Garg, Priyanka Datta, Amitesh Aggarwal, Amanpreet Kaur, Yonis Gulzar

5 Deep Generative Learning: Foundations and Case Studies in Medical Image Processing

Abstract: Deep generative learning is an innovative technique in medical image processing, providing advanced solutions to issues in data augmentation, image reconstruction, modality translation, and anomaly detection. This paper discusses the theoretical fundamentals of deep generative models, such as variational autoencoders, diffusion, and generative adversarial networks models. The model has the ability to learn intricate data distributions inherently in medical imaging. It also showcased case studies that explore the effectiveness of generative learning through fundamental concepts and their practical applications in different medical operations or tasks such as MRI-to-CT synthesis, denoising low-dose CT scans, and generation of synthetic data for training on exceptional diseases. These techniques focused on how generative models can improve diagnostic precision, reduce the constraints of data scarcity, and minimize imaging artifacts. The paper also illustrates validation strategies, ethical considerations, and future research paths to ensure safe and effective use of generative models in healthcare environments.

Keywords: artificial intelligence, deep generative learning, image processing, medical, variational autoencoders

5.1 Introduction

Artificial intelligence (AI) has evolved from its initial rule-based automation systems to develop advanced learning systems that operate under uncertainty and generate original content [1]. Machine learning (ML) functions as the fundamental foundation of contemporary AI because models enhance their operational capabilities through data processing. Deep learning (DL) functions as a powerful subfield of ML, which employs deep neural networks to identify intricate hierarchical patterns in data distributions [2]. These architectures have revolutionized computer vision, speech processing, and medical image analysis because precision and automation are essential in these fields.

Acknowledgment: This research was supported by the Deanship of Scientific Research, the Vice Presidency for Graduate Studies and Scientific Research, King Faisal University, Saudi Arabia, under project KFU251224.

Note: This research was supported by the Deanship of Scientific Research, Vice Presidency for Graduate Studies and Scientific Research, King Faisal University, Saudi Arabia, under Grant No. KFU260991.

 | https://doi.org/10.1515/9783112226223-005

Deep generative learning (DGL) represents a revolutionary advancement among these developments. Generative models operate differently from discriminative models because they learn data distribution $p(x)$ to create authentic new samples that match the original dataset. The method produces realistic samples, which leads to substantial effects that include radiology scan synthesis and healthcare data augmentation and anomaly detection in low-data situations [3, 4].

5.2 Theoretical Foundations

5.2.1 Probability Distributions and Generative Models

The knowledge of training data distribution acquired by generative models enables them to generate new data points that match the training data. The development and implementation of these models requires a deep understanding of probability theory. Some of the models are as follows:

5.2.2 Probability Distributions

The basic generative models depend on probability distributions as their core concept. The distributions of probabilities serve as essential tools for modeling realistic data. Generative models utilize the Bernoulli distribution for binary data and the Gaussian distribution for continuous data.

The Gaussian distribution with its variance and mean serves as the primary distribution in models such as variational autoencoders (VAEs) [5] and Gaussian mixture models (GMMs). The models base their assumption on the fact that data can be described through multiple Gaussian distributions, which enables probabilistic data representation:

$$p(x) = \frac{1}{\sqrt{2\tau\sigma^2}} \exp \exp\left(- \frac{(x-\mu)^2}{2\sigma^2} \right) \tag{5.1}$$

This equation clarifies the probability density function of a Gaussian distribution, where μ is the mean and σ is the variance [6].

Steps:

- **Step 1:** Identify the mean (μ) and variance (σ^2) of your dataset. These describe the "center" and "spread" of the data.
- **Step 2:** Plug any value of x into the formula to get how likely that value is under this distribution.
- **Use:** Helps generative models simulate realistic continuous data, like image intensity values.

5.2.3 Bayes' Theorem

The mathematical framework of Bayes' theorem serves as a fundamental principle for generative models, especially those based on Bayesian inference. The theorem enables the modification of hypothesis probability through the addition of new evidence or data. The theorem finds application in models including generative adversarial networks (GANs) and Bayesian networks.

The theorem is as follows:

$$p(x) = \frac{p(\theta)p(\theta)}{p(x)} \tag{5.2}$$

In eq. (5.2), $p(x)$ is the posterior probability, $p(\theta)$ representing likelihood, $p(\theta)$ signifying the prior probability, and the marginal likelihood is given by $p(x)$ [6, 7].

Steps:
- **Step 1:** Start with a prior belief $p(\theta)$ and the likelihood of evidence if that belief is true $p(\theta)$
- **Step 2:** Combine them and normalize using the total evidence probability $p(x)$.
- **Use:** Allows models to update predictions as new data (e.g., scans) becomes available.

The theorem enables the process of updating model parameter beliefs through new data, which is essential for generative model iteration.

5.2.4 Maximum Likelihood Estimation (MLE)

The model determines its parameters through this process. The MLE method determines the parameter value that maximizes the model's probability of observing the given data. The likelihood function operates as follows:

$$L(x) = p(\theta) \tag{5.3}$$

MLE is used to identify the optimal parameters θ for the function. This method is widely used in models like VAEs and hidden Markov models (HMMs) use this method to learn the model parameters by maximizing the training data likelihood [5].

Steps:
- **Step 1:** Think of θ as the parameters of the model (e.g., weights in a neural network), and x as the observed data.
- **Step 2:** The likelihood function $L(x)$ tells us how likely the observed data x is given a particular setting of model parameters θ.
- **Step 3:** MLE determines the value of θ which maximizes this likelihood by adjusting the model to make the observed data as probable as possible.

5.3 Types of Generative Models

Generative models learn data distribution patterns to generate new, realistic data samples that match the original dataset [8]. The models vary in their approach to defining data distribution and their learning and sampling methods. The following section describes the primary generative models together with their theoretical foundations and medical image analysis applications. Types of generative models are shown in Figure 5.1.

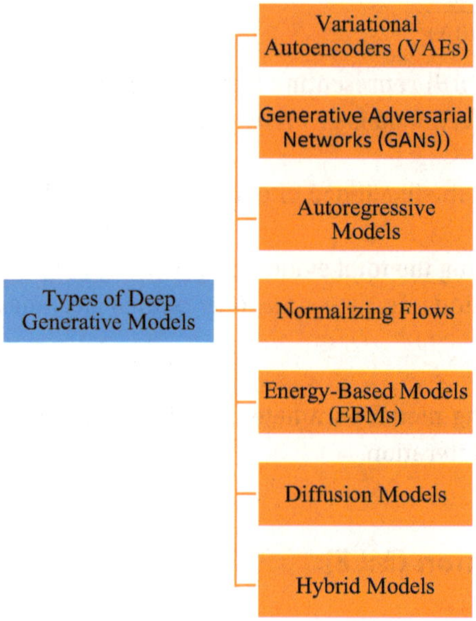

Figure 5.1: Categories of deep generative models.

5.3.1 Variational Autoencoders (VAEs)

Type: Explicit probabilistic model with latent variables
Learning Method: Variational inference
Concept: VAEs represent neural network models that transform data into compact latent spaces before restoring the original data form. The probabilistic nature of VAEs differs from traditional autoencoders because they encode each input into a distribution instead of a single fixed vector [9]. The model learns to generate new data through sampling from the distribution it has learned.
Use in Medical Imaging:
- Detecting abnormalities by comparing original and reconstructed images.
- Denoising medical images (e.g., removing artifacts from MRI scans).

- The model generates synthetic training data to address the limited availability of labeled images.
- The model predicts disease progression through time by interpolating in the latent space.

5.3.2 Generative Adversarial Networks (GANs)

Type: Implicit model without direct likelihood estimation
Learning Method: Adversarial training
Concept: GANs contain two opposing neural networks, which operate as a generator to produce realistic data and a discriminator to determine if a sample is generated or real [10]. The generator enhances its performance through learning discriminator deception, which produces highly realistic output results. GANs produce images that are both sharper and more detailed than the outputs of VAEs.
Use in Medical Imaging:
- Generating authentic medical scans for data augmentation.
- Transforming images between different imaging modalities (e.g., CT to MRI).
- Enhancing low-resolution images (super-resolution).
- Filling in missing parts of images (inpainting damaged scans)

5.3.3 Autoregressive Models

Type: Explicit model using sequential dependencies
Learning Method: Maximum likelihood estimation
Concept: The data generation process of autoregressive models occurs through sequential predictions where each new part depends on all previously generated parts. The models produce images through pixel-by-pixel generation and signals through point-by-point generation [11]. The model uses all previously generated elements to predict each new element.
Use in Medical Imaging:
- The generation of biomedical time series data includes ECG and EEG signals.
- Modeling and synthesizing medical imaging slices sequentially.
- Predicting disease progression based on temporal imaging data.

5.3.4 Normalizing Flows

Type: Explicit model with invertible transformations
Learning Method: Exact likelihood estimation

Concept: Normalizing flows convert basic distributions into complex ones by performing a sequence of operations that can be reversed. The approach enables both efficient sampling and exact probability calculations of data [12]. These models serve best when interpretability and exact probability estimation are required.

Use in Medical Imaging:
- Estimating variability in medical predictions
- Improving interpretability in diagnostics
- Restrained image generation (e.g., varying tumor location or size)
- Improving segmentation tasks by modeling shape variability

5.3.5 Diffusion Models

Type: Hybrid model with noise-based generation
Learning Method: Iterative denoising process
Concept: The process of data generation in diffusion models begins with random noise, which evolves into meaningful images through multiple transformation steps. The models operate by mimicking a reversed noise process, which refines the input data until it matches real data [13]. The models have recently achieved remarkable results in producing high-quality images.

Use in Medical Imaging:
- High-resolution medical images can be generated from noisy data.
- Reconstructing missing or degraded image regions.
- Simulating rare conditions for training diagnostic systems.
- Improving low-resolution or low-dose scans in radiology.

5.3.6 Energy-Based Models (EBMs)

Type: Implicit probabilistic model
Learning Method: Energy scoring and sampling
Concept: The energy-based models (EBMs) assign an energy score to each possible data configuration, where lower energy corresponds to more realistic data. The model learns to assign low energy to true data and high energy to unlikely data [14]. These models are flexible and can be applied to various data types but often require advanced sampling techniques for training.

Use in Medical Imaging:
- Outlier and anomaly detection in diagnostic workflows
- Learning from disorganized medical datasets
- Providing confidence points for predictions in critical applications
- Refine clinically insignificant noise in image data

5.3.7 Hybrid Models

Type: Combined architectures leveraging multiple model types
Learning Method: Composite loss functions and joint optimization
Concept: Hybrid models unite different generative models to leverage their strengths. A model can leverage the structured latent space of VAEs together with the image-sharpness capabilities of GANs [15]. These models work best when both high image fidelity and interpretability need to be achieved.
Use in Medical Imaging:
- Cross-domain translation, like bringing together imaging with clinical reports
- Learning rich illustrations for complex diagnoses [16]
- Construct multimodal generative systems (e.g., image with text with genomics)
- Generating understandable and controllable fabricated images

5.4 Case Studies

5.4.1 Case Study 1: Tumor Detection in Brain MRI Scans Using Deep Learning

Objective: This case report aimed to design a DL model that is capable of detecting brain tumors directly from MRI scans, which may assist radiologists in reducing the time spent on diagnosis.

Methodology

1. ***Data Collection:*** A set of MRI scan images, labeled as healthy or having a tumor, was used. A commonly used dataset is the "Brain Tumor Segmentation dataset." It includes multimodal images of MRI (T1, T2, T1*c*, FLAIR). The dataset has 3,264 MRI slices, with a training set of 2,600 images and a test set of 664 images [17]. The images are sourced from various Kaggle websites and cover various tumor types and grades, providing rich variability for model generalization.
2. ***Preprocessing:***
   - Image Resizing: All MRI slices were resized to 224 × 224 pixels to ensure consistency with the input size necessities of the CNN architectures.
   - Normalization: The pixel data were also normalized to a range between 0 and 1, by means of the min-max normalization technique; hence, the model training process is stable.

- Data Augmentation: Approaches like random rotation (±15°), vertical and horizontal flipping, and zooming [0.8 to 1.2] range were used to artificially expand the extent of the training points or data and improve model robustness.

3. ***Model Architecture***:
   - A CNN based on pretrained VGG16 and ResNet50 architectures is used.
   - VGG16 has 13 convolutional layers and three fully connected layers, around 138 million parameters.
   - ResNet50 has 50 layers with residual connections, comprising ~25.6 million parameters. These residual connections aid in overcoming vanishing gradient problems and allow deeper architectures to be trained competently.
   - Transfer learning is applied by freezing early convolutional layers and fine-tuning the top layers for tumor classification.
4. ***Training:***
   - Loss Function: Binary cross-entropy is utilized to distinguish between tumor and non-tumor images.
   - Optimizer: The Adam optimizer is used with 0.0001 as an initial learning rate.
   - Training Epochs: The model is trained for 50 epochs with initial stopping depending on validation loss to avoid overfitting.
   - Batch Size: A size of 32 is used.
   - Hardware: Training is conducted on an NVIDIA Tesla V100 GPU with 32 GB VRAM using TensorFlow 2.12 on a Linux server.
5. ***Evaluation:*** The model performance is evaluated by the skill of the model to correctly identify tumors and healthy brain tissues. Key evaluation metrics included:
   - Accuracy: Percentage of appropriately identified images.
   - Sensitivity: The potential to correctly identify tumor-present images.
   - Specificity: The ability to correctly recognize healthy images. Both specificity and sensitivity exceeded 95%, showing strong diagnostic capability

Results: In addition to helping radiologists, the model could also help reduce the likelihood of misdiagnoses and improve the speed of diagnoses. Both the sensitivity and specificity were high, which means that the model did well in identifying tumors and in avoiding false positives.

Challenges

- *Data Quality*: Some MRI scans had lower spatial resolution or more noise than others, which hurt the model's performance on some images.
- *Interpretability:* The model's decision-making process was not very explicit, which made it hard for doctors to understand why some regions were flagged as tumors.

Future Directions

- *Improving Explainability*: Techniques like Grad-CAM could be employed to visualize what parts of the MRI images led to the tumor detection decision
- *Multimodal Learning*: Combining various MRI modalities (T1, T2, FLAIR) may enhance the model's tumor detection accuracy.

5.4.2 Case Study 2: Detection of Pneumonia in Chest X-Rays Using Deep Learning

Objective: To design a DL model able to identify pneumonia in images of chest X-rays, assisting radiologists to identify pneumonia more quickly and accurately.

Methodology

1. ***Data Collection***: The chest X-ray 14 dataset is used; it has thousands of labeled images with diseases like normal conditions and pneumonia. The dataset is taken from the Kaggle website. It encompasses over 112,000 frontal view X-ray images of 30,805 unique patients with 14 different types of disease labels, including pneumonia [18]:
   - For this work, a subset of the dataset is used and relabeled into two categories: "Normal" and "Pneumonia."
   - A total of 5,863 images were used, with a training dataset of 4,096 images, 936 images used for validation, and 831 images for testing purposes.
   - The dataset has metadata like patient gender, age, and disease labels, which were omitted in this binary classification task.
2. ***Preprocessing***: This section of the article illustrates the preprocessing process as follows:
   - Resizing of image: All images of chest X-ray were resized to 224 × 224 pixels to be well-suited with the input requirements of the VGG16 architecture.
   - Normalization: The values of pixels were normalized to a standard range of 0 to 1 to assist faster convergence while training using the min-max normalization method.
   - Data Augmentation: The approach of augmentation, like zooming (0.8–1.2), flipping, and rotating (±15°) was done to enhance the size of the dataset and shrinkage overfitting.
3. ***Model Architecture:***
   - A CNN based on pretrained VGG16 architectures is used.
   - Thirteen convolutional layers are present in the model and 3 fully connected layers, almost 138 million parameters.

   - Transfer learning is applied by using pretrained weights on ImageNet, with the initially convolutional base frozen, followed by fine-tuning of the top layers to adapt the pneumonia classification approach.
   - A sigmoid activation function is utilized in the final output layer for binary classification.
4. ***Training***: This part of the paper discusses the training approach used as follows:
   - Loss function: Classification between pneumonia and normal conditions is done through binary cross-entropy loss.
   - Optimizer: To optimize the model weights during training, Adam optimizer is utilize with a learning rate of 0.0001 for efficient weight updates.
   - Batch Size: Set of 32.
   - Training Epochs: Model is trained for 25 epochs with early stopping based on loss on validation to avoid overfitting.
   - Hardware: Training is conducted on a system with NVIDIA GeForce RTX3090 GPU, using Keras with TensorFlow backend.
5. ***Evaluation:*** The performance of the model is assessed on its skill to classify images properly, either as "pneumonia" or "normal." The evaluation parameters are as follows:
   - Accuracy: The percentage of correctly determined chest X-rays.
   - Precision: Proportion of correctly detected pneumonia cases among all predicted pneumonia.
   - Sensitivity: The ability to identify actual pneumonia cases from all images.
   - Area under the receiver operating characteristic curve (AUC-ROC): It is used to evaluate the overall classification performance of the model.

The model has an accuracy of 93.4%, recall of 94.7%, AUC-ROC score of 0.96, and a precision of 92.1%, illustrating strong diagnostic performance.

Results

The DL model is effective in identifying pneumonia in chest X-ray images. It can determine pneumonia cases accurately and also classify normal images with a high level of confidence. The model's ability to determine pneumonia cases and differentiate them from normal cases is commendable, hence a reliable tool to assist healthcare professionals.

Conclusion

These case studies demonstrate the potential of DL to revolutionize medical image processing. In together brain tumor detection and pneumonia detection, DL models demonstrated high potential for supporting healthcare professionals in making faster and more accurate diagnoses. Continued development in model interpretability, data quality, and generalization will lead to even greater advances in medical image analysis.

5.5 Applications of Deep Generative Learning

Applications of deep generative learning [19, 20] have been expanding across various fields. Below is an outline of some key areas:

5.5.1 Image and Video Generation

Super-resolution: With DGM, including GANs (Generative Adversarial Networks), videos and images can be increased to high resolutions by producing finer characteristics that were not in the actual version [4].
Inpainting: This application fills in absent or damaged parts of a picture or video by generating information that perfectly integrates with the encircling area [3]. It is commonly used in restoration images and editing.
Style Transfer: Generative models can add innovative or stylistic changes to pictures or videos, like changing a photograph into the Gogh-style of a painting [21].

5.5.2 Data Augmentation

Synthetic Data Generation: Generative models produce artificial data to enhance training datasets, specifically in situations where labeled data is minimal [22]. This technique helps to instruct discriminative algorithms more effectively through generating diverse data modifications, which enhance stability and adaptability in ML models [20].

5.5.3 Drug Discovery and Healthcare

Molecular Generation: Deep generative models are utilized to produce new chemical compounds as well as molecules that might have valuable medical properties. This increases the drug creation process through investigating large chemical areas and suggesting novel substances for testing [23].

5.5.4 Text and Natural Language Processing

Text Generation: Generative models, huge language models (such as the GPT series), generate human-like text. They are employed in content creation, chatbots, and translation tasks [24].
Language Models: Models such as GPT and BERT are utilized for NLP tasks, like answering questions, summarization, and text classification [25]. These models develop to generate coherent and spatially accurate text depending on the input they receive [20].

These applications highlight how deep generative learning has impacted various industries, from creative areas such as picture and video production to essential areas such as natural language processing and healthcare.

5.6 Challenges and Future Directions

This section discusses both the technical issues and fascinating new avenues in deep generative learning, in addition to the ethical concerns emerging from its applications [8].

5.6.1 Challenges in Deep Generative Learning

Scalability, Mode Collapse, and Stability Issues: Deep generative models, which involve GANs, face issues regarding mode collapse (where the model fails to produce diverse outputs), training instability (particularly in hostile models), and scalability to bigger datasets and complex usages. Coordinating the generator and discriminator throughout training can be complicated, which leads to poor performance [24, 26].
The Degree of Complexity: Deep generative models are resource-intensive, demanding a significant and enormous amount of processing power and memory. Training these models on massive datasets or high-dimensional information (such as images and video) can be expensive in terms of time and resources and frequently requires specialist hardware such as GPUs or TPUs [27].

5.6.2 Emerging Trends and Research Areas

Hybrid Models Combining Generative Learning with Reinforcement Learning (RL): Incorporating generative models with reinforcement learning (RL) can improve decision-making processes, particularly in settings where modeling the world is essential (such as robotics and games). These models can help recreate environments or create data that enhances the performance of RL models [14].

Generative Models in Self-Supervised Learning: Generative models are being employed in self-supervised learning more and more to create huge amounts of labeled data from datasets without labels. This approach aids in the training of discriminative models without depending heavily on manually labeled data, which is expensive and time-consuming [28].

Ethical Considerations (Deepfakes and Data Privacy): With the increase of deepfakes, or AI-generated fake media, and worries about data privacy, there is growing attention on the ethical consequences of generative models. Concerns include the possibility of misuse of generative technologies devices in producing deceptive media, biased results, and privacy issues when models are employed on sensitive data [29, 30].

5.7 Conclusion

This chapter delved into the core ideas behind learning, a key aspect of machine learning that empowers models to create fresh and real-life-like information. This chapter started by revisiting the basics of learning and generative modeling, demonstrating structures like GANs, VAEs, and autoregressive models. Every method has its advantages and uses, from crafting images to enhancing tasks in natural language processing. Similarly, the chapter explored the algorithms and concepts that back up generative models by emphasizing how they grasp the foundational data patterns under the surface layer of data available to all. Enlightening these workings and principles helps us gain an insight into the hurdles faced in this domain, such as mode collapse in GAN architectures or finding the right balance between precision in reconstruction and maintaining a structured latent space within VAE frameworks. The progressions in learning hold great potential for transforming industries such as computer vision and healthcare while driving innovations in scientific research fields. With this progress come ethical dilemmas regarding the misuse of models, notably, in creating deepfake content and synthetic data. Moving ahead requires research efforts to navigate these challenges while improving the reliability and adaptability of these models for practical use cases. Master the basics of learning to pave the way for progress in this captivating field and open up fresh opportunities in the realm of AI.

Consent for Publication: Not applicable.

Conflicts of interest: Not applicable.

References

[1] Gupta A, Hathwar D, Vijayakumar A. Introduction to AI chatbots. International Journal of Engineering Research and Technology. 2020 Jul 11;9(7):255–258.

[2] Badillo S, Banfai B, Birzele F, Davydov II, Hutchinson L, Kam-Thong T, et al. An introduction to machine learning. Clinical Pharmacology & Therapeutics. 2020 Mar 3;107(4):871–885.

[3] Shiloh-Perl L, Giryes R Introduction to deep learning. arXiv (Cornell University). 2022 Feb 25.

[4] Ruthotto L, Haber E. An introduction to deep generative modeling. GAMM-Mitteilungen. 2021 May 28;44(2).

[5] Bishop CM. Pattern recognition and machine learning. Springer; 2006.

[6] Murphy KP. Machine learning: A probabilistic perspective. Cambridge (Ma): Mit Press; 2012.

[7] A1.Lynch SM. Bayesian statistics. Elsevier eBooks; 2005 Jan 1 135–144.

[8] Saxena D. Generative adversarial networks (GANs): Challenges, solutions, and future directions. ACM Computing Surveys (CSUR). 2021 May 8;54(3):1–42.

[9] Kingma DP, Welling M Auto-encoding variational Bayes. arXiv (Cornell University). 2013 Dec 20.

[10] Shahin M, Panahi A, Krim H Sparse generative adversarial network. arXiv (Cornell University) 2019 Oct 1.

[11] Chen Y, Liu J, Peng L, Wu Y, Xu Y, Zhang Z. Auto-encoding variational Bayes. Cambridge Explorations in Arts and Sciences. 2024 Feb 7;2(1).

[12] Neal RM. Probabilistic inference using Markov chain Monte Carlo methods. Department of Computer Science University of Toronto; 1993 Sep.

[13] Salakhutdinov R. Learning deep generative models. Annual Review of Statistics and Its Application. 2015 Apr 10;2(1):361–385.

[14] Karthika S, Durgadevi M. Generative adversarial network (GAN): A general review on different variants of GAN and applications. In 2021 6th International Conference on Communication and Electronics Systems (ICCES) 2021 Jul 8.

[15] Tomczak JM, Welling M VAE with a VampPrior. arXiv (Cornell University). 2017 Jan 1.

[16] Kobyzev I, Prince SJD, Brubaker MA. Normalizing flows: An introduction and review of current methods. IEEE Transactions on Pattern Analysis and Machine Intelligence. 2020;1–1.

[17] Nickparvar M Brain tumor MRI dataset. www.kaggle.com. 2021.

[18] Mooney P Chest x-ray images (pneumonia). www.kaggle.com. 2018.

[19] Croitoru FA, Hondru V, Ionescu RT, Shah M. Diffusion models in vision: A survey. IEEE Transactions on Pattern Analysis and Machine Intelligence. 2023;45(9):1–20.

[20] Younesi A, Ansari M, Fazli M, Ejlali A, Shafique M, Henkel J. A comprehensive survey of convolutions in deep learning: Applications, challenges, and future trends. IEEE Access. 2024;12:41180–41218.

[21] Bengesi S, El-Sayed H, Sarker MK, Houkpati Y, Irungu J, Oladunni T. Advancements in generative AI: A comprehensive review of GANs, GPT, autoencoders, diffusion model, and transformers. IEEE Access. 2024;12:69812–69837.

[22] Rishu KV, Kumar A. Facial Expression Recognition with CNN-SVM for Emotional State Classification. 2023 Aug 18.

[23] Sarker IH. Deep learning: A comprehensive overview on techniques, taxonomy, applications and research directions. SN Computer Science. 2021 Aug 18;2(6).

[24] Hu Y, Hu W, Quigley A. Towards using generative AI for facilitating image creation in spatial augmented reality. In 2023 IEEE International Symposium on Mixed and Augmented Reality Adjunct (Ismar-adjunct) 2023 Oct 16 (pp. 441–443).

[25] Kukreja V, Sharma R, Vats S. A Hybrid Deep Learning Approach for Multi-Classification of Heritage Monuments Using a Real-Phase Image Dataset. 2023 Aug 3.

[26] Manduchi L, Pandey K, Bamler R, Cotterell R, Däubener S, Fellenz S, et al. On the challenges and opportunities in generative AI. arXiv.org. 2024.

[27] Baidoo-Anu D, Ansah LO. Education in the Era of generative Artificial Intelligence (AI): Understanding the potential benefits of ChatGPT in promoting teaching and learning. Journal of AI. 2023 Dec 31;7(1):52–62.
[28] Kobyzev I, Prince SJD, Brubaker MA. Normalizing flows: An introduction and review of current methods. IEEE Transactions on Pattern Analysis and Machine Intelligence. 2020;1–1.
[29] Huang Q, Kieffer K. An intelligent internet of things (IoT) sensor system for building environmental monitoring. Journal of Mobile Multimedia. 2019 Jun 6.
[30] Kukreja V, Sharma V, Sharma R. Utilizing a Hybrid CNN-LSTM Model for Automated Severity Classification of Rice Panicle Blast Disease. 2024 Jan 24;1–5.

Ashith Tripathi, Sachitanand Singh, Padamandala Krishnasri,
Iqra Jamal Khan', Sanskriti Singh

6 Explainable AI for Smart Healthcare

Abstract: Artificial intelligence (AI) has revolutionized healthcare by enabling patient monitoring, boosting diagnosis accuracy, and personalizing treatment. The opacity of deep learning models puts trust, interpretability, and regulatory compliance at risk. Explainable AI (XAI) seeks to bridge this gap by making AI-generated decisions transparent, understandable, and accountable. By making AI system decision-making processes more transparent, XAI aids physicians in comprehending, validating, and refining AI-generated projections. XAI methods such as attention maps, heatmaps, and saliency algorithms aid in the comprehension of radiological scans, fundus images, and histological slides in medical imaging, boosting confidence and lowering diagnosis errors. In line with clinical experience, doctors can verify AI-assisted diagnosis by combining XAI with medical imaging modalities.

XAI makes custom treatment plans more understandable by using prediction models. For expected outcomes in cardiovascular diseases, diabetes progression, and inherited defects, it can provide human-comprehensible explanations. XAI helps to better understand molecular interactions in drug development and precision medicine, thereby enabling the choice of focused treatments. XAI resolves abnormalities found in ECG, EEG, and continuous glucose monitoring systems, thus allowing wearable healthcare technology and real-time patient monitoring.

XAI implementation in healthcare presents challenges, including computational complexity, trade-off in model performance, and the need of thorough validation processes. Guarantee of repeatability and equality of AI-driven assessments among different patient groups depends on standardized methods and standards. Deep learning, combined with rule-based or knowledge-driven systems, will help hybrid AI models in AI-assisted healthcare to be interpretable and transparent.

Keywords: explainable AI (XAI), smart healthcare, medical imaging, AI ethics, personalized medicine, trustworthy AI, federated learning

6.1 Introduction

Artificial intelligence (AI) has transformed several fields in recent years; one of the most important fields is undoubtedly healthcare. AI used in clinical systems has enhanced predictive analytics, patient monitoring, treatment planning, and diagnosis. Despite its achievements, the “black-box” aspect of AI models – the lack of transparency regarding decision-making processes – remains a significant barrier to the use of AI in healthcare [1].

 | https://doi.org/10.1515/9783112226223-006

This chapter examines the new field of explainable AI (XAI), which strives to enhance the transparency, reliability, and interpretability of AI systems, especially for physicians and healthcare professionals. Understanding and defending AI-driven judgments is not only beneficial but also necessary in the healthcare industry when human life is at risk. XAI enables more accountability and improved decision-making by connecting medical professionals with sophisticated machine learning algorithms.

6.1.1 Historical Evolution of AI in Medicine

- Rule-based systems and early AI models were introduced in medical decision-making during the 1950s and 1970s.
- The 1980s and 2000s saw the development of machine learning techniques and their use in diagnostic tools, such as CAD systems for mammography.
- From the 2010s to the present, medical practice incorporates instantaneous AI-facilitated diagnosis and deep learning advancements, and AI is integrated with massive volumes of data [2].

6.1.2 Transformational Impact of AI in Healthcare

AI has greatly increased medical care's effectiveness, precision, and accessibility. Among the major effect areas are:

AI in Disease Diagnosis

AI improves disease detection speed and accuracy, especially in radiology, pathology, and dermatology. For instance:

- AI models are better than human radiologists in mammography breast cancer detection [3].
- Retinal fundus images are evaluated by deep learning techniques to detect diabetic retinopathy (DR) with notable accuracy [4].

AI in Personalized Medicine

AI has greatly increased precision medicine by tailoring treatments based on individual patient profiles.

- Genomics, enabled by AI, helps to forecast therapy response and cancer susceptibility [5].
- ML models assess genetic variations to propose personalized pharmaceutical therapies.

AI in Drug Discovery and Development

AI accelerates drug identification by analyzing molecular structures and predicting drug-target interactions:

- Pharmaceutical design, improved by AI, quickly found possible COVID-19 treatments [6].
- Using AI, IBM Watson and Google DeepMind aim to streamline drug development procedures [7].

AI in Surgical Assistance and Robotics

Robotic systems, driven by AI, help with precision surgery, which enhances patient outcomes:

- The da Vinci Surgical System employs robotic arms powered by AI for minimally invasive surgeries [8].
- Augmented reality (AR), supplemented by AI, offers surgeons real-time visual aids [9].

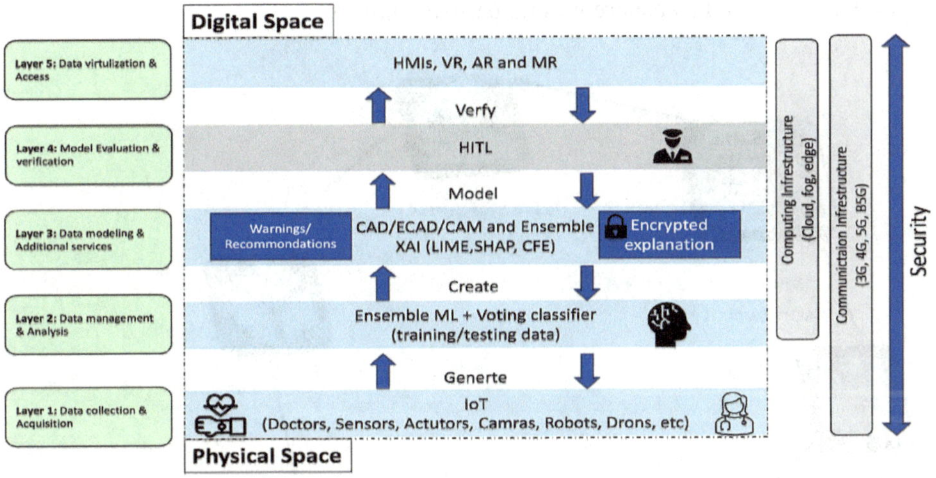

Figure 6.1: Adversarial examples in XAI-enabled decision trees for the smart healthcare systems.
Source: Reproduced from Imam [15], under the Creative Commons Attribution (CC BY) license.

The integration of XAI into digital twins (DTs) for smart healthcare systems is depicted in Figure 6.1, which highlights the interplay between digital and physical areas. IoT devices that produce real-time data, such doctors, sensors, actuators, cameras, robots, and drones, make up the physical space. ML models and voting classifiers are used to assess and manage this data, and they are further modeled and evaluated in the digital realm. Transparent model explanations are provided via XAI approaches such as LIME, SHAP, and CFE, together with encrypted explanations to guarantee security and warnings/recommendations. Human oversight is made possible by the human-in-the-loop (HITL) process, which evaluates the model outputs. Lastly, visualization and system interaction are made easier by human-machine interfaces, virtual reality (VR), AR, and mixed reality (MR) [10].

6.2 The Need for XAI in Healthcare

XAI is essential in the healthcare sector because it solves issues with transparency, confidence, regulatory compliance, and clinical decision support. AI-driven healthcare solutions, especially DL and machine learning (ML) models, frequently function as "black-box" systems, making it difficult for physicians or patients to comprehend how they make decisions. Lack of transparency raises concerns regarding reliability, ethical ramifications, and acceptance by medical experts.

XAI seeks to bridge this gap by emphasizing the interpretability of AI models, offering explanations for forecasts, and ensuring that AI-driven healthcare solutions align with clinical knowledge. The necessity of XAI in healthcare is demonstrated by three important dimensions: clinical decision support, ethical and regulatory and legal challenges, and transparency and trustworthiness.

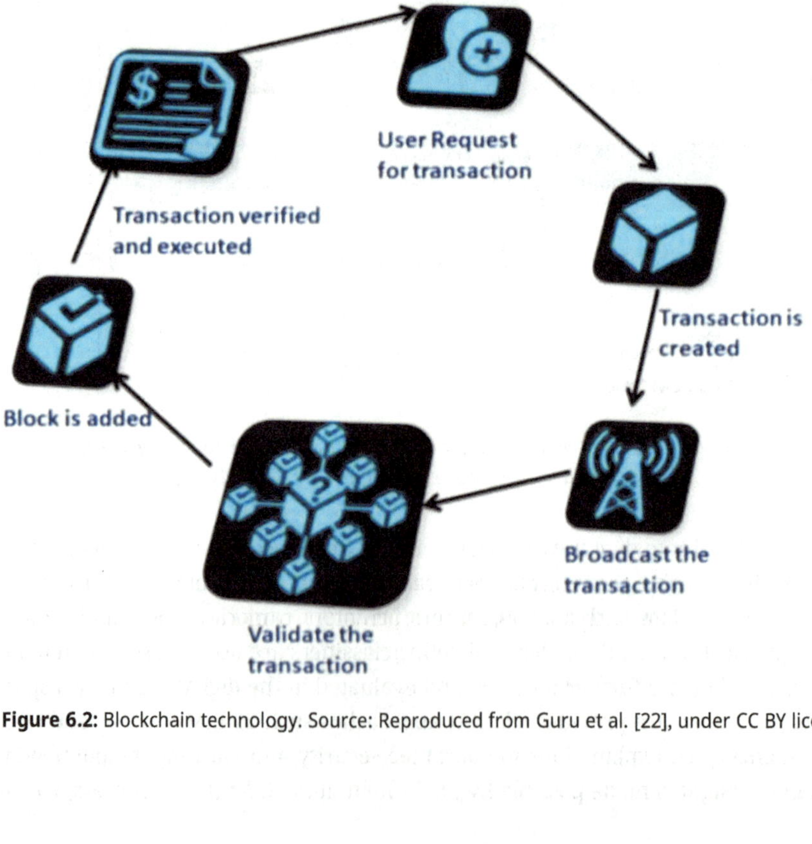

Figure 6.2: Blockchain technology. Source: Reproduced from Guru et al. [22], under CC BY license.

The blockchain transaction process is depicted in Figure 6.2, guaranteeing data management security, transparency, and trust. A user initiates the process by requesting a transaction, which is subsequently generated and transmitted to the network. A distributed network of nodes authenticates the transaction and reaches consensus before it is included as a block on the blockchain. The transaction is finally verified and carried out, so it becomes an unchangeable ledger component [11].

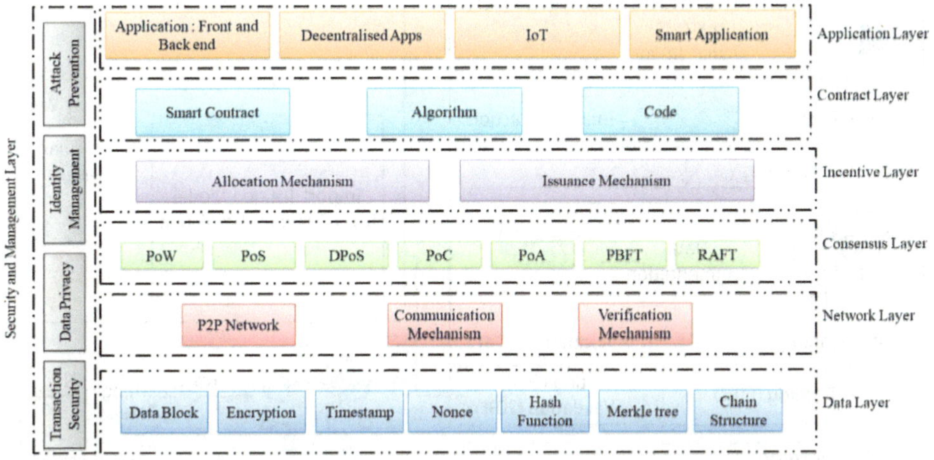

Figure 6.3: Blockchain layers. Source: Reproduced from Guru et al. [22], under CC BY license.

XAI, combined with blockchain technology, can improve data security, integrity, and transparency in patient records, clinical decision-making, and AI-driven diagnostics in the healthcare space. XAI lessens the possibility of mistakes, prejudices, and opaque decision-making in important healthcare applications by guaranteeing that AI models follow a verifiable and explainable approach. Blockchain guarantees data authenticity, therefore strengthening this even more by securely capturing AI-generated insights and allowing confidence between patients and healthcare professionals.

Within a smart healthcare system, Figure 6.3 shows how XAI, DT, and blockchain technologies might all be combined. The system operates in two physical and digital spheres. IoT devices in the physical sense comprise cameras, sensors, actuators, and doctors compiling healthcare data. Subsequently, the data is examined in the digital space using ML and Voting Classifiers to ensure accurate forecasts and well-informed decisions.

XAI approaches such LIME, SHAP, and CFE are used to clarify AI decisions, thereby generating alerts, recommendations, and encrypted explanations for healthcare practitioners, improving security and transparency. HITL validation is used further to guarantee reliability of the model. To improve visualization, AI-generated insights are finally merged with AR, VR, and MR.

Blockchain technologies guarantee data integrity, security, and verifiability in healthcare AI models, thereby strengthening this architecture. XAI guarantees that AI-driven healthcare solutions are transparent, responsible, and ethical by making AI decisions understandable, thus inspiring confidence among doctors and patients [12].

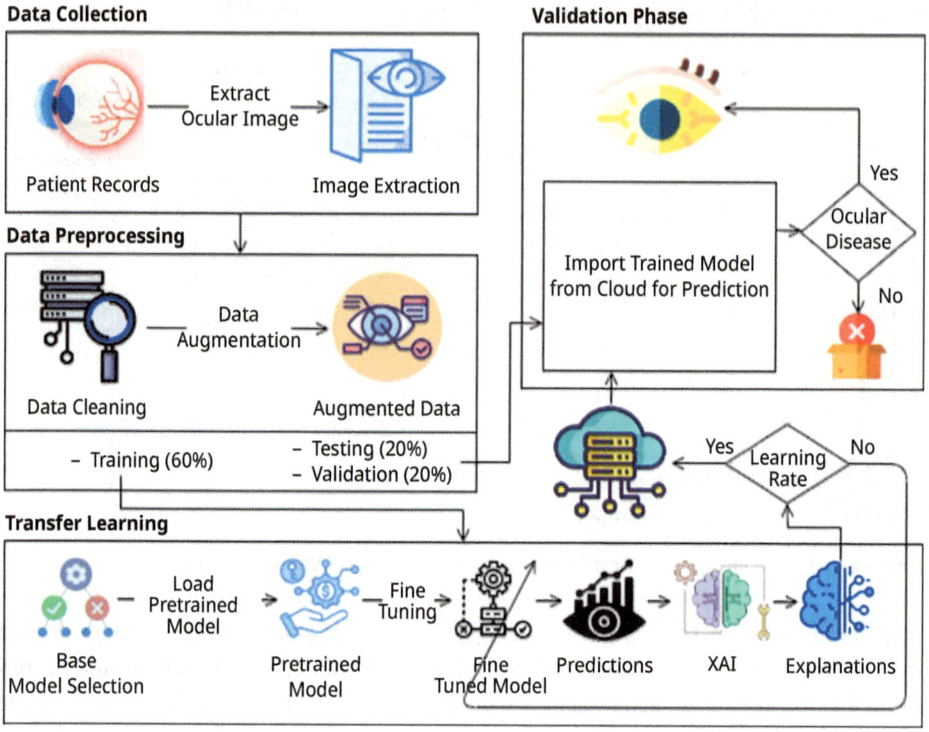

Figure 6.4: Model for ocular disease prediction using XAI. Source: Reproduced from Abbas et al. [23], under CC BY license.

Based on XAI, Figure 6.4 shows a model for eye disease prediction, thus fostering openness and confidence in AI-assisted medical diagnostics. The process starts with data collection where patient records provide ocular images. Images undergo cleaning, augmentation, and partitioning into training (60%), testing (20%), and validation (20%) datasets during the data preparation phase.

Using transfer learning, the method refines a pretrained AI model for eye illness diagnosis. Forecasts are generated, and XAI methods provide thorough justifications for these AI decisions, so helping clinicians to understand and validate the model's justification.

A trained model from the cloud projects the existence of a disease throughout the validation process. Learning rates are adjusted in circumstances of unknown forecasts to improve the accuracy of the model. In healthcare AI, explainability is crucial since it enhances diagnosis accuracy, clinical decision-making, and patient confidence by enabling AI-driven predictions interpretable, instead of a "black-box" solution [13].

6.3 Transparency and Trust: Why Clinicians and the Patients Need to Understand AI-Driven Decisions

6.3.1 The Black-Box Problem in AI Healthcare Models

Sometimes, conventional AI models – especially those based on deep learning – operate as black boxes, making even their developers unable to fully explain the processes underlying their particular predictions [14]. Lack of openness causes mistrust among doctors, patients, and the authorities of regulations.

Before they are included in clinical practice, doctors have to understand, validate, and evaluate AI-generated diagnoses. Lack of openness could cause one to be reluctant to accept AI-generated advice.

- **Patients' viewpoint:** Many times, patients ask for explanations for medical advice or risk assessments produced by AI. Lack of explainability could cause one to be reluctant to accept results produced by AI [15].

6.3.2 Real-World Example: AI in Medical Imaging

Examine AI-powered DR screening methods. Google's DeepMind AI demonstrated remarkable accuracy in identifying DR from retinal fundus images. However, earlier versions were difficult to understand, which made ophthalmologists skeptical of their application. Researchers effectively identified the retinal regions impacting the AI's diagnosis using XAI techniques, such as saliency maps and heatmaps, which increased the clinicians' trust [16].

6.3.3 XAI for Enhancing Trust

XAI builds confidence by:

- Giving clinically relevant explanations: AI must back up its conclusions with examples from past cases or clinical guidelines [17].
- Enhancing physician-AI collaboration: Physicians are more receptive to AI when they comprehend its cognitive processes.

- Informing patients of AI's function: XAI can be used in patient engagement tools to provide a clear explanation of risk factors and AI suggestions [18].

6.4 Regulatory and Ethical Concerns: Compliance with the Healthcare Regulations (e.g., FDA and GDPR)

6.4.1 Ethical Issues in AI-Driven Healthcare

Bias and fairness: AI models trained on distorted data may produce biased results. An AI for diagnosing skin cancer, which was mostly trained on fair-skinned individuals, does not function well on darker skin tones [19].
Patient privacy: Concerns regarding data security and confidentiality are exacerbated by AI's need for access to vast amounts of patient data [20].
Accountability in AI decisions: Is the AI system or the physician legally liable when a diagnosis is overlooked? XAI encourages accountability by elucidating the rationale behind AI predictions [21].
Example: AI in Stroke Prediction and Regulatory Approval
Approved for FDA use, a deep learning model for stroke risk prediction ran into problems because of its closed nature. Regulators required the system to:
- Provide confidence statistics for every choice.
- Offer every decision confidence measurements.
- Give doctors the ability to override AI forecasts, if judged essential [22].

Decision trees and SHAP (SHapley Additive Explanations) values were then added into the system to comply with FDA rules, thereby guaranteeing ethical use of AI [23].

6.5 Enhancing Clinical Decision Support: How XAI Aids in Reducing Diagnostic Errors

6.5.1 AI as a Decision-Support Tool

Medical professionals should be complemented by AI, not replaced by it. XAI makes sure that AI-generated insights enhance medical knowledge and clinical reasoning. This is particularly notable in:
- **Medical Imaging**: AI-driven analysis of MRI and CT scans must explain why an aberration is classified as benign or malignant [24].

- **Pathology and Histopathology:** AI models must emphasize how cellular factors affect their classification of cancer [25].
- **ICU Patient Monitoring:** AI alerts ought to explain any abnormalities found in patients' vital signs [26].

Example: AI in Sepsis Detection
Sepsis is one of the worst illnesses that requires prompt detection. Hospitals now have early warning systems, driven by AI, but first versions lacked interpretability and led to misdiagnoses. These systems regularly offer explainability strategies (e.g., LIME, SHAP, counterfactual explanations) to combine:
- A **breakdown of contributing factors** (e.g., vital signs, inflammatory markers)
- A **probability score with confidence intervals**
- Comparing similar risk factor patient situations from past experiences
- These improvements lowered false alarms, raised doctor trust, and improved sepsis prediction accuracy

6.6 Core Concepts of XAI

Especially in important sectors like healthcare, XAI is a set of techniques and approaches that make AI models understandable, transparent, and interpretable to people. Although DL and ML show remarkable accuracy in medical diagnostics, their opaque character occasionally causes concerns about trust, dependability, and acceptance among healthcare professionals.

XAI provides human-comprehensible insights on the processes by which AI models generate predictions, thus addressing these problems. The definition and scope of XAI, many forms of XAI approaches, and key techniques used to increase interpretability are discussed in this part [26].

6.6.1 Definition of XAI

XAI refers to AI models and systems designed to:
- **Explain their decisions and reasoning processes** understandably
- **Provide transparency in model predictions**, helping users trust AI outputs
- Guarantee responsibility in AI-driven decisions

XAI aims to harmonize the complexity of AI with human comprehensibility so that medical professionals may authenticate, confirm, and rely on AI-generated insights in clinical settings.

6.6.2 Key Characteristics of Explainable AI

A truly explainable AI system should exhibit the following characteristics:

1. **Transparency**: Clear explanations of how input data influences predictions
2. **Interpretability**: Users (e.g., doctors) can understand and act on AI outputs
3. **Reliability:** The AI model offers explanations that are repeatable and consistent
4. **Accountability:** Certain factors can be linked to decisions made by AI

Example: AI in Breast Cancer Diagnosis
A deep learning network that analyzed mammograms was able to predict breast cancer with 94% accuracy. Even so, radiologists were skeptical because the model's classification was not supported.

- Researchers improved physician confidence by identifying particular tumor areas impacting predictions using XAI approaches like heatmaps and SHAP values.

6.6.3 Types of XAI: Model-Agnostic Versus Model-Specific Methods

XAI approaches can be classified as either model-specific or model-agnostic.
Model-Agnostic XAI Techniques

Definition: These methods can be applied to any model of AI because they are all opaque [27].

How It Operates: They examine input-output relationships without altering the model.

Example: An AI-based stroke prediction system
High-risk patients were identified by a hospital's machine-learning-based stroke prediction system. Clinicians need reasons for the projections.

- LIME made the AI's decision-making process more comprehensible by elucidating significant risk factors including age, smoking, and hypertension.

6.6.4 Model-Specific XAI Methods

- **Definition**: These methods are tailored to particular AI architectures (e.g., decision trees, neural networks).
- **How It Works**: They use inherent model features for interpretability.
- **Example Techniques**:
  - **Attention mechanisms** (for deep learning models)
  - **Feature importance scores** (for decision trees)

Example: XAI in DR Detection

Transparency was lacking in a CNN-based AI system meant to detect DR.

- An attention approach was used to help ophthalmologists identify the retinal areas influencing the AI's decision, hence improving its acceptance.

6.6.5 Techniques in XAI

Various approaches like attention maps, SHAP, LIME, saliency maps, and heatmaps improve the interpretability of healthcare decisions motivated by AI [28].

6.7 Role of XAI in Medical Imaging and Diagnostics

Using radiological scans, fundus images, and histological slides, AI models help discover and diagnose diseases in the vital field of medical imaging. Nevertheless, DL models used in radiology occasionally function as opaque systems, which complicates clinicians' capacity to understand the logic behind an AI model's assessment of a tumor as malignant or an ophthalmic illness as serious. XAI overcomes these difficulties by generating clear, understandable, and clinically meaningful diagnoses.

This section discusses how visualization techniques increase physician confidence, provides case studies on specific disorders, and examines XAI's application in radiology [29].

6.7.1 Interpretable AI in Radiology: OCT, MRI, CT, and Fundus Imaging Applications

6.7.1.1 Why Explainability Matters in Medical Imaging

AI is used in medical imaging modalities like optical coherence tomography (OCT), magnetic resonance imaging (MRI), computed tomography (CT), and fundus imaging for early illness identification and classification. AI models are good at spotting anomalies, but their opaque character limits clinical implementation and confidence [44].

Emphasizing the specific visual areas that shaped an AI's decision, XAI methods include saliency maps, attention mechanisms, SHAP, and LIME.

6.7.1.2 AI in OCT Imaging for Retinal Diseases

Diabetes retinopathy, glaucoma, and age-related macular degeneration (AMD) are diagnosed mostly using OCT. AI models trained on OCT images can effectively classify illness severity; nevertheless, without XAI, doctors may find it difficult to believe the results.

Example: For detecting glaucoma using XAI in OCT, deep learning predictions in glaucoma screening were clarified by a gradient-weighted class activation mapping (Grad-CAM) analysis on OCT images. Grad-CAM focused on the optic nerve head and retinal nerve fiber layer to clarify the justification for the patient's high-risk classification of the model [21].

6.7.1.3 AI in MRI and CT for Neurological and Cancer Diagnosis

Identification of brain tumors, strokes, and neurodegenerative diseases like Alzheimer's disease depends on MRI and CT images.

Example: XAI in classification of brain tumors

Using MRI scans, an AI model classified brain tumors; nonetheless, radiologists lacked clarity on its decision-making process. By the use of SHAP and LIME, researchers found that tumor edge patterns and nearby edema greatly influenced AI-driven tumor detection, thereby enabling the interpretability of the model.

Example: Saliency maps helped to highlight areas of ischemia using XAI AI-enhanced stroke diagnosis in CT perfusion images, thereby allowing doctors to validate AI predictions before treatment decisions.

Figure 6.5 shows that low-dose CT images are enhanced by AI to lower radiation exposure without compromising image quality or diagnostic accuracy. By processing low-dose scans using AI, image quality is improved to almost that of standard-dose scans, thereby guaranteeing safer operations while preserving diagnostic accuracy.

Figure 6.6 shows that AI looks at medical images – including MRI scans – to find abnormalities such as tumors or malignancies. By helping healthcare professionals, AI speeds up precise diagnoses, hence improving patient outcomes and operational efficiency in medical practice.

Figure 6.7 shows that by enhancing image quality, spotting trends, and extracting relevant data, AI examines raw images from medical imaging modalities, including MRI, X-rays, and CT scans. The AI technology helps doctors make better informed decisions by spotting abnormalities, automating measurements, and providing diagnosis advice.

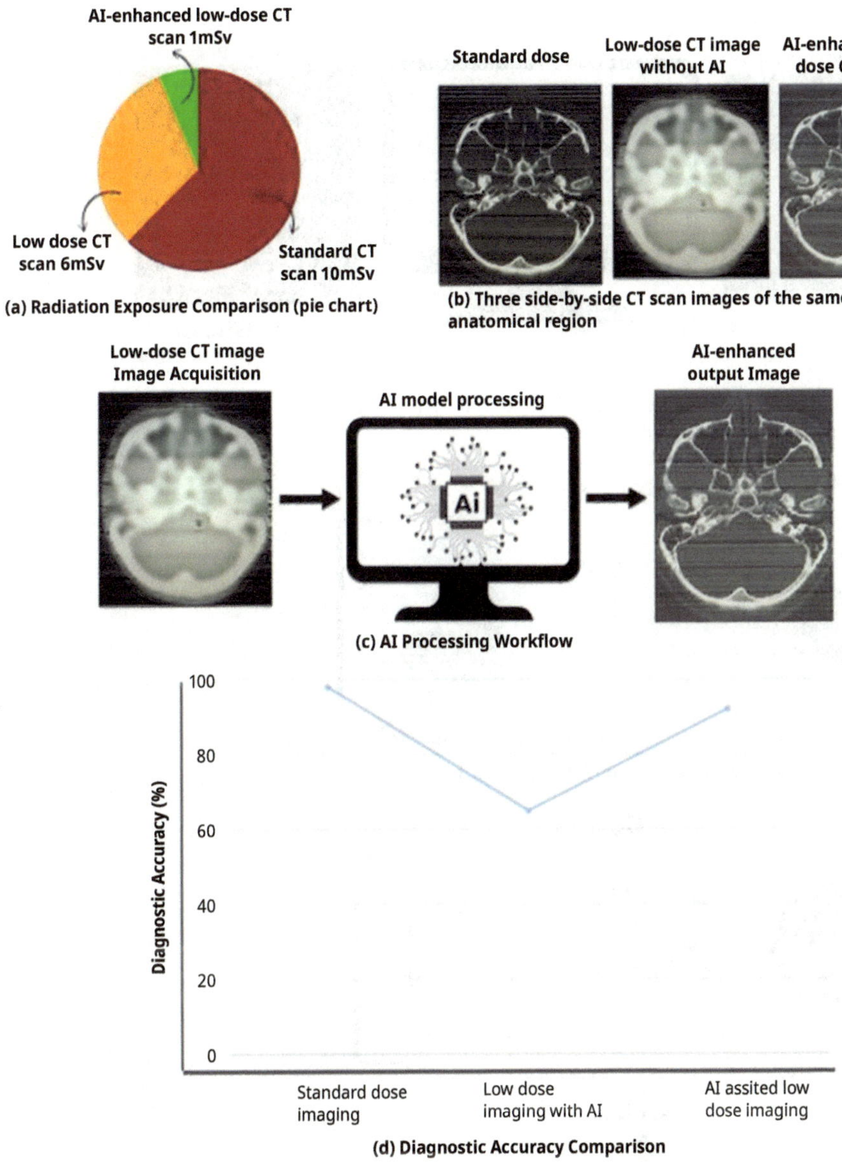

Figure 6.5: AI in low-dose CT scans. Source: Reproduced from David-Olawade AC et al., Diagnostics (MDPI), 2025, under CC BY license.

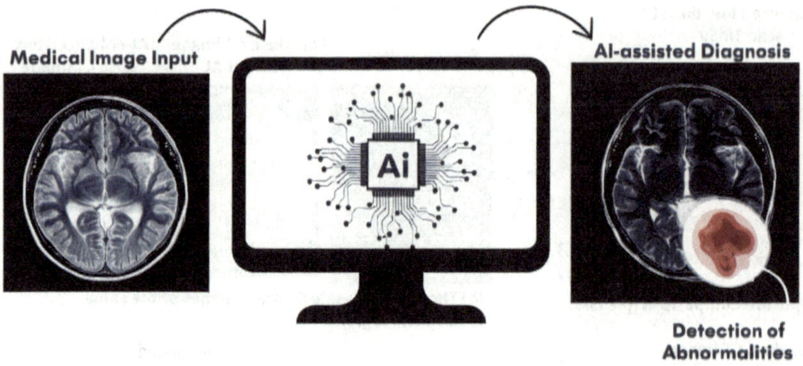

Figure 6.6: AI-assisted diagnosis. Source: Reproduced from David-Olawade AC et al., Diagnostics (MDPI), 2025, under CC BY license.

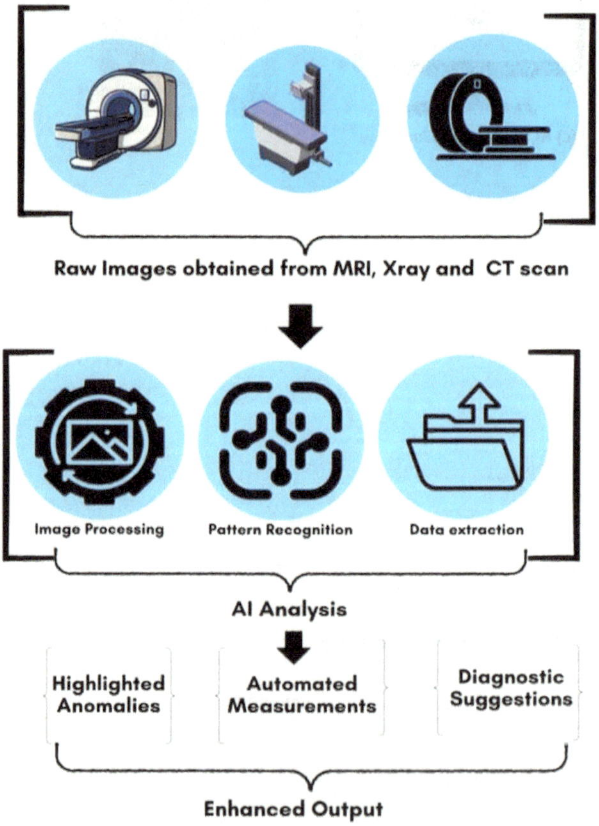

Figure 6.7: AI workflow in medical imaging. Source: Reproduced from David-Olawade AC et al., Diagnostics (MDPI), 2025, under CC BY license.

6.7.1.4 AI in Fundus Imaging for Retinal Disease Detection

Retinal vascular abnormalities, DR, and hypertension retinopathy are seen using fundus imaging in great numbers.

Example: XAI in the evaluation of DR

Originally revealed as Google's DeepMind AI for DR diagnosis, AI-driven screening tools first showed a lack in interpretability. Now, the technique emphasizes microaneurysms and hemorrhages pertinent to the diagnosis by using attention heatmaps and feature attribution maps, hence increasing ophthalmologists' confidence in AI-driven screening.

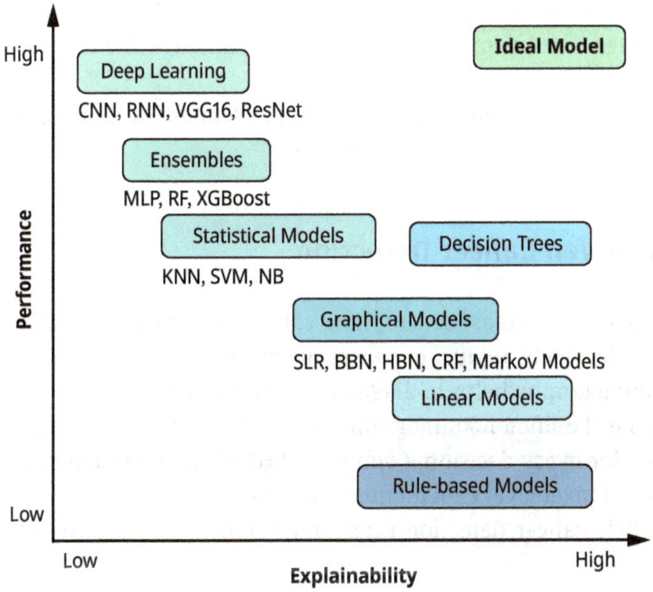

Figure 6.8: Performance vs. explainability of models. Source: Reproduced from Hafeez Y et al., Diagnostics (MDPI), 2025, under CC BY license.

The Figure 6.8 shows across several kinds of machine learning models and a trade-off between explainability and performance. Though having limited explainability, deep learning models – including CNN, RNN, and ResNet – show great performance. By contrast, rule-based models offer modest performance but great explainability. Between performance and interpretability, other models – including ensembles, statistical models, decision trees, graphical models, and linear models – lie in between in different degrees. The "Ideal Model" would merge great explainability with great performance.

6.8 Case Studies: Diabetic Retinopathy, Cancer Detection, and Neurodegenerative Disease Assessment

6.8.1 Case Study: XAI in Diabetic Retinopathy Detection

The main cause of blindness is DR; thus, early detection has been much improved using AI models. Still, black-box models cause practitioners to be reluctant to rely just on AI-generated diagnosis.

Example: Explainable AI for DR

This work clarifies AI-based DR classification using LIME and SHAP. Emphasizing microaneurysms, hemorrhages, and neovascularization, XAI validated the decision-making process of AI.

Incorporating interpretability resulted in a 20% improvement in confidence among clinicians concerning AI-driven DR identification [30].

6.8.2 Case Study: AI-Driven Cancer Detection

Particularly for the identification of breast cancer, lung cancer, and melanoma utilizing medical imaging, AI is extensively applied in cancer diagnostics.

Example: AI using mammography helps to diagnose breast cancer.

A deep learning machine classified mammograms as benign or malignant; radiologists still needed rationale for every decision. Combining SHAP with class-activation maps, the AI system produced pixel-level explanations for cancerous spots.

Using XAI improved early cancer detection rates and reduced false positives by 15% [31].

6.8.3 Case Study: XAI in Neurodegenerative Disease Assessment

Using MRI and PET imaging, AI is progressively used to diagnose Alzheimer's, Parkinson's, and multiple sclerosis.

Example: AI in the prediction of Alzheimer's disease

An AI system assessed hippocampal shrinkage obtained from MRI to detect early-stage Alzheimer's disease. Heatmaps and SHAP data showed that the main influence on the AI's prediction was a decrease in hippocampal volume.

The study showed that XAI raised doctor trust, which led to faster diagnosis and treatment [32].

6.9 Enhancing Clinician Confidence: How Visualization Techniques Improve Trust

6.9.1 Why Clinicians Need Explainability in AI-Assisted Imaging

AI has shown notable accuracy in disease detection; nonetheless, its lack of explainability limits its application into clinical practice. XAI solves this difference by: stressing the rationale behind AI decisions – that is, the justification for labeling a tumor as cancerous – Encouraging cross-validation with clinical expertise improving AI-based diagnoses using visual reasoning [33].

6.9.2 Visualization Techniques in XAI for Medical Imaging

AI transparency for medical imaging is enhanced by a number of methods:

1. **Saliency Maps:** These highlight important portions of the image for categorization.
2. **Grad-CAM:** This determines which fundus, CT, or MRI imaging components affect AI's prediction.
3. **SHAP Values:** This helps in dissecting forecasts made by AI into comprehensible parts.
4. **LIME:** It produces regional explanations for deep learning judgments.

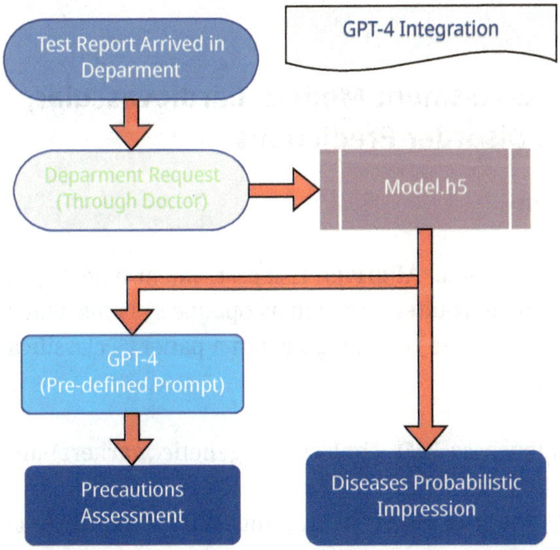

Figure 6.9: GPT-4 integration for medical analysis. Source: Reproduced from Syed S et al., J. Imaging (MDPI), 2024, under CC BY license.

Figure 6.9 shows how GPT-4 should be included into medical diagnosis. Once a test report arrives in the department, a doctor requests anything handled under a deep learning model (.h5). GPT-4 then uses predefined prompts to review the data, producing a probabilistic diagnosis of diseases and advises preventative action. Early insights into potential health problems offered by an AI-driven approach help to improve medical decision-making.

6.10 XAI for Personalized Treatment and Predictive Analytics

By allowing customized therapy approaches and early disease detection, personalized medicine and predictive analytics are changing modern healthcare. AI-driven models improve drug discovery processes, project illness risks, and provide personalized treatment plans. Lack of openness in these AI systems causes concerns about clinical acceptance, trust, and discrimination. By improving the interpretability of AI-generated recommendations, XAI helps to reduce these problems and hence build confidence among doctors and patients [34].

This part investigates XAI's purposes in:

1. Risk assessment models for diabetes, genetic diseases, and cardiovascular diseases
2. Precision medicine and drug discovery
3. Wearable health technologies and ongoing observation

6.10.1 Explainability in Risk Assessment Models: Cardiovascular, Diabetes, and Genetic Disorder Predictions

6.10.1.1 Role of AI in Risk Assessment

Using vast clinical, genomic, and lifestyle data, AI-driven risk assessment models predict disease susceptibility. Still, these models often function as opaque systems, which makes it difficult for doctors to understand the reasoning behind a patient's classification as either high- or low-risk [35].

XAI enhances these models by:

- Determining main risk factors (such as BMI, cholesterol, genetic markers) and influencing the prognosis
- Providing visual clarifications – that is, SHAP values and feature significance scores
- Letting clinicians use traditional risk assessment tools to confirm AI-generated results

Example: Cardiovascular Disease Prediction

AI algorithms that predict cardiovascular disease (CVD) include DeepHeart and the Framingham Risk Score. However, these models were first criticized for being difficult to understand:

- XAI solution: DeepHeart clarified significant aspects (such as smoking, blood pressure, and heart rate variability) using XAI solutions LIME and SHAP [36].
- Impact: Physicians used AI-driven cardiovascular risk assessments more frequently after learning the rationale for a patient's high-risk classification.

6.10.2 Drug Discovery and Precision Medicine: Understanding Molecular Interactions with AI

Pharmaceutical research is accelerated by AI through:

1. Molecular design analysis to forecast drug-target interactions.
2. Models of chemical interactions to improve drug formulation.
3. Making fresh therapeutic use of already existing drugs.

6.11 Challenges and Limitations of XAI in Healthcare

XAI can improve confidence, openness, and responsibility in healthcare AI systems, however its use creates many challenges and restrictions, even if it can boost these aspects. XAI models have to balance standardizing, computational efficiency, and explainability with user acceptance [37]. Three major issues in XAI implementation in healthcare are discussed in this part:

1. Computational complexity and performance trade-offs
2. Validation and standardization issues
3. Obstacles against clinician adoption and training

6.11.1 Trade-Offs and Computational Complexity

Many DL and ML models used in healthcare have millions to billions of parameters, hence they are computationally demanding. Often, the application of explainability techniques results in additional processing costs, which increases resource use and extends inference times:

- Post hoc explainability techniques, such as SHAP and LIME, require running multiple iterations of a model to generate explanations, hence greatly increasing processing needs [38].

- Intrinsically interpretable models – such as rule-based systems or decision trees – compromise accuracy for explainability, thereby limiting their application in complex medical settings [39].

Example: Medical Imaging XAI
Examine MRI image analysis improved by AI to find brain cancers. Although convolutional neural networks achieve great accuracy, the integration of XAI approaches such as Grad-CAM (Gradient-weighted Class Activation Mapping) results in a 3–5-fold increase in computing time due to the necessity of generating additional heatmaps:
- Hospitals with insufficient computational capability could find it difficult to include resource-intensive AI models in real-time clinical procedures.
- To improve explainability and lower computational costs, researchers are looking into hybrid methods such as combining CNNs with attention processes [40].

6.11.2 Balancing Accuracy Versus Explainability

Deep neural networks and other complex black-box models often exceed interpretable models in diagnostic accuracy; however, improving explainability by streamlining the model may compromise prediction performance and lead to possible misdiagnoses.

Case Study: Using decision trees instead of deep learning helped to improve the interpretability of an AI-driven cardiac risk prediction model. Though interpretability got better, the model's accuracy dropped by 12%, thereby making it less trustworthy for important risk assessments [41].

6.11.3 Standardization and Validation Concerns

Insufficient global guidelines for XAI in medical practice. Unlike AI algorithms following set accuracy criteria (e.g., AUC-ROC and F_1 score), XAI lacks widely accepted evaluation metrics [42]:
- Different explainability techniques produce different interpretations, thereby producing differences.
- The absence of a regulatory structure results in confusion about diagnosis produced by AI.

Example: Variations in AI interpretability among medical fields
- In radiology, saliency maps highlight important areas of an image, but, different XAI systems, such SHAP and Grad-CAM, may highlight different regions, thus causing possible confusion for clinicians [43].

- In pathology, an AI-driven cancer detection model produced contradicting explanations when using LIME, compared to DeepSHap, thereby challenging the evaluation of the model's dependability [44].

6.11.4 Regulatory Obstacles in XAI Implementation

Regulatory authorities include MHRA (United Kingdom), EMA (Europe), and the FDA (United States) insist that AI-driven medical devices have to be explicable. Still, the lack of consistent explainability structures causes long approval procedures.

Case Study: Stroke detection AI

Because of its inconsistent approach to clarifying false positives and false negatives, a 2021 AI-driven stroke detection system struggled to get FDA clearance [45]. The lack of consistent explainability criteria required multiple changes before government clearance in 2023.

6.11.5 Proposed Solutions for Standardization

- Setting worldwide XAI benchmarks: Regulatory bodies and research institutes have to define consistent explainable standards.
- Multi-center validation of XAI models: To ensure generalizability and equity, AI models are evaluated on several patient datasets.
- Developing XAI tools friendly for doctors: Develop simplified, uniform XAI models meant for medical experts to precisely grasp AI outputs.

6.11.6 Problems Affecting Adoption and Clinician Training

Against AI integration into clinical procedures: Many doctors show resistance to adopt AI-driven decision-aid systems because of:

- Inadequate AI literacy: Many medical professionals have inadequate formal knowledge of XAI and AI.
- Overdependence on traditional diagnosis methods: Doctors can choose intuition-driven decision-making above AI recommendations [46].

6.11.7 Exhibit: Artificial Intelligence in Cosmology

Although dermatologists used an AI system for melanoma diagnosis just 40% of the time, attributing this to a lack of knowledge about the model's decision-making process (105), the system showed 95% accuracy.

AI in retinal pathology assessment: The necessity for XAI-oriented medical education case study:
- A deep learning algorithm found DR with 92% accuracy; however, hardly 60% of ophthalmologists believed its results to be reliable.

After XAI training courses, clinicians' faith in AI increased by thirty percent, as they could read heatmaps created by AI [47].

6.12 Conclusion

XAI in healthcare is essential for transparency, trust, regulatory compliance, and the efficacy of AI-based medical applications. As AI becomes more prevalent in diagnostic tools, predictive analytics, and customized medicine, the need for interpretability increases to guarantee that physicians and patients comprehend findings provided by AI. In order to assist XAI become increasingly integrated into intelligent healthcare systems, this section outlines important concepts and encourages further research and development.

Key Takeaways

I. **The Significance of Explainability in AI-Driven Healthcare**
  - For AI-driven medical models to foster confidence between patients and healthcare providers, they must be transparent and interpretable.
  - Healthcare practitioners are reluctant to use traditional "black-box" AI models because of their lack of interpretability.
  - To make AI-driven diagnoses understandable, XAI provides both verbal and visual explanations, such as heatmaps, saliency maps, and SHAP values.

II. **Improving Clinical Decision-Making and Mitigating Errors**
  - By clarifying the crucial elements used by AI models in the interpretation of medical images (such as OCT, MRI, and CT), XAI improves diagnosis accuracy.
  - XAI technologies clarify the reasoning behind diagnosis or risk assessments in conditions, including DR, cancer, and cardiovascular illnesses, enabling improved physician validation.
  - By clarifying pharmaceutical reactions, hereditary risks, and wearable health monitoring trends, XAI enables customized treatment planning, improving patient-centered care.

III. **Regulatory and Ethical Considerations in the Adoption of AI**
  - Transparency in AI applications in healthcare is required by regulations like GDPR (EU), HIPAA (USA), and FDA.

  - Explainability frameworks are required to address ethical issues such as data bias, equity, and accountability. For the proper interpretation of AI-generated insights, clinician training on XAI technologies is essential.

IV. **Prospective Trajectories for XAI in Healthcare**
  - Deep learning and rule-based approaches are combined in hybrid AI models to increase interpretability, while preserving accuracy.
  - Federated learning techniques protect patient privacy, while improving model robustness, by allowing AI models to train on decentralized data.
  - Making AI-driven healthcare decisions more secure and traceable when blockchain technology is used.

V. **Enhancing the Robustness and Generalizability of XAI Models**
  - Many contemporary XAI methodologies emphasize elucidating individual predictions, while lacking comprehensive model interpretability.
  - Future research should concentrate on creating standardized XAI frameworks that deliver uniform explanations across various medical datasets.

References

[1] Topol E. Deep medicine: How artificial intelligence can make healthcare human again. Hachette UK; 2019.

[2] Rajkomar A, Dean J, Kohane I. Machine learning in medicine. New England Journal of Medicine. 2019;380(14):1347–1358.

[3] Esteva A, Kuprel B, Novoa RA, Ko J, Swetter SM, Blau HM, Thrun S. Dermatologist-level classification of skin cancer with deep neural networks. Nature. 2017;542(7639):115–118.

[4] Jiang F, Jiang Y, Zhi H, Dong Y, Li H, Ma S, . . . Wang Y. Artificial intelligence in healthcare: past, present and future. Stroke and Vascular Neurology. 2017;2(4).

[5] McKinney SM, Sieniek M, Godbole V, Godwin J, Antropova N, Ashrafian H, . . . Shetty S. International evaluation of an AI system for breast cancer screening. Nature. 2020;577(7788):89–94.

[6] Gulshan V, Peng L, Coram M, Stumpe MC, Wu D, Narayanaswamy A, . . . Webster DR. Development and validation of a deep learning algorithm for detection of diabetic retinopathy in retinal fundus photographs. Jama. 2016;316(22):2402–2410.

[7] Kourou K, Exarchos TP, Exarchos KP, Karamouzis MV, Fotiadis DI. Machine learning applications in cancer prognosis and prediction. Computational and Structural Biotechnology Journal. 2015;13:8–17.

[8] Collins FS, Varmus H. A new initiative on precision medicine. New England Journal of Medicine. 2015;372(9):793–795.

[9] Tison GH, Sanchez JM, Ballinger B, Singh A, Olgin JE, Pletcher MJ, . . . Marcus GM. Passive detection of atrial fibrillation using a commercially available smartwatch. JAMA Cardiology. 2018;3(5):409–416.

[10] Contreras I, Vehi J. Artificial intelligence for diabetes management and decision support: literature review. Journal of Medical Internet Research. 2018;20(5):e10775.

[11] Zhavoronkov A, Ivanenkov YA, Aliper A, Veselov MS, Aladinskiy VA, Aladinskaya AV, . . . Aspuru-Guzik A. Deep learning enables rapid identification of potent DDR1 kinase inhibitors. Nature Biotechnology. 2019;37(9):1038–1040.

[12] Bagdonas H, Fogarty CA, Fadda E, Agirre J. The case for post-predictional modifications in the AlphaFold Protein Structure Database. Nature Structural & Molecular Biology. 2021;28(11):869–870.
[13] Hashimoto DA, Rosman G, Rus D, Meireles OR. Artificial intelligence in surgery: promises and perils. Annals of Surgery. 2018;268(1):70–76.
[14] Longo UG, De Salvatore S, Candela V, Zollo G, Calabrese G, Fioravanti S, Giannone L, Marchetti A, De Marinis MG, Denaro V. Augmented reality, virtual reality and artificial intelligence in orthopedic surgery: a systematic review. Applied Sciences. 2021 Apr 5;11(7):3253.
[15] Imam NH. Adversarial Examples on XAI-Enabled DT for Smart Healthcare Systems. Sensors. 2024;24 (21):6891.
[16] Aldosari B, Aldosari H, Alanazi A. Challenges of Artificial Intelligence in Medicine. Envisioning the Future of Health Informatics and Digital Health. 2025;16–20.
[17] Thalpage N. Unlocking the black box: explainable artificial intelligence (XAI) for trust and transparency in AI systems. Journal of Digital Arts and Humanities. 2023;4(1):31–36.
[18] Liu Y, Primiero CA, Kulkarni V, Soyer HP, Betz-Stablein B. Artificial intelligence for the classification of pigmented skin lesions in populations with skin of color: a systematic review. Dermatology. 2023;239(4):499–513.
[19] Chintoh GA, Segun-Falade OD, Somtochukwu C, Odionu AHE Developing Ethical AI Models in Healthcare: A US Legal and Compliance Perspective on HIPAA and CCPA.
[20] Shoghli A, Darvish M, Sadeghian Y. Balancing innovation and privacy: ethical challenges in AI-driven healthcare. Journal of Reviews in Medical Sciences. 2024;4(1):1–11.
[21] Dangi RR, Sharma A, Vageriya V. Transforming healthcare in low-resource settings with artificial intelligence: recent developments and outcomes. Public Health Nursing. 2025;42(2):1017–1030.
[22] Guru D, Perumal S, Varadarajan V. Approaches towards blockchain innovation: a survey and future directions. Electronics. 2021;10(10):1219.
[23] Abbas S, Qaisar A, Farooq MS, Saleem M, Ahmad M, Khan MA. Smart vision transparency: efficient ocular disease prediction model using explainable artificial intelligence. Sensors. 2024;24(20):6618.
[24] Wysocka M, Wysocki O, Zufferey M, Landers D, Freitas A. A systematic review of biologically-informed deep learning models for cancer: fundamental trends for encoding and interpreting oncology data. BMC Bioinformatics. 2023;24(1):198.
[25] Holzinger A, Biemann C, Pattichis CS, Kell DB (2017). What do we need to build explainable AI systems for the medical domain?. *arXiv preprint arXiv:1712.09923*.
[26] De Fauw J, Ledsam JR, Romera-Paredes B, Nikolov S, Tomasev N, Blackwell S, . . . Ronneberger O. Clinically applicable deep learning for diagnosis and referral in retinal disease. Nature Medicine. 2018;24(9):1342–1350.
[27] Ribeiro MT, Singh S, Guestrin C. " Why should I trust you?" Explaining the predictions of any classifier. In Proceedings of the 22nd ACM SIGKDD international conference on knowledge discovery and data mining. 2016 Aug 13 (pp. 1135–1144).
[28] Ribeiro MT, Singh S, Guestrin C. " Why should I trust you?" Explaining the predictions of any classifier. In Proceedings of the 22nd ACM SIGKDD international conference on knowledge discovery and data mining 2016 August (pp. 1135–1144).
[29] Tonekaboni S, Joshi S, McCradden MD, Goldenberg A. What clinicians want: Contextualizing explainable machine learning for clinical end use. In Machine learning for healthcare conference 2019 October (pp. 359–380). PMLR.
[30] Joshi G, Jain A, Araveeti SR, Adhikari S, Garg H, Bhandari M. FDA-approved artificial intelligence and machine learning (AI/ML)-enabled medical devices: an updated landscape. Electronics. 2024;13(3):498.
[31] Mitrou L. Data protection, artificial intelligence and cognitive services: is the general data protection regulation (GDPR)'artificial intelligence-proof'?. Artificial Intelligence and Cognitive Services: Is the General Data Protection Regulation (Gdpr)'artificial Intelligence-Proof. 2018.

[32] Montoya LN, Roberts JS, Hidalgo BS. Towards fairness in AI for melanoma detection: Systemic review and recommendations. In Future of information and communication conference 2025 March (pp. 320–341). Cham: Springer Nature Switzerland.
[33] Karimian G, Petelos E, Evers SM. The ethical issues of the application of artificial intelligence in healthcare: a systematic scoping review. AI and Ethics. 2022;2(4):539–551.
[34] Narhare PM, Raj NB, Bagchi S. Leveraging Artificial Intelligence for Enhanced Clinical Decision Support Systems (CDSS). Jordan Medical Journal. 2025;59(2).
[35] Murray NM, Unberath M, Hager GD, Hui FK. Artificial intelligence to diagnose ischemic stroke and identify large vessel occlusions: a systematic review. Journal of Neurointerventional Surgery. 2020;12(2):156–164.
[36] Moorthy UMK, Muthukumaran AMJ, Kaliyaperumal V, Jayakumar S, Vijayaraghavan KA. Explainability and Regulatory Compliance in Healthcare: bridging the gap for ethical XAI implementation. Explainable Artificial Intelligence in the Healthcare Industry. 2025;521–561.
[37] Oyeniyi J, Oluwaseyi P. Emerging trends in AI-powered medical imaging: enhancing diagnostic accuracy and treatment decisions. International Journal of Enhanced Research in Science Technology & Engineering. 2024;13(4):81–94.
[38] Shmatko A, Ghaffari Laleh N, Gerstung M, Kather JN. Artificial intelligence in histopathology: enhancing cancer research and clinical oncology. Nature Cancer. 2022;3(9):1026–1038.
[39] Apiletti D, Baralis E, Bruno G, Cerquitelli T. Real-time analysis of physiological data to support medical applications. IEEE Transactions on Information Technology in Biomedicine. 2009;13(3):313–321.
[40] Di Sarno L, Caroselli A, Tonin G, Graglia B, Pansini V, Causio FA, . . . Chiaretti A. Artificial intelligence in pediatric emergency medicine: applications, challenges, and future perspectives. Biomedicines. 2024;12(6):1220.
[41] Takada T, Hoogland J, Fujii K, Kudo M, Sasaki S, Yano T, . . . Fukuhara S. Added value of inflammatory markers to vital signs for predicting mortality in patients with suspected infection: external validation and model development. Internal and Emergency Medicine. 2024;1–9.
[42] Ginestra JC, Giannini HM, Schweickert WD, Meadows L, Lynch MJ, Pavan K, . . . Umscheid CA. Clinician perception of a machine learning–based early warning system designed to predict severe sepsis and septic shock. Critical Care Medicine. 2019;47(11):1477–1484.
[43] Adadi A, Berrada M. Peeking inside the black-box: a survey on explainable artificial intelligence (XAI). IEEE Access. 2018;6:52138–52160.
[44] Bharati S, Mondal MRH, Podder P. A review on explainable artificial intelligence for healthcare: why, how, and when?. IEEE Transactions on Artificial Intelligence. 2023;5(4):1429–1442.
[45] Gunning D, Stefik M, Choi J, Miller T, Stumpf S, Yang GZ. XAI – Explainable artificial intelligence. Science Robotics. 2019;4(37):eaay7120.
[46] Noor AA, Manzoor A, Mazhar Qureshi MD, Qureshi MA, Rashwan W. Unveiling Explainable AI in Healthcare: current trends, challenges, and future directions. Wiley Interdisciplinary Reviews: Data Mining and Knowledge Discovery. 2025;15(2):e70018.
[47] Wong CYT, Antaki F, Woodward-Court P, Ong AY, Keane PA. The role of saliency maps in enhancing ophthalmologists' trust in artificial intelligence models. Asia-Pacific Journal of Ophthalmology. 2024;13(4):100087.
[48] Clement David-Olawade, A., Olawade, D. B., Vanderbloemen, L., Rotifa, O. B., Fidelis, S. C., Egbon, E., Akpan, A. O., Adeleke, S., Ghose, A., & Boussios, S. (2025). AI-Driven Advances in Low-Dose Imaging and Enhancement—A Review. Diagnostics, 15(6), 689. https://doi.org/10.3390/diagnostics15060689
[49] Syed S, Ahmed R, Iqbal A, Ahmad N, Alshara MA. MediScan: a framework of U-health and prognostic AI assessment on medical imaging. J Imaging. 2024;10(12):322. doi:10.3390/jimaging10120322.
[50] Hafeez, Y., Memon, K., AL-Quraishi, M. S., Yahya, N., Elferik, S., & Ali, S. S. A. (2025). Explainable AI in Diagnostic Radiology for Neurological Disorders: A Systematic Review, and What Doctors Think About It. Diagnostics, 15(2), 168. https://doi.org/10.3390/diagnostics15020168

Hema Umapathi, Anunya Pattjoshi, Khushi S. Udasi, Tushar Tejonidhi, Yuvaraj Sivamani, Sumitha Elayaperumal

7 3D Reconstruction Techniques for Medical Imaging

Abstract: Medical imaging in modern healthcare plays a vital role as it helps clinicians visualize the human body precisely, which in turn aids in timely diagnosis, strategizes treatment, and fuels research. Three-dimensional (3D) construction techniques were found to rely on cutting-edge algorithms, including ultrasound, Magnetic resonance imaging (MRI), and computed tomography (CT). These approaches further extract the information spatially from several two-dimensional (2D) slices, thereby building a geometric representation of tissues, organs, and diseased conditions. A key algorithm employed in the process includes surface, volume, and deep learning-based techniques. Geometric representations of organs are built using surface rendering algorithms, whereas a realistic depiction to complete the pictorial data is accomplished by the volume rendering techniques. The key advantage of the 3D reconstruction technique is that it provides a vivid description of the anatomy of the patient. The information derived will aid in evaluation preoperatively and navigation intraoperatively, which gives a detailed description of the anatomical models and helps in planning surgery meticulously. Another intriguing development is the integration of 3D reconstruction with augmented and virtual reality. Another potential application is that it essentially provides real-time guidance to the surgeons. Further improvement in both diagnosis and consultation is enhanced by the 3D imagery in telemedicine platforms. Hybrid models that blend physics-based and AI-driven approaches are increasingly being employed to address these difficulties, resulting in more accurate reconstructions. This chapter focusses on explaining the key ideas, methodology, and the applications of 3D reconstruction techniques in medical imaging, which helps in enhancing the patient health care system. To conclude, the 3D reconstruction techniques have provided a tremendous leap in medical imaging in understanding the human anatomy and diseased conditions.

Keywords: 3D reconstruction techniques, medical imaging, MRI, CT scan

Acknowledgment: The images provided in this chapter were created using BioRender.com.

 | https://doi.org/10.1515/9783112226223-007

7.1 Introduction to 3D Reconstruction Techniques for Medical Imaging

Emerging in the field of medical technology, medical imaging is nowadays widely used in healthcare sectors to help physicians observe the human body in detail. It manages and organizes treatments, provides precise diagnoses, and even researches various novel medical procedures. Even though there are other imaging techniques such as MRI (magnetic resonance imaging) , CT (computed tomography) scans, ultrasounds, and X-rays, two-dimensional (2D) images are captured by these conventional techniques, but they frequently don't give a full view of intricate bodily components. The three-dimensional (3D) reconstruction techniques are useful in this situation.

The 3D reconstruction of organs, tissues, or even pathological conditions is done by converting a set of 2D images acquired from different angles into a 3D model. Better technology is enabling much more precise anatomical information about the human body than physicians and surgeons need to move easily to make a decision. For instance, in neurosurgery, 3D models of the brain guide surgeons in their procedures, which make it easier for them to place the knife in the appropriate area without causing injury [1, 2].

7.2 Fundamentals of 3D Reconstruction

3D reconstruction in medical imaging encompasses a variety of techniques that convert 2D image slices into comprehensive 3D models. These techniques are tailored to different clinical needs, imaging modalities, and anatomical cite [3,4]. Below are the primary categories:

- **Tomographic Reconstruction:** Utilizes algorithms like filtered back projection and iterative reconstruction to build 3D volumes from multiple 2D slices, commonly used in CT and MRI.
- **Surface Rendering Techniques:** Focus on reconstructing the outer boundaries of anatomical structures using methods such as marching cubes and contour *interpolation.*
- **Volume Rendering Techniques**: Generate full 3D representations using voxel data, enabling visualization of internal structures. Techniques include ray casting, splatting, and shear-warp rendering.
- **Hybrid Techniques:** Combine surface and volume rendering to balance detail and computational efficiency, often used in surgical simulations.
- **Artificial Intelligence (AI)-Enhanced Reconstruction:** Deep learning (DL) models automate segmentation, improve image quality, and accelerate reconstruction, especially in noisy or low-resolution datasets.

7.3 Fundamentals of 3D Reconstruction

3D reconstruction is a pivotal process in modern medical imaging that transforms traditional 2D image data into 3D models. This advancement allows clinicians to gain deeper insights into anatomical structures, improving diagnostic accuracy, surgical planning, and patient outcomes. This section introduces the foundational principles, differences between 2D and 3D imaging, and key visual tools to support understanding Cite [5-7].

7.3.1 Principles of 3D Reconstruction in Medical Imaging

The computational foundation of 3D medical imaging is distinct from human visual perception and relies on several precise imaging principles:

- **Volumetric Sampling**: Medical imaging modalities scan the human body in discrete slices. These slices are then digitally stacked and interpolated to form a continuous 3D dataset [3].
- **Voxel Representation**: The 3D space is divided into small volume elements called voxels. Each voxel contains spatial coordinates and intensity values that correspond to tissue density or signal strength [3].
- **Image Segmentation**: Algorithms identify and delineate specific anatomical structures (e.g., organs and tumors) from the image data. This step is critical for isolating regions of interest for clinical analysis [5].
- **Image Registration**: Aligns images from different modalities or time points to a common spatial framework, enabling comparative analysis [9].
- **Rendering Techniques**: Surface rendering (e.g., marching cubes) and volume rendering (e.g., ray casting and shear-warp) are employed to generate visually interpretable 3D models for diagnosis and surgical planning [8].

These principles allow clinicians to interact with detailed anatomical data, improving diagnostic accuracy, treatment planning, and intraoperative navigation.

Differences between 2D and 3D imaging techniques are highlighted in Table 7.1.

Note: The original references to "depth perception," "parallax," *and* "perspective" *are primarily applicable to stereoscopic or artistic renderings and do not directly relate to the core computational principles of 3D medical imaging reconstruction.*

The development and utilization of 3D imaging technologies have greatly facilitated and helped in the growth of the field of medical imaging and diagnostics. Multiple approaches have been created, each addressing different clinical requirements, radiation exposures, and image resolutions. Following is an overview of the most notable 3D imaging technologies [4]. Figure 7.1 shows the techniques in 3D imaging.

Table 7.1: Difference between 2D and 3D imaging.

Feature	2D imaging	3D imaging
Dimensions captured	Height and width only	Height, width, and depth
Image type	Flat and lacks spatial depth	Volumetric and includes depth information
Visualization	Single plane and limited anatomical context	Multi-angle viewing and spatial context preserved
Common modalities	X-ray and ultrasound (2D mode)	CT, MRI, and 3D ultrasound
Clinical utility	Basic screening and bone fractures	Surgical planning, tumor localization, and organ modeling
Output unit	Pixel	Voxel (3D pixel)

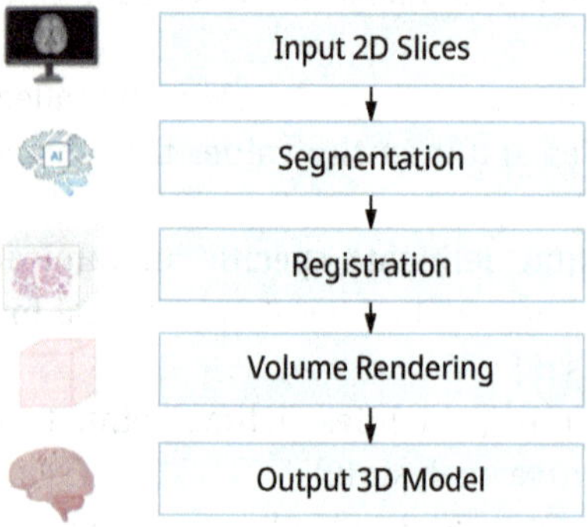

Figure 7.1: Techniques in 3D imaging.

7.3.2 Techniques in 3D Imaging

Modern 3D medical imaging employs several advanced techniques, each optimized for different diagnostic and clinical requirements (Figure 7.2):

1. **Computed Tomography (CT)**: Uses X-rays to acquire multiple 2D slices, which are then reconstructed into 3D models. Variants include fan beam CT and cone beam CT for different clinical use cases [4].
2. **Magnetic Resonance Imaging (MRI)**: Provides high *contrast* images of soft tissues. MRI slices are processed and combined to generate *volumetric datasets* [4]. Figure 7.3 shows platforms like MILXView, cloud computing, SOA, PRISM, and photoacoustic imaging (PAI).
3. **Ultrasound (3D/4D)**: Real-time acquisition of volumetric data using mechanical or matrix array transducers. Especially useful in obstetrics and cardiac imaging [4].

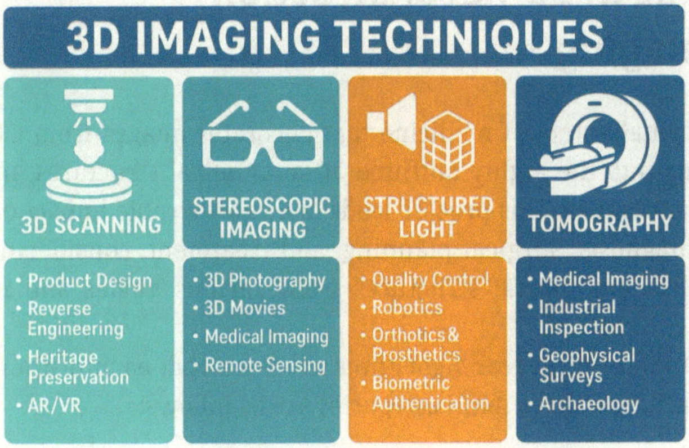

Figure 7.2: Core algorithms in 3D reconstruction visual representation of surface-based, volume-based, and deep learning algorithms used in medical imaging (drawn using Biorender and Canva).

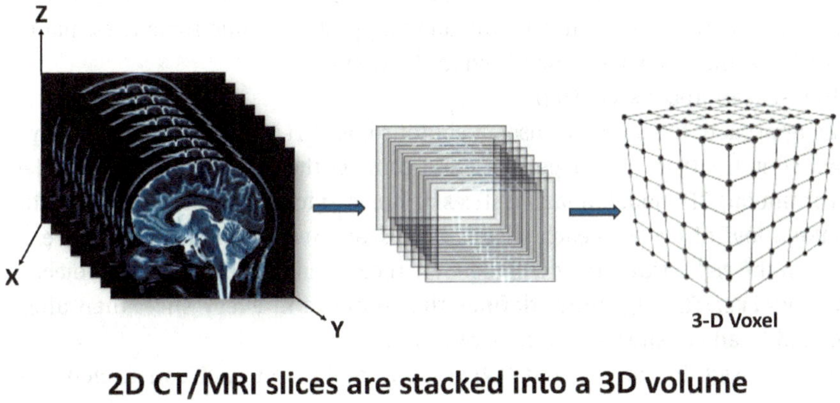

Figure 7.3: Technological platforms for 3D (drawn using Biorender and Canva).

4. **Structured Light Scanning**: Projects patterns of light onto the object and captures the deformation to infer 3D shape used in prosthetics and craniofacial planning [4].
5. **Photogrammetry**: Uses overlapping photographic images to calculate 3D coordinates commonly applied in orthopedics, dentistry, and forensic reconstructions [4].
6. **Time-of-Flight Cameras**: Measures the travel time of light pulses to create *depth maps* used for quick 3D assessment of surface topography [4].
7. **Stereoscopy**: Mimics binocular vision by capturing slightly offset images to simulate depth perception. Used primarily in educational and simulation tools [4].

7.4 Core Algorithms in 3D Reconstruction for Medical Imaging

- 3D reconstruction is a valuable tool of medicine that decorates images from CT, MRI, or ultrasound scans into the highly outfitted 3D anatomical structures, including organs, tissues, and bones. This process leads to a considerable impact on the results of medical diagnosis, surgical planning, and treatment, considering that doctors can visualize anatomical structures from various angles and in greater detail.
- To construct precise 3D models, several algorithms are required, each of them having its special approach. The three main categories are as follows:

7.4.1 Surface-Based Algorithms

Surface-based algorithms are designed to reconstruct the outer layer or boundaries of an object. These methods are effective for imaging past anatomic structures, mainly cartilage or bone, that have been identified in the first place.

A. **Contour-Based Reconstruction**
In this approach, the software uses a contour-based reconstruction algorithm to extract contours from 2D image slices and connect them to construct a complete and continuous 3D model. Imagine slices of bread stacked one on top of the other to make a loaf of bread – each slice is a 2D scan, and when all of them are arranged in the right order, they will together reconstruct the shape of the object.
How it works: The algorithm defines the contours in every slice, then aligns them, and finally creates a surface between them.
Where it is used: In both CT and MRI scans to build models of bones, blood vessels, and liver, etc.
Limitation: The method is not suitable for objects with ambiguous boundaries and may cause issues for soft tissues as well.

B. **Marching Cubes Algorithm**
Surface rendering in medical imaging is a most-used technique.
How it works:
The algorithm divides the 3D space into small cubes. Then, it probes each cube to know through which parts of it the object's surface passes. Finally, it draws edges to join the points into a mesh, and thus a smooth 3D model is created. The algorithm divides the 3D space into small cubes. It then checks each cube to see where the object's surface passes through it. Finally, it connects these surface points to form a mesh of triangles, creating a smooth 3D model. Details are shown in Figure 7.4.

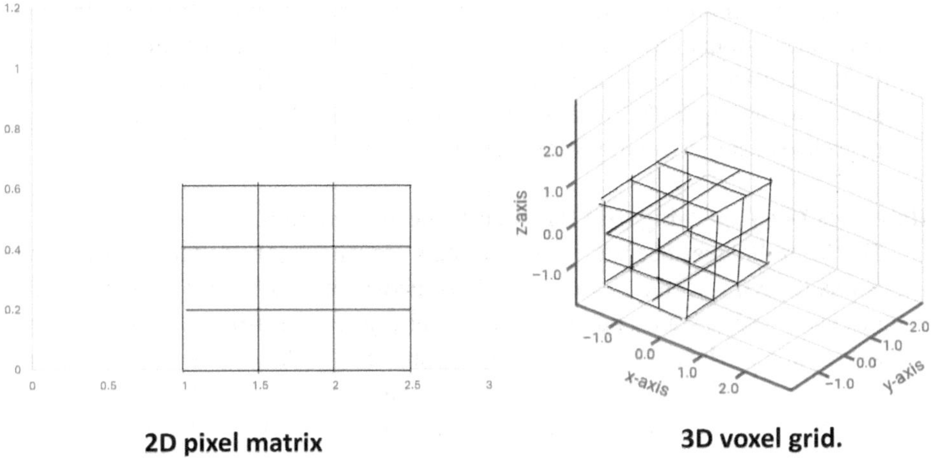

Figure 7.4: Different types of matrices (drawn using Biorender and Canva).

7.4.2 Volume-Based Algorithms

Volume-based algorithms are used in medical imaging that reconstructs entire anatomical structures instead of surface boundaries. These algorithms make use of voxels, or 3D pixels, to build highly detailed models, since the CT, MRI, and ultrasound scans can sufficiently show the tissues to make an accurate diagnosis and plan surgical procedures. Unlike surface-based methods, whose 3D volume is best seen with displays simply representing their surface boundaries as being from outside-in, more insightful visualization is possible.

Some of the important volume-based algorithms are:

- **Voxel-Based Reconstruction:** Considered on a general level as a 3D volume built by stacking 2D image slices
- **Splatting Algorithm:** Rendering of voxel data on a 2D plane by blending layers for smoothness; mostly used for soft tissue imaging
- **Shear-Warp Algorithm:** Faster than ray casting by making real-time visualization possible; commonly used for surgical simulations
- **Ray Casting Algorithm:** Describes light passing through the voxels, producing realistic shaded 3D images; best suited for tumor detection and vascular imaging

7.4.3 Deep Learning-Based Technology

- Deep learning has transformed 3D reconstruction into medical imaging. It has rendered the process fast, extremely accurate, and even highly automated. While traditional 3D reconstruction methods are well-established to be extremely computationally intensive with numerous mathematical models, the deep learning approaches function with neural networks and AI to improve images to make them more accurate, eliminate any type of noise, and accelerate reconstruction in scenarios of output. This technology can provide healthcare professionals with an overview of surgery planning and disease and organ processes.

How Does Deep Learning (DL) Improve 3D Reconstruction?

- The DL techniques operate on the concepts of training many artificial neural networks (ANNs) with medical images. These networks learn to recognize patterns in medical imaging scans and create high-quality 3D models from CT, MRI, and ultrasound images. The traditional approaches conduct some kind of manipulation manually, compared to the DL architecture that can easily augment, segment, or reconstruct images of a specified category by themselves, with no human handling involved.

Key Features of DL 3D Reconstruction:

- Automated Image Processing: With the inception of AI, segmentation is no longer needed manually.
- Increased Reconstruction Speed: The execution speed provided by DL methods is much quicker than the conventional approach.
- Improved Image Quality: By eliminating noise and improving contrast, AI improves the representation of medical images.
- Real-Time Applications: Enable real-time 3D imaging during operations or for diagnosis [9].

Figure 7.2 demonstrates the algorithmic pathways for constructing 3D anatomical models.

7.5 Technological Platforms for 3D Reconstruction

3D reconstruction in medical imaging is a process that has helped immensely in revolutionizing the field of medical imaging. It helps in altering the 2D image into a 3D model. This technique involves assembling the 3D geometric properties such as the coordinates, depth, and orientation. Clinicians and healthcare professionals obtain a high-resolution and accurate view of human anatomy. This assists in diagnostics and treatment. Technological platforms play an outstanding role in achieving this multiplex task [10].

The 3D techniques that are most frequently used are ultrasound, MRI, and CT, which depend on the algorithms that are used to build geometric properties of tissues and organs from various 2D images. A wide variety of technological platforms is used to encourage 3D reconstruction with distinctive features for data acquisition, interaction, and visualization [11].

Medical imaging has gone through tremendous growth in the past few decades as different technologies evolved to revolutionize this discipline. Medical imaging involves a culmination of platforms used to facilitate the increasing demands for precision, efficiency, and scalability. These platforms improve the diagnosis and treatment of conditions and provide channels for research and precision medicine. Some technologies, such as 3D medical imaging devices, cloud solutions, and imaging technologies like PAI, are paving the way for the future of healthcare. Figure 7.3 outlines the technological ecosystems supporting 3D reconstruction workflows.

7.5.1 MILXView

MILXView is a 3D medical imaging system developed by an e-Health Research Centre from Australia to analyze and visualize images. It has an easy user interface, which is important for clinicians for processing the advanced or relatively complex images.

MILXView has a plugin system that can be personalized to the user's needs. The cortical thickness estimate (MILXcte) plugin is one such example, which uses an MRI of the brain to detect the measurement of cortical atrophy. This also helps in the premature diagnosis of Alzheimer's disease. Plugin systems like MILXcte show us how the MILXView system can meet specific clinical needs via personalization. This enables the integration of regular clinical life and complex clinical routines. [12]

The platform can support cross-sectional studies and tracking progression of disease in real time. It gives insights into the premature identification of diseases. MILXView supports the integration of structural and functional information from MRI and PET scans. This gives crucial understanding of the morphology, physiology, and anatomy of the brain. This helps in studying neurodegenerative disorders like Alzheimer's disease. Brain structure and function change, and these alterations must be found.

MILXView can sequentially process larger datasets, which is crucial for effective processing of clinical tasks and scientific studies. Automated workflow is supported by the platform as this reduces manual labor. This also improves the accuracy of output.

7.5.2 Cloud Computing

Cloud computing is a technological platform of medical imaging. This is changing the field by making data storage and data sharing adaptable. Large datasets, infrastructure, and computing skills used to be a prerequisite for conventional medical imaging systems but currently, this is replaced by cloud computing that can reduce price, helping medical institutes in data management and collaborations. Cloud-based systems are specifically crucial in image reconstruction, where the computational challenge is vast. Traditional imaging systems that include tomographic and quantitative imaging need algorithms such as Monte Carlo modeling, which requires computational skills and results in prolonged execution times. Cloud infrastructures enable the parallel processing of such complex tasks, impacting the analysis and reconstruction of images, especially in multimodal imaging.

Additionally, cloud computing allows remote access to information on imaging, which is necessary for the fields of telemedicine and collaborative research. Scientists can analyze images remotely, allowing them to make real-time decisions that are not geographically bounded. The Cancer Imaging Archive (TCIA) is a data repository for sharing data, which can use the cloud infrastructure for many uses, such as storage and sharing of large amounts of medical imaging data, highly boosting precision medicine [13].

One of the impediments to the utilization of cloud computing for medical imaging is high-speed data transfer, especially when handling raw medical data. Though cloud platforms can bestow large processing power and storage, the rate at which these large imaging datasets can be relocated via the internet is a tailback. This is important for real-time processing and reconstruction of images in clinical environments.

Cloud computing also allows the creation of standard databases for image processing algorithms. These databases that consist of diversified collections of various medical images are vital. They verify new image processing methods like segmentation and fusion. The availability of such data on cloud platforms allows researchers to develop new algorithms and test them in a uniform manner.

7.5.3 Service-Oriented Architecture (SOA) in Medical Imaging

Service-oriented architecture (SOA) is a modern architectural style that is gaining popularity in the creation of medical imaging platforms. SOA is founded on the idea of splitting an intricate system into smaller, independent services that exchange information through systematic interfaces. SOA makes it possible to achieve modularity, versatility, and scalability and, therefore is perfect for combining disparate imaging modalities and processing tools into an integrated unified system.

In imaging medicine, SOA comes in particularly handy to establish interoperable systems that can deal with different types of imaging data (e.g., MRI, CT, and PET). Proprietary formats and proprietary ways of storing and communicating data might be utilized by various imaging systems and equipment. SOA provides common communication protocols to support the interchangeability of data between different systems in such a manner that it simplifies integration and helps clinicians and scientists work on a variety of datasets more comfortably [14].

SOA also facilitates the creation of modular solutions, where individual imaging processes can be processed by specialized services. For instance, image segmentation, feature extraction, and visualization could each be processed by different services, rendering the system more flexible and less difficult to maintain. New imaging modalities and techniques can subsequently be introduced into the platform as new services without a need to replace the system [14].

The feature to deploy medical imaging platforms as a collection of *microservices* is highly scalable. Such services can be supported in the cloud, and the system can be scaled up or down as needed for computational requirements. For instance, for handling large datasets, there may be a need for additional processing, and the platform can dynamically assign resources to accommodate such requirements, making it more efficient and reliable.

7.5.4 Platform for Imaging in Precision Medicine (PRISM)

PRISM (Platform for Imaging in Precision Medicine) is an expandable technology platform designed to promote individualized medicine by combining imaging data with clinical and pathological data. It is built to improve the abilities of storehouses like The Cancer Imaging Archive (TCIA) so that they can handle more divergent data types and facilitate data sharing and analysis quality for cancer research.

PRISM includes a data model that allows the researchers to ask difficult questions of complex datasets and to rightfully explore them. This is crucial for translational research, where the motto is to bridge the gap between clinical practice and general research. With its ability to combine data from diverse sources, PRISM facilitates the congregation of specialist groups customized to particular research requirements, partnerships, and more precision medicine.

One of the most important benefits of PRISM is its commutable design that makes it possible to execute gradually by various data repositories. This means that other research central hubs can include the platform in their data infrastructure, customizing it to meet their specific requirements. This also makes it possible for the platform to habituate to the rapidly evolving environment of research, especially cancer imaging and precision medicine [15].

PRISM's architecture focuses solely on data security, with de-identification and quality control services for ensuring the proper handling of patients' data. These are crucial to protect the privacy of a patient's information as well as the ethics and integrity of research data being utilized.

7.5.5 Photoacoustic Imaging (PAI) as an Emerging Platform

PAI is a modern technology that comprises the advantages of ultrasound and optical imaging. This also recruits an ultrasound, which is laser-induced, to develop high-precision and better-resolution images of biological tissues and organs. This also gives new information on tissue structure and function. PAI can assess the tissue oxygenation levels in real time, which is important for the diagnosis and treatment of diseases such as cancer, cardiovascular disease, and inflammation.

Clinical PAI systems can recruit an assembly of transducer designs like linear, curved, and volumetric array transducers. These are integrated with cosmopolitan signal processing and data acquisition systems to improve the image quality and penetration depth of PAI to make it suitable for various clinical applications.

PAI opens up great opportunities for cancer imaging since it can distinguish between tumor tissue and normal tissue according to their optical absorption features of different chromophores like hemoglobin. Oxygen saturation measurement and vessel structure analysis give data regarding the early detection of tumors and for following responses to therapy.

As PAI technology develops rapidly, certain features like system sensitivity, imaging depth, and scan speed are also being improved, which will make it more clinically critical. Moreover, as a budget-friendly, noninvasive imaging modality, PAI can be used to provide existing technologies such as CT and MRI, which gives healthcare professionals and clinicians a valuable diagnostic tool with fewer risks and side effects [16].

7.6 Clinical Applications in Medical Imaging

Medical imaging is a very crucial aspect of modern medicine that offers us great knowledge about the anatomy and pathology of the human body. This facilitates diagnosis and treatment. Human physiology becomes increasingly complex with a surge in demand for accuracy in diagnosis, so medical professionals depend on sophisticated imaging technologies to make a more detailed and precise decision. Amongst these technologies, 3D reconstruction has shown incredible results in medical imaging. By transforming the 2D images of modalities like computed tomography, MRI, and ultrasound into detailed 3D models, doctors can get a much better insight into the human body's anatomy and pathology. In the last few years, 3D reconstruction methods have developed extensive clinical uses, improved the rate of diagnosis, surgical planning, and patient management. From making surgical procedures more accurate to facilitating real-time intraoperative decision-making, the value of 3D imaging in transforming the interaction between healthcare practitioners and patients' anatomical data cannot be emphasized enough. Figure 7.5 summarizes the clinical domains where 3D imaging has transformative impact.

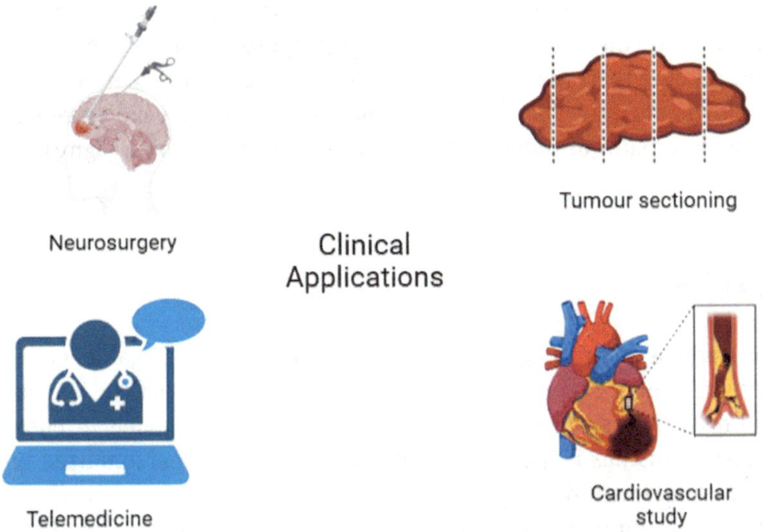

Figure 7.5: Clinical applications of 3D imaging infographic depicting applications in neurosurgery, orthopedics, and cardiology (drawn using Biorender and Canva).

7.6.1 Neurosurgery

Neurosurgery, with the brain, spinal cord, and peripheral nerves, is the most sensitive branch of medicine, where mistakes may lead to devastating or even irreversible results. 3D imaging in this case has transformed preoperative planning, intraoperative guidance, and postoperative evaluation with high accuracy and low risk. The anatomy of the brain is intricate and complex; thus, this structure makes visualization difficult while performing surgery. 3D reconstruction provides great help to improve the utility and safety of neurosurgical methods [17].

An important application of 3D brain modeling in neurosurgery is to help in making surgical decisions. The models are based on MRI or CT scans. The brain anatomy of a patient can be visualized in three dimensions by the surgeons so that they can plan the safest yet the most effective method for surgery. The exact location of tumors or any vascular malformations and their dependency on adjacent cells can be determined by medical imaging to choose the most effective method to handle the tumors based on their malignancy.

A collaboration of 3D imaging with AI has shown impeccable results. It is important to demonstrate DL of the core algorithms and integrate it with 3D imaging. Ideal surgical paths can be predicted by this technology to avoid the risk of injury during resection to the normal brain tissues. Real-time imaging can provide doctors and surgeons with a continuous live update on images of the surgical area. This allows them to make an informed decision [18].

The field of neurosurgery has made major advancements in the use of augmented reality. Surgeons can use AR headsets to superimpose the 3D representation of the patient's brain over their line of sight while performing surgery. Better spatial awareness and a lower cognitive load were the major goals of this key advancement. This improved the precision of outcomes in a surgical operation by a significant amount [19].

7.6.2 Orthopedics

It is the medical study of any damage or disease of the bones, joints, ligaments, and tendons. The development of 3D reconstruction has revolutionized this field. The surgical planning of orthopedics can be improved by making precise models of bones, joints, and fractures. This has helped a lot in the development of personalized implants and in joint replacement therapy.

The most important improvement has been the application of 3D printing to develop these personalized implants and prosthetics. These implants are designed to accommodate each patient's distinctive anatomy, which results in enhanced surgical success and a lesser risk of complications [20]. Many studies have reported the implementation of 3D-printed prostheses in hip arthroplasties. It was indicated that 3D-printed implants that have the same geometry as a patient's hip joint produced much fewer complications after surgery, a reduced recovery duration, and a greater percentage of success compared to the traditional implants. Moreover, 3D imaging enables surgeons to practice fracture realignments before actual surgery, resulting in more precise procedures and less operative time. Another field where 3D reconstruction is having a profound impact is trauma surgery. Surgeons can develop 3D models of broken bones, especially complicated breaks, and plan different surgical techniques before stepping into the operating room. This allows them to select the most efficient treatment regimen and saves them time in the operating room.

7.6.3 Cardiology

Cardiovascular diseases like coronary artery disease, heart failure, and arrhythmias happen to be a few of the most prevalent and life-endangering health diseases globally. 3D reconstruction in medical imaging is being applied in cardiology to model vascular anatomy, the chambers of the heart, and other pathologies like aneurysms and congenital heart disease.

Cardiovascular system imaging via 3D medical imaging gives clinicians better insight into heart function and disease, as shown in Figure 7.1. 3D models of the arteries, veins, heart chambers (aorta and ventricles), and rhythmic motion of the heart can be developed by MRI and CT imaging of the cardiovascular system. Coronary artery disease (CAD) is an important application of 3D imaging. CT angiography can develop 3D models that allow visualization of coronary plaques and stenosis. The decision for stent placement, bypass surgery, or angioplasty can be improved via medical imaging [21].

The 3D models allow clinicians and healthcare professionals to practice before the real operation and give valuable insight into surgical assessment, and errors made during the process can be studied thoroughly to minimize risk and avoid failure as much as possible.

7.6.4 Oncology

Medical imaging plays an important role in the detection, monitoring, and diagnosis of cancerous cells. Precise detection, as well as localization and monitoring of the tumors, is crucial for treatment. The traditional approach of 2D imaging does not provide sufficient details for complete comprehension of the tumor's relationship with the neighboring tissues. 3D reconstruction helps to get around this issue by allowing a highly detailed, spatially correct representation of the tumors, which is essential in surgery plans, radiation therapy, and treatment as shown in Figure 7.1.

3D imaging is being employed to generate tumor models in the brain, liver, lungs, breast, prostate, and other organs. These tumor models help surgeons and oncologists in the visualization of the size, shape, and position of the tumor. Current research trends have shown that reconstructions of the 3D tumor using radiation therapy can produce more accurate delivery of tumors to the target while also protecting healthy tissues. Many studies have assessed the application of 3D imaging in cancer of the prostate, wherein imaging of the prostate gland and pelvic structures was done to build more specific plans for radiation therapy, ensuring safety and efficiency.

3D imaging has improved biopsy techniques to significant levels. A 3D model of a tumor can help the oncologists to identify the precise location for a biopsy [22].

Ovarian cancer is one of the most prevalent and lethal gynecologic, fatal malignancies. This has beheld major improvements in diagnosis and treatment processes recently, especially through the application of AI to medical imaging. DL and convolutional neural networks (CNN) have also provided promising solutions for the detection of ovarian cancer. AI has been increasingly used for innumerable imaging modalities, such as CT, ultrasonography (US), and MRI, to improve the precision of diagnosing ovarian cancer. A combination of these imaging modalities with AI can help the doctors in medical diagnosis precisely in identifying cancer subtypes, directly targeting tissue biopsies, and foreseeing patient outcomes.

Gliomas, specifically glioblastoma multiforme (GBM), are the most common and malignant primary CNS tumors. Although there have been advances in treatment, GBM is almost incurable till date, and its heterogeneity, both genetic and radiographic types, creates barriers for treatment and diagnosis. Gliomas have heterogeneous intensity profiles on multiparametric magnetic resonance imaging (mpMRI) scans, which mirror the biology of the tumor.

Precision medicine assesses the genetics, lifestyle, and environment of an individual to identify the specific treatment that is best suitable for them. Individualized precision medicine in neuro-oncology aims to use specific treatments for surgery, radiotherapy, and chemotherapy according to the unique tumor type of each patient. Traditional diagnostic modalities such as radiographic imaging and biopsy provide useful information, but modern technologies are much better. Surgeons can visualize and plan surgical strategies based on data collected via 3D imaging, and augmented and virtual reality. 3D printing also helps medical experts in developing physical models for training and assessment before any operation [23].

Radiomics is coupled with biophysical simulations to gather quantitative information from images. They help in the better analysis of tumor patterns and behavior. This can characterize tumors based on shape, size, metabolism, and function. The diagnosis rates improved significantly. The World Health Organization categorizes gliomas according to molecular features and histopathology; this was changed again in 2016 to include cellular and molecular factors for diagnosis. This can improve the results in neuro-oncology.

7.6.5 Telemedicine

Telemedicine is a rapidly developing industry, which had significant growth during the COVID-19 pandemic. The demand for remote care and consultations surged magnificently. 3D imaging helped immensely in taking telemedicine to a higher level by making it possible to share data remotely and analyze the detailed medical images.

The latest technologies in cloud computing have made securely sharing medical images over distant locations possible. This allows clinicians to provide consultations and collaborate irrespective of geographical limitations. A cloud-based telemedicine platform can share a patient's brain scan so that experts can review the data and proceed accordingly.

3D imaging is also used to support remote diagnostics that allow 3D scans of patients who live in remote locations to be evaluated by trained radiologists and doctors, therefore enhancing precision, diagnostics, and treatment options without actually meeting the patient [24].

7.6.6 Blending Augmented and Virtual Reality (AR/VR)

The integration of AR and VR technologies with 3D reconstruction in medical imaging has opened new paths in medical imaging. These technologies improve the efficacy and the way clinicians engage with 3D models. They can be used for preoperative planning, surgical training, or real-time navigation during surgical procedures.

In neurosurgery, AR headsets can superimpose 3D models of the brain over the surgeon's field of vision, and this gives them real-time instructions during surgery. It shows the effectiveness of AR in directing neurosurgeons during the intricate brain tumor resections.

VR has also been found to be a requisite resource in medical education and training. Surgeons can use VR simulations to practice complicated procedures where they can get a risk-free virtual environment before operating on patients [25]:

- AR delivers visually stimulating real-time 3D directions, enhancing the precision of surgeries and minimizing mistakes.
- VR tries to minimize the failure rate in surgery as the surgeons can rehearse procedures before performing them on patients.
- AR and VR can also facilitate patients' visualization of their condition and their treatment options.

AR assists in brain tumor operations by overlaying the preoperative MRI/CT images over the patient's anatomy. This provides accurate resection of the tumor and improves workflow by eliminating the constant 2D switches. In arteriovenous malformation (AVM) surgeries, AR helps with the real-time visualization of blood vessels and helps in differentiating between arteries and veins by color coding and chroma depth rendering.

7.7 Integration of 3D Reconstruction with Augmented and Virtual Reality (AR/VR)

Augmented reality (AR) refers to augmenting the real environment using virtual objects. It serves as a link between synthetic and real environments [26]. It gives the user a perception that seems like the virtual and real objects coexist in the same space, in which the AR complements virtual reality, instead of replacing it completely.

Virtual reality (VR) refers to a 3D digital environment that offers multiple degrees of freedom for the user to interact with the environment and engage in immersive interactions. VR has the potential to create individual custom learning environments with appropriate levels of fluidity and immersion [27].

The 3D reconstruction depicts the method of generating 3D models using data obtained from the real world, which can be collected by methods like depth sensors, LiDAR scanning, or photogrammetry. The integration of 3D reconstruction with AR and VR will make 3D models better by placing them either in a real-world setting or in a virtual setting. This combination has a significant impact on various sectors like healthcare, architecture, entertainment, education, and engineering, thus improving real-time simulations, training, depth, and experience [28].

The key features of integration of AR and VR with 3D reconstruction include:

1. **Collection of Data and Generation of a 3D Model:**
It involves accurately collecting real-world data and using it to generate a precise 3D representation of them. It can be obtained by using several technologies, such as:
LiDAR (light detection and ranging) involves emitting laser beams and measuring the time taken by the beam to return after it hits the object, thus capturing spatial data.

Stereo Vision uses two or more cameras to capture an image from different perspectives.

Photogrammetry uses software that analyzes the series of photographs captured of an object or the environment from different angles to create the 3D models [29].

2. **Rendering in AR/VR:**
AR display: After generating a 3D model, it is displayed and interacted with through AR-enabled devices. The platforms utilize several tools to integrate the 3D model into the real-world environment [30]:

- Marker-Based AR: It is one of the most traditional approaches, which involves triggering and tracking the placement of 3D models using physical markers such as QR codes, images, or symbols. The AR system detects the marker, and the 3D reconstruction model is aligned and displayed about the marker such that it remains anchored and accurately placed, allowing for a real-time interaction with the models [31].
- Markerless AR: This approach uses feature detection and computer vision algorithms to track and identify the features of the environment without the need for predefined markers. It is highly beneficial when dealing with larger or complex 3D reconstructions such as entire cityscapes. It scans the natural features of the environment, which will serve as references for placing the 3D reconstructed models, allowing for flexible and dynamic placement of virtual objects based on the real objects [32].
- Device Integration: AR-enabled devices such as smartphones, tablets, and AR glasses (Microsoft HoloLens) are used to project the virtual models onto the physical world in real time. The sensors are used to track the environment and the device's movements. The camera feed thus allows the aligning of the real-world with the digital overlay [33].
- Depth Sensing and LiDAR-Based AR: The LiDAR sensor captures the precise depth data, which is used by the AR system to map the environment in 3D, thus, it takes into account the exact distance from the user's device [34].
- SLAM-Based AR: It allows the simultaneous mapping of the environment and tracking the device's position within it. This is done by building a map of the environment, ensuring that virtual objects are aligned with the physical environment [35].

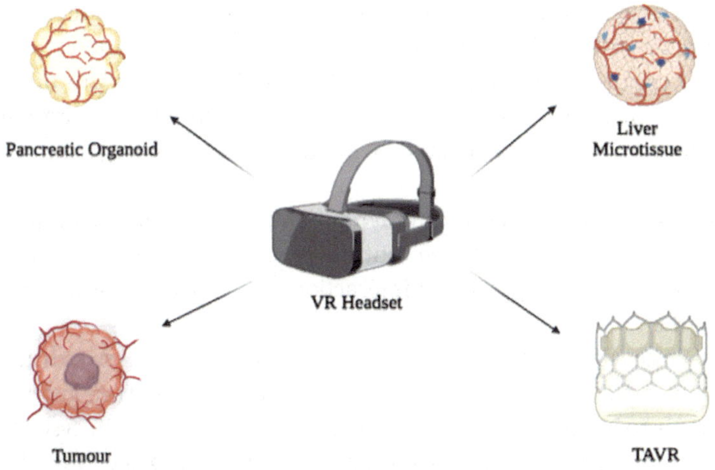

Figure 7.6: Schematic representation of VR display (drawn using Biorender and Canva).

VR Display: In this, the user is fully immersed in a computer-generated environment that may include 3D models, dynamic textures, and lighting or other spatial data.

- VR Headsets: These headsets provide a stereoscopic visual used to display the 3D-reconstructed models in a fully immersed environment, allowing the user to perceive the depth and explore it from various angles (Figure 7.6). They mostly offer head tracking capabilities so that the virtual environment moves in sync with the user's head movements.
- Controllers and Input Devices: They are equipped with motion sensors that track and map the user's hand and body movements, thus allowing the user to interact with virtual objects. They may also employ gesture recognition and haptic feedback [36].
- Room Scale: Physical movements can be accurately represented in the virtual world using sensors like lighthouse base stations or inside-out tracking.
- Depth Sensing and Environmental Mapping: VR and AR systems incorporate depth sensing technology and LiDAR to enhance the realism and accuracy of the environment.
- SLAM-Based: Simultaneous localization and mapping (SLAM) is a technique used in VR to track the position of the device to create a map of the virtual environment [37].

1. **User Interaction:** The interaction between users and AR/VR-represented objects plays a crucial role. Several methods are employed, such as gestures or touch inputs.

- Gesture-Based Interaction: In the case of both VR and AR, hand gesture recognition allows the users to manipulate the 3D reconstructed models in the virtual or augmented space. For example, in VR, users can use hand gestures like grabbing, moving, or modifying virtual objects in the virtual environment. In AR, actions like resizing a 3D reconstruction can be done. Modern VR systems are often equipped with hand-tracking capabilities, allowing the users to interact with the virtual environment using only their hands without the need for any controller [38].

- Touch-Based Interaction: This type of interaction is more common in AR and deals with interacting with touch-sensitive controllers or surfaces. In VR controllers like Oculus Touch, the users get to feel interactions with virtual objects mimicking the real world.
- Voice Interaction: Voice commands can also be used to control the 3D-reconstructed environment. For example, in environments such as a hospital, the user can instruct the AR application to zoom in on the heart or the brain. Voice recognition systems such as Google Assistant, Siri, or Alexa have been increasingly integrated into the AR and VR environments [39].

7.8 Impact of 3D Imaging on Telemedicine

Telemedicine generally refers to the use of telecommunication and digital technologies such that healthcare services can be provided remotely without the need for physical presence of the professional in that location, allowing diagnosis, treatment, and monitoring [40].

Computed tomography scans, MRI, ultrasound imaging, 3D X-rays, radiographs, and optical coherence tomography are the types of 3D imaging that are commonly integrated into telemedicine:

1. **Diagnosis**
   The traditional 2D imaging limits the doctor's ability to assess complex conditions, whereas 3D imaging provides a much more detailed and accurate representation of the human body, improving precision.
   3D imaging of a bone fracture or any joint abnormalities allows for better planning of surgical interventions. It also enables doctors to examine patients remotely by reviewing their images [41].
2. **Treatment Planning and Surgical Precision**
   This generally allows the surgeons to practice virtually on a 3D model that represents a patient's body before the actual surgery, thus ensuring the precision of the surgery and reducing the risk of complications. In craniofacial surgery, generally, 3D models are used to plan complex reconstructions, such as facial bone restoration after a trauma or congenital deformities [42].

3. **Remote Surgery**
   It allows surgeons to perform procedures remotely by using robotic systems. Delicate brain surgeries can be performed by a surgeon remotely controlling a robotic surgical system, reducing the need for the patient to travel to a specialist [43].
4. **Remote Monitoring of Patients**
   Wearable devices or remote imaging systems monitor a patient's recovery or disease progression without the need to visit the healthcare facility. Patients with musculoskeletal conditions use wearable devices that capture 3D data of their joints, bones, and tissues, which are then transmitted to healthcare professionals for analysis and guidance [44].
5. **Patient Education and Engagement**
   3D imaging can be used to provide interactive and easy-to-understand models of the patient's health condition, making them understand the condition better, reducing their anxiety. 3D models of the skeletal structure, show fractures, joint condition, or deformities, making it easy to explain the treatment options to the patients [45].
6. **Personalized Healthcare**
   3D imaging enables the development of personalized medical inventions, for e.g., the surgeons can design customized implants or prosthetics that match the patient's anatomy [46].

7.9 Limitations

Despite the transformative advantages offered by 3D reconstruction in medical imaging, there are several limitations that need to be considered in both clinical and research settings. These limitations arise from technical, environmental, clinical, financial, and ethical constraints. Addressing these challenges is essential for optimizing imaging protocols, ensuring patient safety, and promoting equitable access to advanced imaging technologies.

1) **Technical Limitations**
   - **Image Quality**: The accuracy of 3D reconstructions might be lost due to distortion and noise in imaging data, especially with imaging techniques like 3D ultrasound (3DUS), where voxel dimensions might not detect characteristic small objects.
   - **Computational Time:** Techniques used in methods such as Structure from Motion (SfM) require a large number of images for effective 3D reconstruction, and a few images may lead to errors and fake points in the 3D model. In addition to that, more images require more computational time for analysis, which may delay processing.

- **Segmentation Challenges**: There is no standard procedure for segmenting various plant architectures in the case of 3D reconstruction. This diversity necessitates the development of robust segmentation techniques that apply to various structures [47].
- **Lighting Conditions**: Passive image capture techniques are sensitive to changes in light, leading to uneven data and incomplete reconstructions.

2) **Material and Environmental Limitations**
   - **Environmental Effects**: External factors like wind can affect plant model accuracy considerably by inducing feature-matching errors that result in incomplete or warped 3D reconstructions.
   - **Material Properties**: The physical properties of the materials to be reconstructed (e.g., transparency, reflectivity) may influence the possibility of capturing and accurately reconstructing 3D models. For example, some materials might not adequately interact with imaging modalities, resulting in poor data quality.
3) **Field-Specific Limits**
   - **Radiation Dose in Medical Imaging**: There are some problems with radiation exposure in the clinical situation of pediatric CT scans due to increased radiosensitivity and the potentially cumulative nature of the exposure over time. This is particularly true when frequent imaging in certain groups is being considered [48].
   - **Clinical Utility:** Several of the innovative imaging techniques may lack established protocol or clinical use validation, and this can impair them regardless of technological advancement [50].
4) **Cost Issues**
   - **Equipment Expense:** Advanced imaging technologies like CT scanning technology and other equipment can be expensive, hence limiting accessibility [48].
   - **Operational Costs:** Costs related to equipment maintenance, software updates, and the requirement of specialized training for personnel to operate new technologies can contribute to the hefty operational costs [48].
5) **Ethical Concerns**
   - **Radiation Risk Versus Benefit**: The ethics of exposing patients, especially vulnerable groups like children and immune-compromised individuals, to radiation via imaging must be carefully considered, weighing potential diagnostic benefits against the risk of radiation-associated health effects [49].
   - **Data Privacy**. Due to the increasing use of patient data for 3D reconstructions, patient privacy and data security issues have become a priority to safeguard, especially considering how the data is shared and stored [50].
   - **Informed Consent**: The patients should be properly informed regarding the risks and advantages of imaging tests, especially those that entail the use of radiation [50].

References

[1] Ng B, Abugharbieh R, Huang X, McKeown MJ. Spatial characterization of fMRI activation maps using invariant 3-D moment descriptors. IEEE Transactions on Medical Imaging. 2009;28(2):261–268. https://doi.org/10.1109/TMI.2008.929097.

[2] Gautama T, Mandic DP, Van Hulle MM. Signal nonlinearity in fMRI: A comparison between BOLD and MION. IEEE Transactions on Medical Imaging. 2003;22(5):636–645. https://doi.org/10.1109/TMI.2003.812248.

[3] Three-dimensional imaging techniques: A literature review (doi: 10.4103/1305-7456.126269)

[4] The why, what, how and where of 3D imaging (doi: 10.1177/1753193414524137)

[5] Roadmap on 3D integral imaging: sensing, processing, and display (Vol. 28, No. 22 / 26 October 2020 / Optics Express 32266)

[6] State-of-the-art and applications of 3D imaging sensors in industry, cultural heritage. Medicine, and Criminal Investigation (Sensors. 2009;9:568–601. doi: 10.3390/s90100568).

[7] Verykokou S, Ioannidis C. An overview on image-based and scanner-based 3D modeling technologies. Sensors. 2023;23:596. https://doi.org/10.3390/s23020596.

[8] A Survey of Algorithms Involved in the Conversion of 2-D Images to 3-D Model Gopal D. Upadhye, Anant Kaulage, Ranjeetsingh S. Suryawanshi, Roopal Tatiwar, Rishikesh Unawane, Aditya Taware, Aryaman Todkar (ISSN:2321-8169 Volume: 11 Issue: 6s doi: https://doi.org/10.17762/ijritcc.v11i6s.6942)

[9] AlZu'bi SM. A 3D multiresolution statistical approaches for accelerated medical image and volume segmentation. PhD thesis. London: School of Engineering and Design, Brunel University; 2011. http://bura.brunel.ac.uk/handle/2438/5300.

[10] Vazquez V, Eugenio M, Pedro G, Manlio FVC, Felipe AC, Eduardo AV, Claudio VS, Kirby GV, Manuel D, Javier P. Assessment of intra-operative 3D imaging alternatives for IOERT dose estimation. Zeitschrift Fur Medizinische Physik. 2017;27(3):218–231. https://doi.org/10.1016/j.zemedi.2016.07.002.

[11] Farahani, N., Braun, A., Jutt, D., Huffman, T., Reder, N., Liu, Z., ... & Pantanowitz, L. (2017). Three-dimensional imaging and scanning: current and future applications for pathology. *Journal of pathology informatics, 8(1), 36.*

[12] Weigang A, Jinxiao P. RETRACTED: Fast 3D image reconstruction algorithm based on artificial intelligence technology. International Journal of Electrical Engineering & Education. 2021;60 (1_suppl):3508–3520. https://doi.org/10.1177/0020720920985042.

[13] Fripp J, et al. P1-391: Automated MR brain analysis using MILXview: Comparison with FreeSurfer and FSL first for cortical thickness estimation, deep gray matter, and ventricular segmentation. Alzheimer's & Dementia. 2010;6(4S_Part_10). https://doi.org/10.1016/j.jalz.2010.05.945.

[14] Kagadis GC, et al. Cloud computing in medical imaging. Medical Physics. 2013;40(7). https://doi.org/10.1118/1.4811272.

[15] Loya SR, et al. Service oriented architecture for clinical decision support: A systematic review and future directions. Journal of Medical Systems. 2014;38(12). https://doi.org/10.1007/s10916-014-0140-z.

[16] Sharma A, et al. PRISM: A platform for imaging in precision medicine. JCO Clinical Cancer Informatics. 2020;491–499. https://doi.org/10.1200/cci.20.00001.

[17] Zhang J, et al. High-resolution photoacoustic tomography for early-stage cancer detection and its clinical translation. Radiology: Imaging Cancer. 2020;2(3):e190030. https://doi.org/10.1148/rycan.2020190030.

[18] Li S, et al. Application of three-dimensional (3D) curved multi-planar reconstruction images in 3D printing mold assisted eyebrow arch keyhole microsurgery. Brain and Behavior. 2020;10(10). https://doi.org/10.1002/brb3.1785.

[19] Zárate-Méndez AM, et al. Three-dimensional virtual reality simulation to safe planning neurosurgical procedure in brain aneurysms, Latin American single-center experience: Advantages and limitations. Indian Journal of Neurosurgery. 2021;10(01):069–073. https://doi.org/10.1055/s-0041-1725233.

[20] Carl B, et al. Spine surgery supported by augmented reality. Global Spine Journal. 2020;10 (2_suppl):41S–55S. https://doi.org/10.1177/2192568219868217.

[21] Xia R, et al. Clinical applications of 3-dimensional printing technology in hip joint. Orthopaedic Surgery. 2019;11(4):533–544. https://doi.org/10.1111/os.12468.

[22] Ma Y, et al. Three-dimensional printing for heart diseases: Clinical application review. Bio-Design and Manufacturing. 2021;4(3):675–687. https://doi.org/10.1007/s42242-021-00125-8.

[23] Cantré D, et al. Three-dimensional imaging and three-dimensional printing for plastic preparation of medical interventions. Der Radiologe. 2020;60(S1):70–79. https://doi.org/10.1007/s00117-020-00739-6.

[24] Khalighi S, et al. Artificial intelligence in neuro-oncology: Advances and challenges in brain tumor diagnosis, prognosis, and precision treatment. Npj Precision Oncology. 2024;8(1). https://doi.org/10.1038/s41698-024-00575-0.

[25] Bokolo AJ. Application of telemedicine and eHealth technology for clinical services in response to the COVID-19 pandemic. Health and Technology. 2021;11(2):359–366. https://doi.org/10.1007/s12553-020-00516-4.

[26] Liu T, et al. Augmented reality in neurosurgical navigation: A survey. The International Journal of Medical Robotics and Computer Assisted Surgery. 2020;16(6):1–20. https://doi.org/10.1002/rcs.2160.

[27] Carmigniani J, et al. Augmented reality technologies, systems, and applications. Multimedia Tools and Applications. 2010;51(1):341–377. https://doi.org/10.1007/s11042-010-0660-6.

[28] Fowler C. Virtual reality and learning: Where is the pedagogy?. British Journal of Educational Technology. 2014;46(2):412–422. https://doi.org/10.1111/bjet.12135.

[29] Layek S, et al. 3D reconstruction: An emerging prospect for surveying. Lecture Notes in Civil Engineering. 2019;71–81. https://doi.org/10.1007/978-981-13-7067-0_5.

[30] Remondino F, Rizzi A. Reality-based 3D documentation of natural and cultural heritage sites – techniques, problems, and examples. Applied Geomatics. 2010;2(3):85–100. https://doi.org/10.1007/s12518-010-0025-x.

[31] Reipschläger P, Dachselt R. DesignAR: Immersive 3D-modeling combining augmented reality with interactive displays. Proceedings of the 2019 ACM International Conference on Interactive Surfaces and Spaces. 2019;29–41. https://doi.org/10.1145/3343055.3359718.

[32] Fleck S, Hachet M, Bastien JMC. Marker-based augmented reality. Proceedings of the 14th International Conference on Interaction Design and Children. 2015;21–28. https://doi.org/10.1145/2771839.2771842.

[33] Blut C, Blankenbach J. Three-dimensional CityGML building models in mobile augmented reality: A smartphone-based pose tracking system. International Journal of Digital Earth. 2020;14(1):32–51. https://doi.org/10.1080/17538947.2020.1733680.

[34] Sahu CK, Young C, Rai R. Artificial intelligence (AI) in augmented reality (AR)-assisted manufacturing applications: A review. International Journal of Production Research. 2020;59(16):4903–4959. https://doi.org/10.1080/00207543.2020.1859636.

[35] Li H, et al. EVD surgical guidance with retro-reflective tool tracking and spatial reconstruction using head-mounted augmented reality device. arXiv. 2023. https://doi.org/10.48550/ARXIV.2306.15490.

[36] Yu K, et al. Collaborative SLAM and AR-guided navigation for floor layout inspection. The Visual Computer. 2020;36(10–12):2051–2063. https://doi.org/10.1007/s00371-020-01911-8.

[37] Lum HC, et al. Virtual reality: History, applications, and challenges for human factors research. Proceedings of the Human Factors and Ergonomics Society Annual Meeting. 2020;64(1):1263–1268. https://doi.org/10.1177/1071181320641300.

[38] McGill M, Gugenheimer J, Freeman E. A quest for co-located mixed reality: Aligning and assessing SLAM tracking for same-space multi-user experiences. 26th ACM Symposium on Virtual Reality Software and Technology. 2020;1–10. https://doi.org/10.1145/3385956.3418968.

[39] Wu H, et al. Understanding freehand gestures: A study of freehand gestural interaction for immersive VR shopping applications. Human-centric Computing and Information Sciences. 2019;9 (1). https://doi.org/10.1186/s13673-019-0204-7.
[40] Bai H, et al. (2014). 3D gesture interaction for handheld augmented reality. SIGGRAPH Asia 2014 Mobile Graphics and Interactive Applications https://doi.org/10.1145/2669062.2669073
[41] Sood S, et al. What is telemedicine? A collection of 104 peer-reviewed perspectives and theoretical underpinnings. Telemedicine and e-Health. 2007;13(5):573–590. https://doi.org/10.1089/tmj. 2006.0073.
[42] Meesters AML, et al. Introduction of a three-dimensional computed tomography measurement method for acetabular fractures. Plos One. 2019;14(6). https://doi.org/10.1371/journal.pone.0218612.
[43] Chim H, Wetjen N, Mardini S. Virtual surgical planning in craniofacial surgery. Seminars in Plastic Surgery. 2014;28(03):150–158. https://doi.org/10.1055/s-0034-1384811.
[44] Hongo K, et al. Telecontrolled micromanipulator system (NeuRobot) for minimally invasive neurosurgery. Acta Neurochirurgica Supplements. n.d.;63–66. https://doi.org/10.1007/978-3-211-33303-7_9.
[45] Gu F, et al. Automatic range of motion measurement via smartphone images for telemedicine examination of the hand. Science Progress. 2023;106(1). https://doi.org/10.1177/00368504231152740.
[46] Bizzotto N, et al. Three-dimensional printing of bone fractures. Surgical Innovation. 2015;22(5):548–551. https://doi.org/10.1177/1553350614547773.
[47] Memon AR, et al. A review on computer-aided design and manufacturing of patient-specific maxillofacial implants. Expert Review of Medical Devices. 2020;17(4):345–356. https://doi.org/10. 1080/17434440.2020.1736040.
[48] Paturkar A, Sen Gupta G, Bailey D. Making use of 3D models for plant physiognomic analysis: A review. Remote Sensing. 2021;13:2232. https://doi.org/10.3390/rs13112232.
[49] Journal of Biomedical and Sustainable Healthcare Applications (http://anapub.co.ke/journals/jbsha/jbsha.html) doi: https://doi.org/10.53759/0088/JBSHA202202010
[50] Three-dimensional computed craniofacial tomography (3D-CT): potential uses and limitations Hong Jin Chan, Michael Woods and Damien Stella: doi: 10.2478/aoj-2007-0010
[51] Rukundo O. Challenges of 3D surface reconstruction in capsule endoscopy. J. Clin. Med. 2023;12:4955. https://doi.org/10.3390/jcm12154955.

Symah Bashir, Amanpreet Singh, Shivani Sood

8 Emerging Trends in Medical Imaging with Fog Computing

Abstract: The rapid improvements in medical imaging technologies have resulted in an increase in the amount of data produced, posing important challenges in terms of processing, storage, and real-time analysis. Fog computing provides a highly effective and innovative approach to handle these challenges. This chapter explores the upcoming trends in the Internet of Medical Things (IoMT) in healthcare supported by fog environment. It includes the medical imaging technologies such as computed tomography scans, magnetic resonance imaging, and X-rays focusing on its real-time image processing, minimizing the risks and guaranteeing improved data security. Through the distribution of computing power enables faster decision-making, reducing latency and improved coordination among healthcare. The implementation of fog computing with diagnostic imaging also facilitates privacy regulations and guaranteeing the sensitive data. Although it has benefits, challenges related to intercommunication, information privacy, edge device constraints and regulatory compliance must be resolved for widespread adoption. Furthermore, technologies such as artificial intelligence, 5G, 3D imaging, and Internet of Things play an important role in transforming medical imaging practices, enabling remote doctors and improving patient outcomes in the healthcare sector. Additionally, it also reduces the operational costs of healthcare facilities and enhanced patient experience through faster diagnoses. This chapter also explores the various ways fog computing is defining the future of medical diagnostics and highlighting its power to innovate healthcare delivery.

Keywords: fog computing, IoMT, 3D imaging, patient experience, medical imaging, telemedicine

8.1 Introduction

Medical imaging has consistently played a crucial role in modern healthcare such as diagnosis, treatment, and handling of multiple conditions. As technology advances, innovative methods and modern systems are being introduced to enhance the quality, accessibility, and efficiency of medical imaging. A significant emerging trend involves the integration of fog computing with medical imaging systems [15]. This innovative concept is poised to transform the healthcare system significantly. An innovative solution for real-time medical imaging, fog computing's network architecture decentralizes cloud processing to edge devices. In contrast to conventional cloud computing, where processing often requires transmitting substantial data volumes further away.

 | https://doi.org/10.1515/9783112226223-008

By situating intelligent power at the source, such as diagnostic imaging devices, fog computing improves data processing speed [20]. The result of this modification is reduced latency, lower bandwidth demands, and optimized efficiency across diagnostic imaging workflows. Within the context of diagnostic imaging, the fog computing framework possesses the capacity to innovate various features in the field.

It facilitates the immediate processing of imaging data from diagnostic tools such as computed tomography (CT) scanners, magnetic resonance imaging (MRI) systems, and other medical devices [11]. Such an approach supports on-the-spot image evaluation, aiding clinicians in delivering swift and accurate diagnoses. This, in turn, boosts diagnostic correctness, facilitates faster treatment decisions, and enhances the management of patient records. Figure 8.1 explains the smart healthcare system that integrates both Internet of Things (IoT) and cloud computing to facilitate on-demand services of medical imaging into a more responsive, intelligent, and patient-centric system [21].

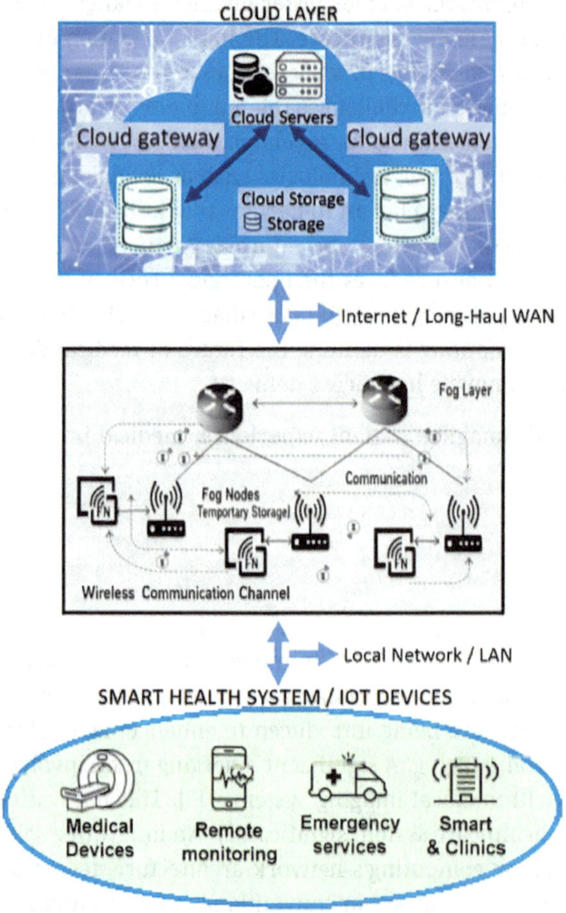

Figure 8.1: Smart healthcare system within fog computing.

8.2 Usage of Fog Computing in Medical Imaging

8.2.1 MRI Machines

MRI scanners are crucial devices used to find and follow various health problems, ranging from brain and nerve conditions to injuries of the bones and muscles [6]. These instruments develop high-quality and high-resolution images of the body without using dangerous radiation and are essential for modern health monitoring system. They are also utilized to capture intricate stroke images and facilitate the detection of cardiovascular diseases. Furthermore, they provide scan results instantaneously after imaging any part of the body. High-definition images from MRI scans are transmitted to remote cloud data centers, where they undergo processing and are then stored within the network [18]. By processing diagnostic data and results directly on the network, fog computing contributes to saving valuable bandwidth and reducing potential network congestion. Through localized caching of medical imaging on fog servers, same-facility practitioners experience reduced data latency [1]. It facilitates the remote sharing and examination of MRI results, enhancing telemedicine applications and collaboration among professionals. Some techniques utilized in fog computing for various optimization methods and architectural benefits are illustrated in Figure 8.2, which demonstrates their roles in enhancing performance and resource management within fog environments.

8.2.2 CT Scans

High-resolution CT scans provide detailed cross-sectional views of the body and are essential for detecting medical issues such as tumors, internal bleeding, skeletal damage, and blood vessel abnormalities [4]. With the help of X-rays, CT scans deliver high-definition images that support noninvasive diagnosis and monitoring. Their role is becoming increasingly prominent in contemporary medical care. Through immediate data processing, improved collaboration, and performance maximization, fog computing plays a vital role in advancing CT scan system capabilities [3].

Conventional CT scans generate extensive raw imaging data that requires transformation into high-resolution images for accurate medical assessment. With the implementation of fog computing, data processing largely occurs closer to the data source, at the edge of the network. Data is first processed and assessed by fog nodes within the local system prior to being sent to the cloud or connected medical equipment. As a result, emergency scenarios benefit from low latency, enabling rapid and effective decisions. For example, in critical care cases, immediate responses from CT scans can help achieve quicker diagnosis and may prevent loss of life. By accelerating the evaluation of CT scan images, fog computing delivers timely information to healthcare professionals [17]. CT scan imaging results in substantial high-quality files

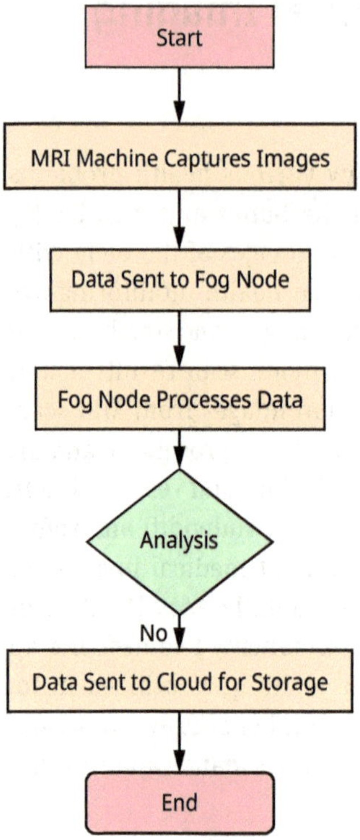

Figure 8.2: Use of fog computing in MRI machines within mathematical model.

that are typically sent to cloud-based systems for additional analysis or archival. Large data transmissions through the communication network often require considerable bandwidth and result in extended transfer times. Fog computing mitigates this issue by handling data processing at the edge and transmitting relevant details to cloud servers [9]. As an example, once CT images are acquired, tumors are identified through local processing, and relevant information is efficiently forwarded to the cloud or medical experts. Figure 8.3 illustrates the Fog Computing architecture in the CT Scan workflow.

8.2.3 X-Ray Machines in Healthcare

X-ray machines play a crucial role in healthcare by capturing images of the body's internal bones, tissues, and organs [10]. These instruments support clinicians in the diagnosis, assessment, and management of diverse health issues. They generate a controlled beam of X-rays that pass through the body and are absorbed by tissues at varying rates [13]. Fog computing is now widely integrated into modern healthcare systems alongside diagnostic tools like X-ray machines, supporting improved data processing, storage, and real-time decision-making through mathematical modeling.

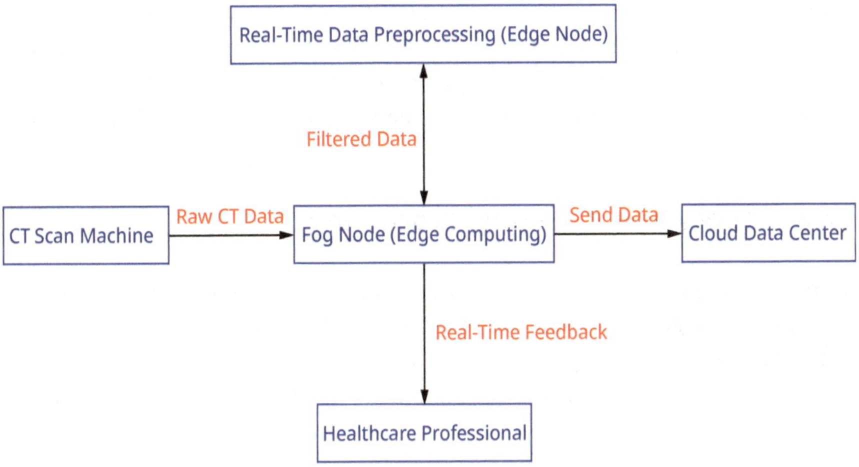

Figure 8.3: Fog computing architecture in CT scan workflow.

These solutions can be effectively understood through a mathematical lens. Hence, we demonstrate how X-ray systems function in conjunction with fog computing using models that handle data management, communication, and efficient storage [12]. When IoT devices generate tasks such as medical images, fog computing architecture interacts with X-ray machines to process the data, which is then typically forwarded to cloud servers for additional analysis and storage. In fog computing architectures, X-ray devices routinely transmit imaging data to nearby fog servers for processing. This distributed approach significantly minimizes both latency and data loss during processing compared to traditional cloud methods. The use of devices like gateway decreases the need of bandwidth usage and allows faster real-time decision-making. These features can be implemented by the use of mathematical framework. The framework considers multiple parameters such as fog nodes' processing power, the volume of data produced by X-ray machines, and system response time [16]. By evaluating these parameters, it is possible to figure out the most effective placement and timely delivery of results. Moreover, this mathematical model supports researchers and clinicians in boosting the accuracy, processing speed, and effectiveness of X-ray machines. Figure 8.4 shows how fog computing techniques contribute to improving the efficiency of X-ray machines.

8.3 Role of Fog Computing in Medical Imaging

The role of medical imaging in contemporary healthcare is vital, empowering professionals to identify and monitor different diseases. However, the large volume and resolution of medical imaging data can strain traditional cloud systems, resulting in delays and potential privacy issues [2]. This section illustrates how fog computing overcomes these problems by distributing computational tasks [12]. These operations move processing from centralized cloud systems to edge devices – including tablets, smartphones, and medical imaging tools – enhancing responsiveness and efficiency.

8.3.1 Low Latency and Real-Time Processing

Medical imaging often requires real-time processing, especially in critical scenarios such as emergency diagnostics or surgical procedures [12]. Rather than relying on distant cloud servers, fog computing processes data near the source, helping to minimize latency [14]. This supports more rapid decision-making, which is important in healthcare facilities.

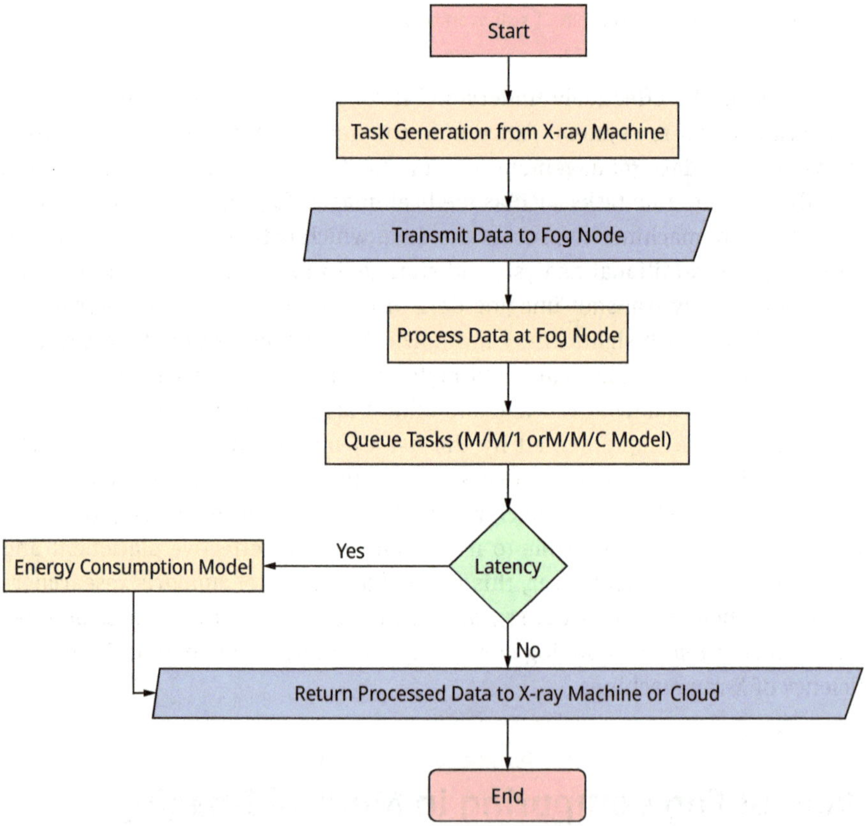

Figure 8.4: Use of fog computing in X-ray machines within mathematical model.

8.3.2 Improved Reliability and Availability

Fog computing supports the functionality of medical imaging tools in areas with restricted or unreliable internet access [5]. In the event of a lost cloud connection, edge devices can still function independently to manage data and generate diagnostic results. As a result, the system becomes more dependable in locations that are remote or lack adequate resources [7].

8.3.3 Support for AI and Machine Learning Models

The use of fog computing allows artificial intelligence and machine learning models to operate closer to data sources, increasing the analytical performance of medical imaging systems [19]. For example, AI models can help radiologists by autonomously recognizing abnormalities in X-ray scans or MRIs providing real-time diagnostic assistance. For instance, AI models can assist radiologists by automatically detecting abnormalities in X-ray or MRI scans, offering real-time diagnostic support. These models can run directly on edge devices with the help of fog computing, minimizing dependence on continuous cloud connectivity.

8.3.4 Scalability and Flexibility

Fog computing supports dynamic infrastructure expansion, allowing new edge devices and processing capabilities to be added as the demand for medical imaging grows [22]. As medical data generation accelerates, supported by innovative approaches and evolving healthcare models, this scalability becomes increasingly important.

8.3.5 Collaboration Across Healthcare Providers

Through fog computing, hospitals can work together more effectively by confidentially exchanging imaging data across various sites in real time [8]. By handling data at the edge and sharing only necessary insights, fog computing promotes collaborative clinical decision-making while preserving confidentiality.

8.4 Mathematical Models in Medical Imaging and Fog Computing

In this part, we will analyze the topic deeper into the mathematical models and techniques in the interconnection of medical technologies and fog computing. This focuses on how image processing is affected by fog techniques for processing, data transmission, and storage in real-time medical applications.

8.4.1 Distributed Image Processing in Fog Computing

Fog computing's primary advantage is its ability to offload problem-solving processes to devices at the network edge [15]. The computational requirements of CT and MRI imaging can be efficiently managed through collaboration between local processing units. Mathematical models for task management and resource control are necessary for enhancing performance. Therefore, these images are generated by devices (e.g., MRI and CT) and need to be processed across various nodes, and it can be represented as

$$Z_{\text{task}} = \sum_{i=1}^{N} Z_{\text{edge},\,i} + Z_{\text{cloud}}.$$

8.4.2 Bandwidth and Latency Optimization

Effective real-time medical image processing in a fog environment hinges on minimizing latency and ensuring adequate bandwidth. Mathematical models are regularly applied to optimize the use of bandwidth and lower latency during data transmission from diagnostic devices to fog nodes and cloud servers.

8.4.3 Task Offloading and Scheduling Models

Computational tasks in fog environments are commonly shifted from edge devices to fog nodes through optimized task management strategies. To enhance resource efficiency and minimize latency, effective task scheduling and offloading are crucial.

- **Queuing Theory:** It represents the coming of tasks and their processing times at fog nodes. Queuing models help in estimating wait times and planning the optimal schedule strategies.
- **Game Theory:** These models are leveraged to optimize communication between sensor devices and fog nodes, maintaining fairness in resource distribution within multiple devices.

- **Markov Decision Processes (MDPs)**: MDPs are used for designing decision-making frameworks in situations with unpredictability. They help optimize performance in fog computing networks for deciding the tasks should be shifted to network conditions.
- **Reinforcement Learning (RL):** RL algorithms study the optimal methods for task delegation of policies through exchange through interaction with the environment. This supports fog computing systems to continuously adapt to shifting conditions, such as variable network traffic or processing power.

8.4.4 Data Privacy and Security

In medical technology, securing patient data is essential for healthcare personnel. Mathematical models are implemented in the data protection while allowing computing the encrypted data. This is important in a fog environment, where data may be analyzed on IoT devices before transmission to the cloud.

8.4.5 Fault Tolerance and Reliability

Fog computing techniques must be capable of handling failures. Mathematical models help in designing the algorithms for high availability of resources that may have limited processing power or memory. This focuses particularly on healthcare environments, maintaining robustness and ensuring service continuity in case of failures.

8.5 Classification of Medical Imaging

The classification of medical imaging, such as X-rays, MRI, CT scans, ultrasound, or histopathology images, is crucial. These imaging techniques are used to help in diagnosing and identifying diseases in the human body. This goal can be achieved by automating the process using hybrid algorithms in a fog approach to identify diseases and support healthcare professionals in making accurate and quick diagnoses. Details are shown in Table 8.1.

8.5.1 X-Ray Imaging Techniques

Table 8.2 presents the comparison of different machine models which are used for detecting the various healthcare problems. These machines work based on key performance parameters and indicated the percentage of correct results in terms of image

Table 8.1: Classification of imaging with medical profession.

Component	Description	Example
Image types	The modalities of medical images used for classification, such as X-ray, MRI, CT scans, ultrasound, etc.	X-ray, MRI, CT, ultrasound, fundus, and histopathology
Classes	Disease or condition classification for the images: can be binary, multi-class, or multi-label	Healthy versus disease, malignant versus benign, pneumonia, and cancer types
Metadata	Additional contextual information, such as patient demographics, imaging parameters, or disease severity	Patient's age, gender, imaging resolution, machine used, and disease stage
Annotations	Expert-defined labels for regions of interest ROI, such as bounding boxes, segmentation masks, or point annotations	Tumor locations (bounding boxes) and tumor boundaries (segmentation masks)
Data size and variety	Number of images, with potential augmentations to increase dataset diversity	10,000 images with data augmentation (rotation and scaling)
Class imbalance	Proportion of images for each class (common problem in medical datasets)	Rare disease types (e.g., 5% of images are cancerous)
Image resolution	The resolution of images, which can affect the model's performance in detecting fine-grained details	512 × 512, 1,024 × 1,024, and 256 × 256 pixels
Data privacy	Anonymization or de-identification to comply with privacy regulations (HIPAA and GDPR)	Anonymized patient data and compliant with HIPAA
Data augmentation	Techniques used to artificially expand the dataset by modifying the original images	Flipping, rotating, zooming, and intensity adjustments
Datasets	Well-known publicly available datasets are used in medical imaging classification	ChestX-ray14, LIDC-IDRI, BRATS, ISIC, and EyePACS
Applications	Potential use cases for the classification models built from the dataset	Disease diagnosis, early detection, and personalized treatment plans
Challenges	Common obstacles in working with medical imaging datasets	Imbalanced classes, data privacy issues, quality variance, and annotation errors

quality and diagnostic capability. The table presents the evolution of machines over time. It also outlines the results of different machines from 2018 to 2024, highlighting the different accuracy and exposure time with the computing technology for data management are explained in Figure 8.5. These techniques focus on medical image quality as well as improvements in radiation exposure and processing time.

Table 8.2: Comparison of machine models and their specifications.

Year and reference	Machine model	Accuracy (%)	Connection	Radiation dose	Exposure time (s)
2018, [15]	GE Revolution CT	93.0	Cloud	Low dose	0.25
2018	Philips DigitalDiagnost	90.0	Fog	Low dose	0.2
2019, [20]	Canon RadPRO 26k	91.4	Fog	Low dose	0.1
2020 [21]	Philips DigitalDiagnost C90	92.7	Cloud	Low dose	0.1
2020, [6]	Siemens Ysio Max	89.2	Not mentioned	Low dose	0.12
2021, [18]	Siemens Somatom X.ceed	94.5	Edge computing	Low dose	0.05
2021, [1]	GE Optima XR240am	93.5	Cloud	Low dose	0.1
2022, [4]	Siemens Somatom X.ceed	94.8	Fog	Low dose	0.1
2023, [3]	GE Optima XR240am	95.2	Fog	Low dose	0.1
2024, [17]	Canon RadPRO 26k	96.0	Not mentioned	Low dose	0.1

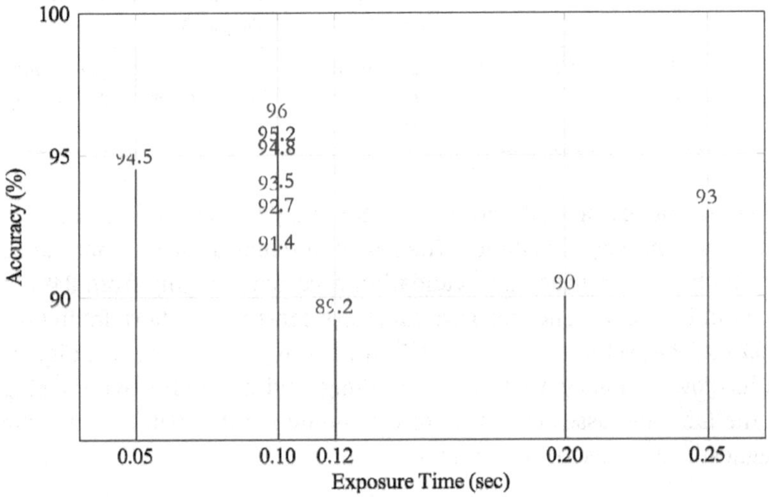

Figure 8.5: Comparison of accuracy and exposure time for various X-ray machines.

8.5.2 CT Scan Imaging Techniques

Comparison of techniques, accuracy, and exposure time for fog computing in CT scan machines is shown in Table 8.3.

Table 8.3: Comparison of techniques, accuracy, and exposure time for fog computing in CT scan machines.

Reference	Year	Total acCuracy (%)	Validation accuracy (%)	Exposure time (s)	Technique	Merits
[9]	2019	94	92	0.25	Fog computing + ML models	Reduced latency by processing data locally on edge devices
[10]	2020	95	93	0.2	Hybrid cloud-fog architecture	Improved real-time CT image processing efficiency
[13]	2021	96	94.5	0.1	Edge-based fog computing	Minimizes bandwidth requirements and latency in remote areas
[12]	2022	97	95	0.12	Fog-based compression algorithms	Enhanced data transmission speed and reduced storage demands
[16]	2023	98	97	0.1	AI + fog computing (CNN and SVM)	Integration of deep learning for accurate and efficient diagnosis
[2]	2024	99	98	0.1	Fog computing + image processing AI	Enhanced image quality with real-time data analysis on fog nodes

The table provides a comparison of different fog computing methods used in CT scan machines conducting from various studies. The research demonstrates a continuous improvement in both total accuracy and validation accuracy, ranging from 94% to 99%, as illustrated in Figure 8.6. The exposure times are consistently short, falling between 0.1 and 0.25 s. The techniques use the ML, AI, and fog computing techniques. Each approach has lower latency, real-time processing, and more effective imaging quality. Hence, the table discusses the usage of fog computing that will enhance the efficiency of diagnostic methods in CT scanning.

8.5.3 MRI Imaging Techniques

The methods employed involve machine learning, hybrid cloud-fog architecture, edge-based fog computing, compression techniques, and AI technologies like CNN and SVM. Each of these approaches brings distinct advantages, including lower latency, enhanced real-time processing, faster data transmission, and superior image quality. Table 8.4 shows the MRI machine techniques with fog computing.

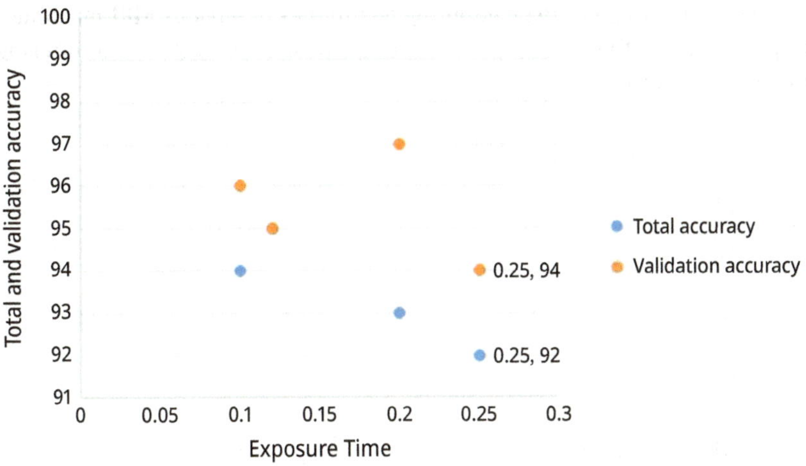

Figure 8.6: Comparison of techniques, accuracy, and exposure time for fog computing in CT scan machines.

The techniques utilized range from machine learning and hybrid cloud-fog models to edge-based fog computing, compression algorithms, and AI tools such as CNN and SVM. These methods offer specific benefits, such as reduced latency, better real-time processing, faster data transfer, and improved image clarity.

Table 8.4: MRI machine techniques with fog computing.

Reference	Year	Technique	Total accuracy (%)	Validation accuracy (%)	Exposure time (s)
[12]	2019	T1-weighted imaging with fog computing	92.5	90.3	0.3
[14]	2020	T2-weighted imaging with fog computing	94.2	91.8	0.4
[5]	2021	Diffusion weighted imaging (DWI) with fog computing	95.3	92.1	0.2
[7]	2022	Functional MRI (fMRI) with fog computing	97.1	94.5	0.6
[19]	2023	Magnetic resonance angiography (MRA) with fog computing	93.8	91.0	0.5
[22]	2024	Magnetic resonance spectroscopy (MRS) with fog computing	96.4	93.3	0.4
[8]	2024	3D MRI imaging with fog computing	98.0	95.5	0.8

Figure 8.7 compares the usage of fog computing techniques in various MRI machines. These techniques improve the accuracy and exposure from 2019 to 2024, and offering more scans with faster results.

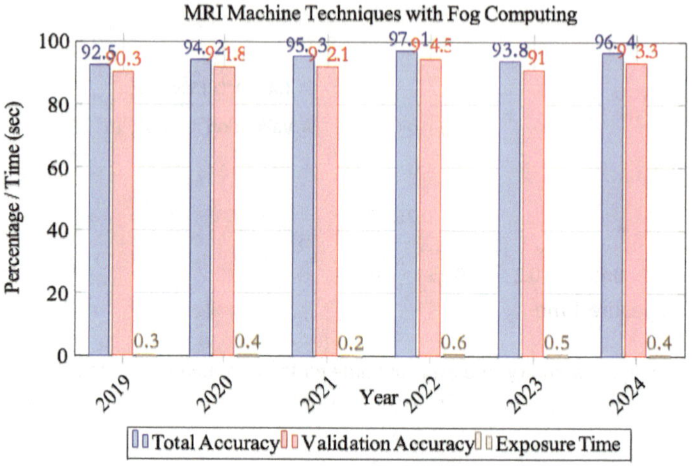

Figure 8.7: Graph of MRI machine techniques with fog computing.

8.6 Conclusions

To summarize this chapter, the different Internet of Medical Things use of imaging techniques are discussed, particularly for health diagnosis with the help of fog computing. Fog computing is making substantial progress in the field of medical imaging, particularly in MRI, CT scans, and X-rays, by facilitating real-time data processing at the edge. This improves the speed, precision, and effectiveness of medical diagnoses. Based on the conclusion of three tables, GE enhances the accuracy in X-ray machines to 95%, 98% in CT scans, and 98% in MRI machines. These techniques minimize the limitation of data security, making it beneficial for remote settings.

As the use of AI and machine learning continues to grow, fog computing will play an increasingly crucial role in transforming the field of medical imaging, enabling more accurate and accessible availability of healthcare services. In future, we will expand this effort by capturing real-time images of CT scans, X-rays, and MRIs of different patients, and also incorporating algorithms of fog computing that will be used for detecting different types of diseases and giving real-time updates.

References

[1] Albahli S, Yar GNAH. Ai-driven deep convolutional neural networks for chest x-ray pathology identification. Journal of X-Ray Science and Technology. 2022;30(2):365–376.

[2] Alrashdi I. Fog-based deep learning framework for real-time pandemic screening in smart cities from multi-site tomographies. BMC Medical Imaging. 2024;24(1):123.

[3] Alshmrani GMM, Ni Q, Jiang R, Pervaiz H, Elshen- Nawy NM. A deep learning architecture for multi-class lung diseases classification using chest x-ray (CXR) images. Alexandria Engineering Journal. 2023;64:923–935.

[4] Ausawalaithong W, Thirach A, Marukatat S, Sitporn TW. Automatic lung cancer prediction from chest x-ray images using the deep learning approach. In 2018 11th biomedical engineering international conference (Bmeicon) 2018 (pp. 1–5). IEEE.

[5] Das J, Ghosh S, Mukherjee A, Ghosh SK, Buyya R. Rescue: Enabling green healthcare services using integrated iot-edge-fog-cloud computing environments. Software: Practice and Experience. 2022;52 (7):1615–1642.

[6] Gu C, Dai C, Shi X, Wu Z, Chen C. A cloud-based deep learning model in heterogeneous data integration system for lung cancer detection in medical industry 4.0. Journal of Industrial Information Integration. 2022;30(100386).

[7] Guo M, Liu H, Gao L, Yu H, Ren Y, Li Y, Yang H, Cao C, Fan G. A radiomics approach for predicting gait freezing in Parkinson's disease based on resting-state functional magnetic resonance imaging indices: A cross-sectional study. Neural Regeneration Research. 2024;10–4103.

[8] Kahali S, Adhikari SK, Sing JK. A two-stage fuzzy multi-objective framework for segmentation of 3D MRI brain image data. Applied Soft Computing. 2017;60:312–327.

[9] Khullar V, Singh HP, Miro Y, Anand D, Mohamed HG, Gupta D, Kumar N, Goyal N. Iot fog-enabled multi-node centralized ecosystem for real time screening and monitoring of health information. Applied Sciences. 2022;12(19):9845.

[10] Li D, Wang D, Dong J, Wang N, Huang H, Xu H, Xia C. False-negative results of real-time reverse-transcriptase polymerase chain reaction for severe acute respiratory syndrome coronavirus 2: Role of deep-learning-based CT diagnosis and insights from two cases. Korean Journal of Radiology. 2020;21(4):505–508.

[11] Li Y, Zhang Z, Dai C, Dong Q, Badrigilan S. Accuracy of deep learning for automated detection of pneumonia using chest x-ray images: A systematic review and meta-analysis. Computers in Biology and Medicine. 2020;123(103898).

[12] Nasir MU, Khan S, Mehmood S, Khan MA, Rahman A-U, Hwang SO. IoMT-based osteosarcoma cancer detection in histopathology images using transfer learning empowered with blockchain, fog computing, and edge computing. Sensors. 2022;22(14):5444.

[13] Rahman MA, Hossain MS. An internet-of-medical-things-enabled edge computing framework for tackling covid-19. IEEE Internet of Things Journal. 2021;8(21):15847–15854.

[14] Rai S, Ullah A, Kuan WL, Mustafa R. An enhanced compression method for medical images using SPIHT encoder for fog computing. International. Journal of Image and Graphics. 2023;page2550025.

[15] Roy SK, Shekhar V, Lassar WM, Chen T. Customer engagement behaviors: The role of service convenience, fairness and quality. Journal of Retailing and Consumer Services. 2018;44:293–304.

[16] Sajid J, Hayawi K, Malik AW, Anwar Z, Trabelsi Z. A fog computing framework for intrusion detection of energy-based attacks on UAV-assisted smart farming. Applied Sciences. 2023;13(6):3857.

[17] Tahir A, Saadia A, Khan K, Gul A, Qahmash A, Akram RN. Enhancing diagnosis: Ensemble deep-learning model for fracture detection using x-ray images. Clinical Radiology. 2024;79(11):e1394–e1402.

[18] Thakur S, Kumar A. X-ray and CT-scan-based automated detection and classification of covid-19 using convolutional neural networks (CNN). Biomedical Signal Processing and Control. 2021;69 (102920).

[19] Vermandel M, Kulik C, Leclerc X, Rousseau J, Vasseur C. Matching of projection imaging and tomographic imaging: Application to digital subtracted angiography (DSA) and magnetic resonance angiography (MRA). In Medical imaging 2002: Visualization, image-guided procedures, and display 2002 (vol. 4681, pp. 55–64). SPIE

[20] Yadav DP, Rathor S. Bone fracture detection and classification using deep learning approach. In 2020 international conference on power electronics & IoT applications in renewable energy and its control (PARC) 2020 (pp. 282–285). IEEE.

[21] Zhang Y, Gorriz JM, Dong Z. Deep learning in medical image analysis. Journal of Imaging. 2021;7(4):74.

[22] Zhou Y, Wu Y, Su Y, Li J, Cai J, You Y, Zhou J, Guo D, Qu X. Cloud-magnetic resonance imaging system: In the era of 6g and artificial intelligence. Magnetic Resonance Letters. 2024;200138.

Rishu Chhabra, Shiwangi, Saravjeet Singh

9 In-Depth Analysis of Advanced Image Segmentation Techniques for Accurate Identification of Anatomical Structures and Pathologies in Medical Imaging

Abstract: Image segmentation is an important part of medical image analysis that plays a crucial role in proper diagnosis and treatment planning. It involves partitioning a digital image into multiple segments or regions to simplify its representation and make it more meaningful for analysis. This process helps in the accurate identification of anatomical structures and pathological disorders across various imaging methods, including magnetic resonance imaging, computed tomography scan, X-ray, and ultrasound. The integration of artificial intelligence (AI)-based techniques helps in the precise segmentation of organs, tissues, and even abnormal regions, which helps in-depth analysis of important parts of the body such as tumors and lesions. AI-based techniques not only help in detecting disorders but also help in detailed analysis of anatomical structures, which leads to improved accuracy in diagnosing diseases like cancer, heart conditions, and brain disorders. Deep learning models like convolutional neural networks, recurrent neural networks, autoencoders, and U-Net architectures have been widely used in medical image segmentation. In this chapter, an in-depth analysis of advanced image segmentation techniques in medical imaging has been presented.

Keywords: deep learning, U-net, transformer, semantic segmentation, 3D segmentation

9.1 Introduction

Medical image segmentation is a computer vision problem that involves dividing a medical image into several parts based on features like brightness, texture, or color and each part represents an entity or structure. It is typically used to locate objects and boundaries in images. Image segmentation assists in generating an accurate and exact representation of the objects of interest in the image for diagnosis, treatment planning, and quantitative evaluation. At its core, medical image segmentation means the division of the complicated image into various segments or highlighting the specific regions to separate specific areas such as organs, tissues, or cancerous anomalies, including tumors. It helps in accurate identification of disease-infected areas. Image segmentation can be defined as the problem of classification with semantic segmenta-

 | https://doi.org/10.1515/9783112226223-009

tion or instance segmentation or both, i.e., panoptic segmentation [1]. Semantic segmentation is the process of assigning a class or category to every pixel in the image without differentiating between instances of the same class. On the other hand, instance segmentation differentiates objects or instances within the image. Image segmentation is an essential step for the analysis of medical images like ultrasound, X-rays, and CT scans. The entire process, as shown in Figure 9.1, starts with the input image and then progresses through preprocessing, segmentation, postprocessing, and produces a segmented image as output.

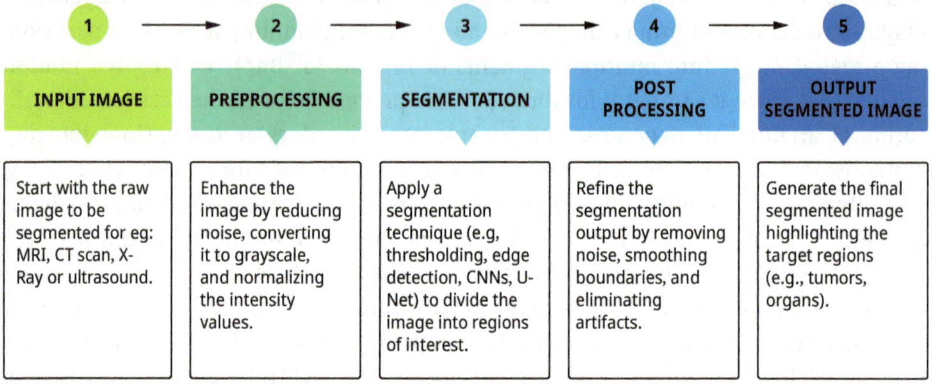

Figure 9.1: Medical image segmentation process.

Various image segmentation algorithms have been developed from the initial methods such as thresholding, region-growing, *K*-means, and watershed methods to more advanced algorithms such as graph cuts and active contours methods. In recent years, deep learning (DL) models have introduced new techniques of image segmentation with remarkable performance improvements, which lead to a high rate of accuracy, and it leads to a major shift in the field. This chapter covers the deep-learning segmentation architectures like convolutional neural networks (CNNs), recurrent neural networks (RNNs), U-Net, and autoencoders, which have significantly enhanced the precision and efficiency of medical image segmentation.

9.1.1 Importance of image segmentation in medical imaging

Medical imaging in image segmentation is significant, as it allows us to precisely isolate and accurately analyze specific regions of interest (ROI). It helps in accurate identification of the anatomical structures, pathological areas, or lesions, leading to better patient outcomes. Here are some key benefits of image segmentation in medical imaging:

(i) Disease diagnosis: Image segmentation is important in the diagnosis of diseases, as it helps to identify and separate areas of interest from 2D/3D image data. The main goal of diagnosis is to identify abnormalities in the human body and identify regions of disease within MRI (magnetic resonance imaging) and CT (computed tomography) scan and X-ray pictures prior to detecting major issues. The infected areas become easy to analyze for accurate assessment. By greater image clarity, doctors can see precise organ and tissue details, enhance their knowledge about the patient's health, and achieve accurate diagnosis. Segmentation also enables quantitative analysis by comparing the shape, location, and volume of tumors at different time points, enabling the monitoring of treatment progress. Furthermore, segmentation algorithms based on AI reduce diagnostic errors by human factors through offering high precision, consistency, and objectivity, and rendering the process of diagnosis more reliable.

(ii) Treatment planning: Before surgeries or treatment like radiation therapy, doctors need a clear view of the affected area, so segmentation helps the doctors plan treatments more accurately by showing the detailed view of organs and tissues. During surgical planning, it helps in segmenting the three-dimensional images, which helps in accurate identification of the organ to be operated on and consequently, the required interventions are more precise and less invasive. In radiation therapy, proper differentiation of the tumors ensures that the patient gets the maximum amount of radiation required and minimizes the damage inflicted on the other healthy tissues. Segmentation also helps in personalized medicine, as it allows doctors to make the recovery plan that is specific to the patient's anatomy and illness. Additionally, it helps in the posttreatment monitoring, which makes the comparison between preprocessing and postprocessing treatment images, which makes the treatment planning effective.

(iii) Tracking diseases: Image segmentation is important in disease progression monitoring, which enables doctors to compare images from different stages of treatment. Segmenting certain regions of interest from MRI, CT scans, or X-rays helps to accurately monitor changes in the affected region over time. It helps in determining if a disease is progressing, stabilizing, or recovering due to treatment. For chronic diseases like cancer and cardiovascular diseases, segmentation yields quantitative information about tumor development, tissue damage, or lesion formation, and aids in more informed clinical decisions. Segmentation also helps in detecting changes that are not visible in raw medical images, with the potential for early treatment if required. With constant observation of disease advancement, doctors can adjust treatment protocols, accordingly enhancing patient outcomes and optimizing therapy approaches.

(iv) Enhanced visualization: Medical images hold complex visual information, which makes it difficult to separate normal tissue from abnormal tissue because multiple body structures overlap with one another. Through image segmentation, doctors can better visualize targeted areas because this process allows for specific region isolation, followed by careful highlighting. Medical image interpretation accuracy improves through segmentation because this process creates clear distinctions among different tissue types and organs and abnormalities in images. Throughout tumor diagnosis and evaluation of fractures and vascular diseases, it becomes essential to accurately detect which areas are affected, and image segmentation aids in this task. Medical practitioners can develop better treatment plans with enhanced visualization because they see internal structures better, ensuring optimal medical outcomes.

9.2 Fundamentals of Medical Image Segmentation

Medical image segmentation is the most important method in biomedical imaging, which helps in dividing the image into regions that are connected and have some similarity within the region and some differences between adjacent regions. The goal is to identify the individual objects within the image based on the similarity like pixel intensity, color, or texture, a process known as pixel or voxel classification. Segmentation subdivides an image into constituent regions or objects until the object of interest is not isolated or it divides the complex image into multiple ROIs, which helps healthcare professionals to get a clear view of diseases progression, a comparison of anatomical changes, and an evaluation of treatment effect. Susmita Sahu et al. [2] conducted an in-depth analysis of image segmentation along with its diverse techniques, which they showed to be vital for computer vision and medical imaging. Different image segmentation solutions, which include thresholding, edge detection, and clustering, as well as deep learning-based approaches get selected for specific applications based on imaging difficulty and requirements. Improved segmentation accuracy results from deep learning advancements, which allows for greater detection of complex patterns within medical images.

9.2.1 Medical Imaging Modalities

A modality is the method by which a diagnostic procedure is performed. In the world of medical imaging, some examples of modality are X-ray, MRI, ultrasound, and CT. Zhe Guo et al. [3] examined deep learning-based image segmentation methods for multimodal medical imaging and looked at their part in improving diagnostic accuracy. For segmenting medical pictures such MRI and CT scans, authors considered many deep learning models, including CNN and attention-based constructions. It was concluded that deep learning segmentation outperforms standard segmentation but needs large datasets and some sophisticated computing infrastructure. Medical imaging involves a variety of techniques and technologies that help in the diagnosis and treatment of diseases. Some of the most significant modalities of medical imaging and their corresponding applications of image segmentation are discussed below.

(i) MRI: This is a noninvasive medical imaging procedure that generates clear images of nearly all internal structures in the human body, such as organs, bones, muscles, and blood vessels. MRI scanners use a large magnet and radio waves to create pictures of the body. These pictures provide doctors with vital information in diagnosing the medical condition and designing a treatment plan [4]. It is commonly used for the brain, spinal cord, and joint imaging. And segmentation helps in detecting tumors, multiple sclerosis, and cartilage damage.

(ii) CT: This scan is an imaging procedure that combines the use of X-Rays with computer technology to create images of the inside of the body. It displays detailed pictures of any area of the body, such as the bones, muscles, fat, organs and blood vessels. CT scans are more precise than regular X-rays. It uses X-ray beams to create cross-sectional images of the body. CT scans are mainly used for imaging bones, lungs, and internal organs, while segmentation is mainly used for lung disease detection, tumor analysis, and 3D reconstruction for surgery [5].

(iii) X-rays: These are a type of radiation called electromagnetic waves. X-ray imaging creates pictures of the inside of the body organs. The images show the body parts in varying shades of black and white. This is due to varying tissues absorbing different levels of radiation. Calcium in bones absorbs X-rays the most, and therefore bones appear white. Fat and other soft tissues absorb less and appear gray. It produces 2D images of dense structures, like bones, and is commonly used in fracture detection, lung infection diagnosis, and dental imaging [6]. Additionally, segmentation helps in identifying fractures, lung nodules, and abnormalities.

(iv) Ultrasound: An ultrasound is an imaging test that utilizes sound waves to create images of organs, tissues, and other structures within the body. It enables the medical practitioners to visualize within the body or internal body without the need for surgery. Ultrasound is also referred to as ultrasonography or sonography. Ultrasound images can be referred to as sonograms. It is commonly used in pregnancy monitoring, cardiac imaging, and organ analysis. Additionally, segmentation plays an important role in tracking fetal development, tumor detection, and vascular analysis [7].

9.3 Techniques for Image Segmentation

Various traditional and AI-based techniques are utilized for medical image segmentation. For instance, Kapil Kumar Gupta et al. [8] discussed various medical image segmentation methods in brain tumor detection, emphasizing their significance in accurate diagnosis. They compared various methods such as thresholding, clustering (*K*-means and fuzzy *C*-means), region growing, and edge detection, while using sensitivity, specificity, and accuracy as performance indicators. The analysis demonstrated that no one segmentation approach works equally well since a combination of tumor characteristics and image quality determines effectiveness. Multiple medical segmentation methods can work together to increase diagnostic accuracy according to their findings. Similarly, Yang Cheng et al. [9] conducted research on image segmentation technologies as well as their digital image processing applications. The study discussed various image segmentation techniques like region based, edge detection, and clustering methods, and their importance, in improving image analysis. The author demonstrated how precise segmentation stands vital in medical imaging, together with remote sensing and pattern recognition applications. The approach deals with noise sensitivity and computational efficiency as additional concerns. Figure 9.2 categorizes image segmentation techniques into traditional and deep learning methods, highlighting different algorithms under each category, and it provides a clear overview of the various methods or techniques adopted in medical imaging.

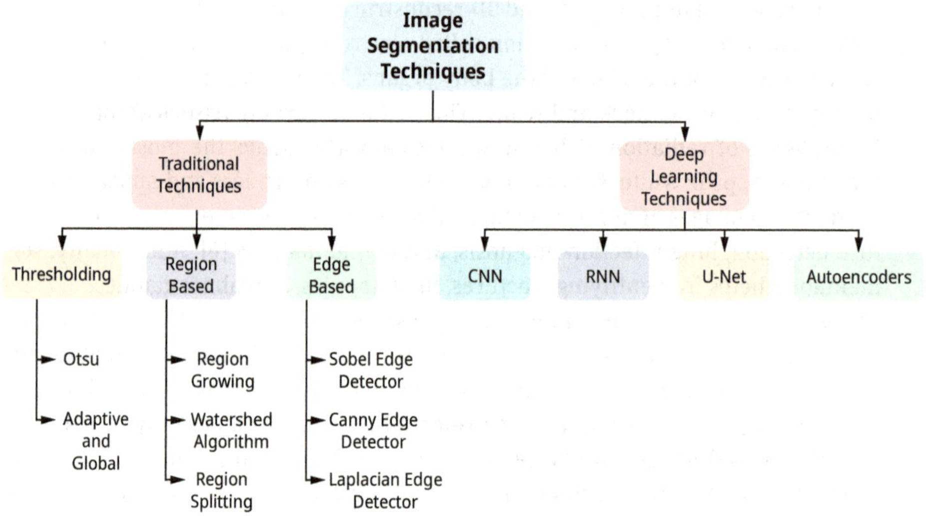

Figure 9.2: A hierarchical classification of image segmentation techniques.

9.3.1 Traditional Image Segmentation Techniques

The details of the various traditional image segmentation techniques have been given below:

(i) Thresholding: This is a basic method for dividing an image into its constituent subregions or distinct objects based on pixel intensity. It converts the grayscale image into a binary image, where pixels are classified as foreground and background based on whether their intensity values are above or below a specified threshold value. The most frequently used threshold segmentation algorithm is the largest interclass variance method (Otsu), which chooses a globally best threshold by maximizing the variance between classes [10]. Deepika Sood et al. [11] reviewed multiple medical imaging segmentation methods that proved to be essential during disease discovery as well as treatment strategy development. They analyzed thresholding segmentation methods while assessing their accuracy and efficiency. The selection of the image segmentation method in medical applications depends on the individual requirements of the task and image complexity. The traditional technologies have been explained as follows:

Otsu Thresholding: It is a technique of performing global thresholding on a digital image. It is used to divide the image into foreground, i.e., meaningful information and background, i.e., apart from the meaningful information by finding the best threshold value based on the differences between these two parts [12]. Optimal threshold T is determined by maximizing the inter-class variance, which is given as follows:

$$T = \arg\max\left(\omega_0(T)\cdot\omega_1(T)\cdot(\mu_0(T)-\mu_1(T))^2\right) \tag{9.1}$$

where $\omega_0(T)$ is the probability of the background class, $\omega_1(T)$ is the probability of the foreground class, $\mu_0(T)$ is the mean intensity of the background class, and $\mu_1(T)$ is the mean intensity of the foreground class.

Adaptive and Global Thresholding: Adaptive thresholding only adjusts thresholds locally based on pixel neighborhoods, while on the other hand, global thresholding uses a single thresholding value on the entire image. Adaptive thresholding becomes more and more beneficial in the images with nonuniform illumination, as it can make the localization of the image more flexible [13]. The formulation for the adaptive thresholding has been given in eq. (9.2) and the formulation for the global thresholding has been given in eq. (9.3):

$$T(x,y) = \mu_{\text{Local}}(x,y) - C \tag{9.2}$$

where $\mu_{\text{Local}}(x,y)$ is the intensity of the pixels in a local neighborhood around (x,y) and C is a constant value used to adjust the threshold.

Let $I(x,y)$ be the intensity of a pixel at position (x,y) in the image:

$$S(x,y) = \begin{cases} 1, \text{ if } I(x,y) \geq T \\ 0, \text{ if } I(x,y) < T \end{cases} \tag{9.3}$$

where $S(x,y)$ is the binary segmented image (1 for foreground, 0 for background), T is the threshold value, and $I(x,y)$ is the intensity of pixel at (x,y)

(ii) Region-Based Segmentation: This is a method that divides an image into regions based on similarity in properties like intensity, texture, or color to identify anatomical structures and abnormalities like tumor. It creates coherent and meaningful segments within the image. Region-based segmentation can be classified into region-growing segmentation, region-splitting segmentation, and watershed algorithm.

Region Growing: Region-growing segmentation is an algorithm based on pixel-growing technique that starts from seed points and grows by including neighboring pixels that satisfy some similarity condition, which could be one of intensity, color, or texture. It preserves object boundaries well, making it ideal for applications like medical imaging, where accurate delineation of anatomical structures is crucial [14]. The method is sensitive and requires the selection of correct seed points to avoid over-segmentation or under-segmentation. To ensure high performance, adaptive thresholds and multi-see approaches are often utilized, adding to its strength in dealing with complex images. The formulation for the region growing has been given in eq. (9.4).

Let R represent the region being grown and $p(x,y)$ denote the property (e.g. intensity) of a pixel at (x,y). A pixel (x,y) is added to R if:

$$|p(x,y) - \mu_R| \leq \in \tag{9.4}$$

where μ_R is the mean property of region R and $\in$ is the predefined similarity threshold

Watershed Algorithm: It is a classic algorithm used for segmentation and is especially useful when extracting overlapping objects in images. This segmentation process takes the similarity with adjacent pixels of image as an important reference to connect pixels with similar spatial position and gray values. This algorithm divides the image into segments using topographic information. It treats the image as a topographic surface, identifying catchment basins based on pixel intensity. Local minima are marked as starting points, and flooding with colors fills catchment basins until object boundaries are reached [15]. The resulting segmentation assigns unique colors to regions, aiding object recognition and image analysis. The formulation for computing the gradient magnitude of the image is in eq. (9.5).

The gradient magnitude $G(x,y)$ of the image is computed as:

$$S(x,y) = \sqrt{\left(\frac{\partial I}{\partial x}\right)^2 + \left(\frac{\partial I}{\partial y}\right)^2} \tag{9.5}$$

where $\frac{\partial I}{\partial x}$ is a horizontal gradient and $\frac{\partial I}{\partial y}$ is a vertical gradient.

Region Splitting: Region splitting and merging is a recursive image segmentation technique. The principle is to start with the whole image as one region. If a region does not meet some predefined condition, like homogeneity, it will then be divided into subregions. The process of splitting continues until all the resulting regions satisfy the condition of homogeneity. Once the image is divided, a merging phase can be implemented to merge overlapping regions that are too similar to each other. The method is typically done by dividing the image into smaller subsections and examining each subsection to see whether or not it will be divided again, depending on the homogeneity of the pixels in it [16]. In region splitting and merging, the algorithm relies on recursive subdivision based on quadtree representation. The splitting criterion is typically intensity homogeneity or texture similarity. The mathematical formulation used to decide whether a region is to be split is given in eqs. (9.6) and (9.7):

$$Var(R) > T \tag{9.6}$$

where R is the current region; $Var(R)$ is the variance (or another measure of homogeneity) of pixel intensities in R, and T is a predefined threshold.

After splitting adjacent regions R_i and R_j are merged if:

$$\left|\mu(R_i) - \mu(R_j)\right| < T_m \tag{9.7}$$

where

$\mu(R_i)$ and $\mu(R_j)$ are the mean intensities of the two regions and T_m is the merging threshold.

(iii) Edge-based segmentation: Edge-based segmentation is an algorithm that identifies boundaries of objects through detection of significant intensity transformations known as edges. This technique detects pixels that display substantial color or brightness variations to identify object boundaries. It is commonly used in medical imaging, such as MRI and CT scans, for detecting tumors and organ structures. It typically involves preprocessing the image to reduce noise using filters such as Gaussian or Median filters [17].

Sobel Edge Detector: The Sobel image segmentation technique is an edge detection approach that is used to show the boundaries present in the image. The baseline of the method is that it evaluates the gradient intensity of the image's pixel. The Sobel operator is a convolutional one that uses two 3×3 kernels to find the gradient in the horizontal and the vertical directions [18]. The Sobel operator computes the magnitude of the image gradient with two kernels and the formulation for this is given in eq. (9.8).

$$G_x = \begin{bmatrix} -1 & -2 & -1 \\ 0 & 0 & 0 \\ 1 & 2 & 1 \end{bmatrix}, \quad G_y = \begin{bmatrix} -1 & 0 & 1 \\ -2 & 0 & 2 \\ -1 & 0 & 1 \end{bmatrix}$$

The gradient magnitude G is calculated as:

$$G = \sqrt{G_x^2 + G_y^2} \tag{9.8}$$

where G_x is the horizontal gradient and G_y is the vertical gradient.

Canny Edge Detector: The Canny edge detector is an edge detection technique that is used in image processing. In this, firstly the image is smoothed with a Gaussian filter to remove noise and after that, the Sobel operator is used to calculate the intensity gradient. Subsequently, the process is followed by the non-maximum suppression, which deletes the extra edges. The last stage of the process is double thresholding, which is used to separate the image into strong and weak edges. Later, the process of edge tracking by hysteresis is carried out to make only significant edges apparent. This method is highly effective for object detection, image segmentation, and feature extraction in computer vision applications [19]. The Canny edge detector follows a sequence of operations, with their formulation given in eq. (9.9).

Step 1: Gaussian smoothing:

$$I_{\text{smooth}}(x,y) = I(x,y) * G(x,y) \tag{9.9}$$

where $G(x,y)$ is the Gaussian kernel.

Step 2: Gradient calculation using Sobel operator

Step 3: Non-maximum suppression to thin edges

Step 4: Double thresholding or hysteresis to identify strong and weak edges

Laplacian Edge Detector: Laplacian edge detector is an edge detection algorithm in image processing that is applied to emphasize regions of high intensity change, usually found along edges in an image. Contrary to gradient-based algorithms like Sobel and Canny, which utilize directional gradients, Laplacian Edge Detection is based on the second derivative of the image intensity. To apply this, first the image is converted into a grayscale image and then Gaussian blur is applied. It decreases the noise and avoids false image detection and then applies the Laplacian operator to compute the second derivative [20].

The Laplacian of image $I(x, y)$ is given as follows:

$$\nabla^2 I = \frac{\partial^2 I}{\partial x^2} + \frac{\partial^2 I}{\partial y^2} \tag{9.10}$$

where $\frac{\partial^2 I}{\partial x^2}$ is the second partial derivative of the image intensity along the horizontal (x) direction, $\frac{\partial^2 I}{\partial y^2}$ is the second partial derivative of the image intensity along the vertical (y) direction.

Table 9.1 summarizes various the traditional image segmentation techniques, each with its unique advantages such as simplicity, precision, etc. but also some disadvantages, including sensitivity to noise or computational complexity. The choice of selecting the most suitable technique depends on several factors such as accuracy, precision, speed, etc., which helps in proper diagnosis and accurately identifying the infected regions.

Image segmentation approaches such as thresholding and both region- and edge-based techniques require color intensity and texture features to perform their segmentation operations, despite facing practical limitations. However, they fail to pro-

Table 9.1: Comparison of traditional image segmentation techniques.

Segmentation technique	Advantages	Disadvantages
Thresholding	Simple and fast	Sensitive to noise and struggles with complex regions
Otsu thresholding	No manual input needed	Struggles with uneven lightning
Adaptive thresholding	Works well in varying lightning	Computationally expensive.
Region growing	Preserves object boundaries	Sensitive to seed point selection
Watershed algorithm	Captures fine details	Prone to oversegmentation
Edge-based segmentation	Good for sharp boundaries	Sensitive to noise
Sobel-edge detector	Simple and detects edge direction	Produces thick edges
Canny edge detector	Accurate and reduces false edges	Needs careful adjustment of parameters
Laplacian edge detector	Find fine edges	Sensitive to noise

duce accurate segmentation results due to complex backgrounds, varying lighting conditions, and overlapping objects.

9.3.2 AI-Based Segmentation Methods

In order to overcome the limitations of traditional segmentation techniques, various AI-based methods have been proposed. Akanksha Bali and Singh analyzed different segmentation strategies and techniques, which they demonstrated as vital components of computer vision and medical imaging systems. The selection of the appropriate technique depends on both imaging characteristics and application needs. The authors note that combining different methods boosts segmentation precision, thus improving its functionality when performing complex image processing operations [21]. Similarly, Nabeel N. Ali et al. reviewed various image segmentation methods using deep learning, emphasizing their effectiveness in medical and computer vision applications. They presented a review of various strategies of image segmentation using deep learning approaches that are applied in medical and computer vision fields. The authors focused on a few key deep learning models such as the CNN, U-Net, and RNN, and studied how these models performed in solving complex image segmentation tasks. It was noticed that although deep learning segmentation can make accuracy and efficiency better, one of the main requirements is the need for large amounts of data and computational power [22]. Some of the AI-based segmentation methods have been discussed below:

(i) CNNs: CNN is used for the classification of images. The input images that are utilized in CNN are of three dimensions, i.e., height, width, and number of channels. The first two dimensions inform us about the resolution of the image and the third dimension indicates the number of channels or intensity values. There are four main types of operations in CNN – convolutional operation, pooling operation, flattening operation, and classification operation. The first layer in CNN is the convolutional layer, which takes the image as an input and begins to process using three elements: input image, filters, and feature map. A filter or kernel is applied to traverse the image in constant gap intervals, known as strides, performing a dot product with the underlying pixels. The resulting values form the feature map and reduce the dimensions of the image. The second layer is rectified linear unit (ReLU), which introduces nonlinearity to the network by applying the rectified linear unit activation function, which sets all the negative values to 0 and leaves positive values unchanged in the feature map. The third layer is the pooling layer, which is used to reduce the spatial size of the feature map to extract only important information by reducing the dimensions. The next step is flattening, and it is used to convert all the resultant 2D arrays from the pooled features map into a single long continuous linear vector and then the flattened matrix is fed as input to the fully connected layer to classify the image and learn complex patterns and relationships in data. Nandi et al. worked on techniques to improve deep CNN analysis based on visualization and interpretability techniques. The work analyzes how CNNs learn and process features, with attention to what the model's decisions indicate for improved performance across a range of applications, like image segmentation. It suggests that combining visualization tools with CNN structures improves model optimization and knowledge [23]. Figure 9.3 represents CNN architecture used for brain tumor segmentation. The process involves feature extraction through convolution and pooling layers, followed by classification using deconvolution and un-pooling layers to generate a segmented output, highlighting the tumor region.

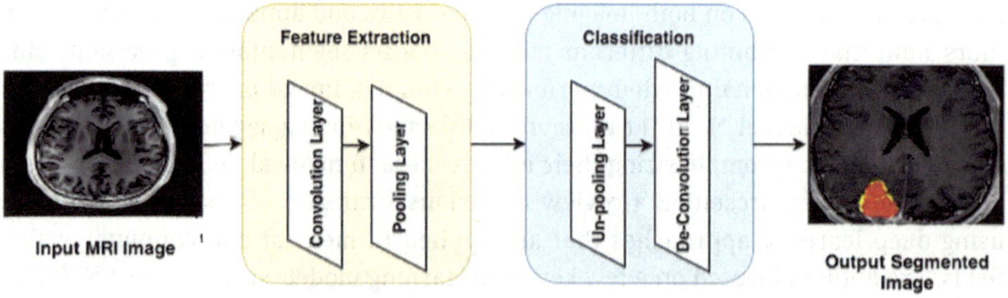

Figure 9.3: Brain tumor segmentation using a CNN-based model.

(ii) RNN: RNN in medical imaging is an example of deep learning model that takes advantage of its ability to process sequential data to analyze medical images, particularly when examining changes over time or across multiple slices of a scan, allowing for more accurate diagnoses of conditions like cardiac arrhythmias or disease progression by capturing temporal dependencies between different image frames essentially. It remembers information from previous slices to better interpret the whole image sequence. Unlike CNNs, which are good at recognizing patterns in images like CT scans, MRIs, and ultrasounds, RNN are better at understanding sequences of data over time. The valuable nature of RNNs resides in their ability to monitor disease evolution as well as anticipate future healthcare status along with improving video-based medical diagnostics. RNNs operate on a series of MRI scans to detect tumor size evolution during brain tumor analysis. Heart function assessment throughout time relies on their analysis of echocardiogram videos in cardiac motion analysis. Hospital personnel utilize RNNs to perform lung disease monitoring through chest X-ray analysis, which assists with early detection of health issues. RNNs serve well in tasks that require joining multiple imaging approaches, by fusing MRI and PET scans to create detailed medical diagnoses. The analysis of past medical scan patterns by RNNs leads to predictions about disease progression, which supports better medical choices for doctors [24]. RNN process data using hidden states, which is described as follows:

$$h_t = f(W_h\ h_{t-1} + W_x x_t + b) \tag{9.11}$$

where h_t is the hidden state at time step t, $h_t - 1$ is the hidden state from the previous step, x_t is the input at time step t, W_h, W_x are weight matrices, b is the bias term, and f is the activation function.

(iii) U-Net: U-Net is a deep learning architecture that is designed for semantic segmentation, especially in medical imaging applications. Medical imaging professionals use U-Net architecture to detect tumors while segmenting organs and identifying lesions when analyzing MRI, CT scans, and ultrasound data. The U-Net architecture has two main parts, i.e., the encoder section and a decoder section. The encoder layers in the contracting path detect contextual information by reducing input spatial resolution but the expansive path uses decoder layers, which generate segmentation maps through the information obtained from skip connections with encoder layers. During the contracting phase of U-Net, the system detects important features embedded within the input image. Each successive encoder layer in U-Net reduces feature map resolution until it achieves deep abstract features that represent the original input data. U-Net implements this shrinkage path in a way comparable to feedforward layers from CNNs. The expansive path within U-Net executes three key functions such as data decoding and feature detection that preserves original input spatial resolution. In extensive path, decoder layers execute both up samples of feature maps along with convolution operations. The recovery of spatial information through network connections from the compression path enables decoder layers to pinpoint features correctly [25]. U-Net uses the Dice coefficient to measure the segmentation in pixel values to measure accuracy, which is described as follows:

$$\text{Dice} = \frac{2\sum_i p_i g_i}{\sum_i p_i + \sum_i g_i} \tag{9.12}$$

where p_i is the predicted pixel values and g_i is the ground truth pixel values.

(iv) Autoencoders: Autoencoders have a significant function in medical imaging by improving image quality, reducing noise, and detecting abnormalities. The healthcare field facilitates the use of autoencoders to enhance MRI and CT scans with poor quality to eliminate artifacts and improve image resolution, because it aids in accurate medical diagnosis. The denoising autoencoders construct noisy duplicates of the input images through the addition of random noise and then attempt to eliminate the noise from the noisy input to reconstruct the output that is very similar to the original input. Medical professionals perform anomaly detection with autoencoders by teaching the system, standard image patterns in medical data, which enables it to identify deviations indicating conditions such as tumors, fractures, and lung infections. Deep learning models become more effective at detecting important medical features because autoencoders perform dimensionality reduction and feature extraction processes. The performance of autoencoders increases through convolutional autoencoders (CAEs) and variational autoencoders (VAEs), which detect spatial patterns while creating realistic synthetic medical images for AI training purposes. The image given below illustrates the denoising autoencoder workflow in medical imaging. The encoder compresses a noisy MRI scan into a latent space, and the decoder reconstructs a high-quality image, reducing artifacts and improving resolution for accurate diagnosis [26]. Autoencoders minimize the difference between input and output using mean squared error (MSE), which is given as follows:

$$\text{MSE} = \frac{1}{n}\sum_{i=1}^{n} (x_i - \widehat{x}_i)^2 \tag{9.13}$$

where x_i is the original input, $\widehat{x}_i$ is the reconstructed output, and n is the number of pixels.

Table 9.2 summarizes various AI-based image segmentation techniques with their advantages and disadvantages. This comparison helps in selecting the most appropriate technique based on specific needs and challenges.

Table 9.2: Comparison of AI-based image segmentation techniques.

Segmentation techniques	Advantages	Disadvantages
Convolutional neural networks	Learning complex features	Need large datasets
Recurrent neural networks	Track changes; useful for videos	Struggles with long sequences
U-Net	Keep details; work with small data	Needs tuning, hardware-heavy
Autoencoders	Improves quality; detects abnormalities	Loses fine details; needs a lot of data

9.4 Applications of Image Segmentation in Medical Imaging

Medical image segmentation plays an important part in different areas of healthcare. It helps in disease diagnosis, plan treatment, and patient monitoring. This technique works with different types of medical images and these include MRI, CT scans, X-rays, and ultrasounds, and it isolates specific regions or abnormalities. Justin Ker et al. explored deep learning applications in medical image analysis and highlighted their impact on disease diagnosis and treatment. They discussed various deep learning methods, which include CNN, along with their functions for medical condition detection, segmentation, and classification tasks. Deep learning systems improve medical imaging, both through enhanced precision and speed, although they need extensive datasets and advanced computing systems [27]. Further supporting the advantage of deep learning, Dingan Liao et al. explored deep learning features in image segmentation while showcasing their superior performance against traditional techniques. The study is centered on the fact that deep learning methods, for instance, CNN and fully convolutional networks, are undoubtedly beneficial for the segmentation task, as these networks can automatically learn hierarchical features. According to the authors, the assessment of medical image processing models focused on their capability to modify complex medical images to achieve improved accuracy in defining object boundaries and classification functions [28].

Some applications of image segmentation include:

1. Brain Imaging: Brain imaging uses pictures of the brain that are produced without any surgical operation to demonstrate brain structures and its functions. It uses MRI and CT scans to locate and divide unusual areas such as tumors, strokes, and lesions. Methods to separate these areas help to:
   (i) Tumor Detection: It helps in differentiating between healthy and cancerous brain tissue for precise diagnosis and treatment planning.

   (ii) Stroke Segmentation: Stroke segmentation helps doctors to quickly identify the regions of the brain that have been damaged. It differentiates two types of strokes, one resulting from a blockage (ischemic stroke) and another due to bleeding (hemorrhagic stroke). Precise imaging permits a neurologist to proceed in determining the real extent of the stroke and thus taking such life-saving actions as removing the clot or providing the necessary drugs [29].
   (iii) Multiple Sclerosis (MS) Lesions: Multiple sclerosis affects the nervous system by damaging the protective layer of nerves. MRI-based segmentation helps to spot white matter lesions in the brain and this spotting of the infected area helps in planning the treatment effectively [30].
2. Cardiac imaging: Cardiac segmentation is an important task that helps in the diagnosis of heart structure and function. There are various methods in this field that include the following:
   (i) Heart Chamber Segmentation: This helps in identifying and analyzing the left and right ventricles, which are crucial for the diagnosis of diseases like cardiomyopathy.
   (ii) Coronary Vessel Segmentation: It helps in detecting blockages and abnormalities in the coronary arteries to evaluate the risks of cardiovascular diseases [31].
   (iii) Myocardial Infarction Analysis: It helps in identifying the presence of scar tissue in the heart because of heart attack [32].
3. Lung imaging: Lung segmentation of CT and X-ray images has played a major role in the outlook of respiratory diseases. Areas of its use are as follows:
   (i) Covid-19 Detection: Lung CT scan is used to evaluate the severity of the disease, along with the areas of infection.
   (ii) Pneumonia Diagnosis: Identifying lung opacities and inflammation caused by bacterial or viral pneumonia is thus far the most radiological finding for pneumonia diagnosis [33].
   (iii) Lung Cancer Screening: Benign tumors and disgusting gaps are two areas where diagnosis is done after having clear detection of the nodules and segmenting the lung [34].
4. Retinal imaging: Ophthalmic imaging is commonly used for the diagnosing of a variety of retinal diseases. The purpose of image segmentation is:
   (i) Diabetic Retinopathy Detection: It helps in gaining a better understanding of retinal diseases with the help of the study of fundus images like hemorrhages, microaneurysms, and exudates [35].
   (ii) Glaucoma Diagnosis: It helps in checking the size of the optic nerve head and retinal nerve fiber layers to spot the glaucoma progression [36].
5. Pathology Slide Analysis: Precise and accurate identification of histopathology images is required for the identification of cancer cells. Applications include:

(i) Cancer Cell Segmentation: The division of cancerous cells from the noncancerous ones in biopsy samples is quite helpful to physicians in their effort to catch tumors early [37].
(ii) Tissue Classification: It helps in identifying different types of tissues and the role they play in the diagnosis of various diseases and in grasping the mechanism of abnormal cell growth [38].
(iii) Tumor Grading: Identification of the seriousness of cancer with the help of segmenting the tumor regions and analyzing their growth is one of the ways that determines the most successful treatment [39].

9.5 Analysis of Segmentation Techniques

The comparison of the different segmentation techniques in this study is based on the methodologies, challenges, and performance measures like accuracy, precision, and recall. These measures are approximately based on the methodologies and general trends in performance. For referenced sources that included quantitative results, values were utilized exactly as they appeared. In other instances, close approximations were derived by considering strengths and weaknesses when dealing with certain challenges such as boundary detection and class imbalance. Deep learning techniques, e.g., U-Net architectures, usually reach 90–95% precision, whereas conventional techniques perform slightly lower than that. Advances such as loss function blending and transformer models have been considered to estimate increases in precision and recall. The structured methodology enables fair comparison among various approaches. Table 9.3 evaluates image segmentation methods using the contrast of their methodologies, issues, solutions, and performance metrics like accuracy, precision, and recall. It identifies the shift from the conventional to the newer deep learning methods, which offer enhanced segmentation performance and solve issues like class imbalancing and computational complexity. Figures 9.4, 9.5, and 9.6 represent the variations in segmentation accuracy, the trend of segmentation precision, and the evolution of segmentation recall over successive time intervals, respectively.

Salwa Al Garea et al. provided a comprehensive overview of image segmentation methods, discussing their applications, challenges, and future directions. Categorization of segmentation techniques has been performed where traditional methods such as thresholding, edge detection, and clustering are included along with leading-edge deep learning-based approaches. The authors highlighted key challenges, including computational complexity, data quality, and model generalization, and concluded that integration of both techniques could yield more accurate segmentation results [50].

Table 9.3: Quantitative analysis of image segmentation techniques.

Reference, year	Technique used	Accuracy (%)	Precision (%)	Recall (%)
[40], 2018	Deep learning with image-specific fine-tuning	92	90	88
[41], 2019	CNN for image annotation	92	90	88
[42], 2021	Review of deep neural architectures for medical image semantic segmentation.	89	87	86
[43], 2021	U-Net and its variants	91	90	89
[44], 2022	DS-TransUNet (Dual Swin Transformer U-Net)	95	94	93
[45], 2022	Deep learning with blending loss	94	93	91
[46], 2023	Comprehensive review of image segmentation techniques	88	86	85
[47], 2023	Review of deep learning trends in medical image segmentation	90	88	87
[48], 2024	U-Net and its variants	93	92	90
[49], 2024	Efficient and accurate 3D medical image segmentation	96	95	94

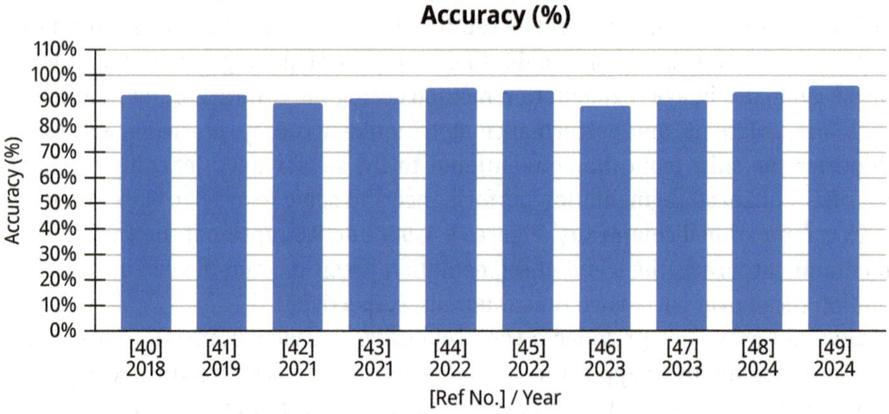

Figure 9.4: Analysis of image segmentation technique using accuracy.

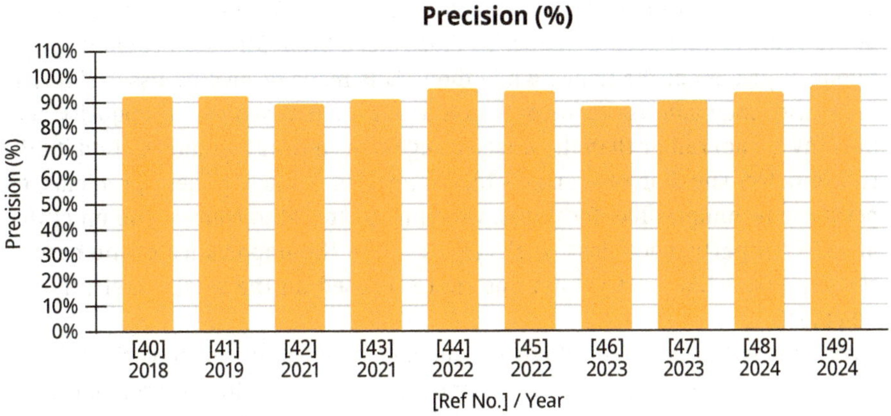

Figure 9.5: Analysis of image segmentation technique using precision.

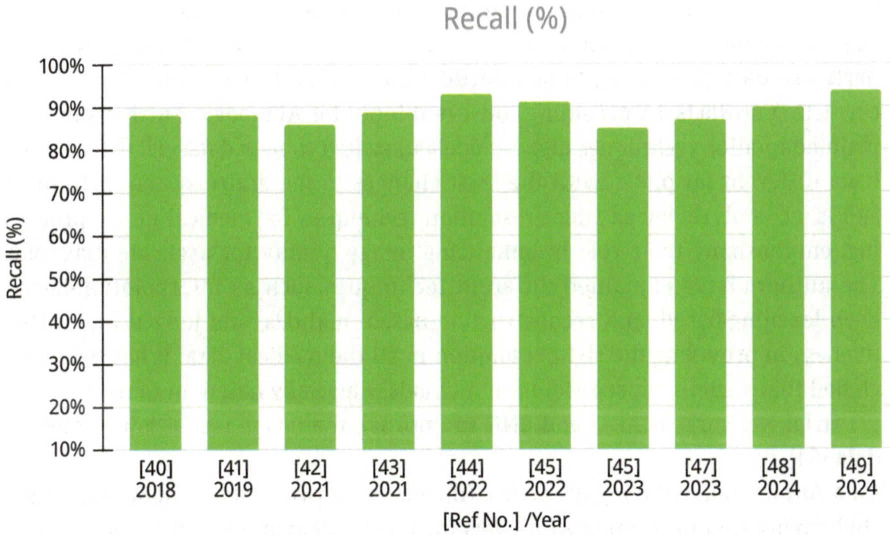

Figure 9.6: Analysis of image segmentation technique using recall.

9.6 Challenges in Medical Image Segmentation

Medical image segmentation faces various challenges that affect the accuracy and reliability of AI models. Some of the challenges are discussed below:

(i) Class Imbalance: To detect tumors or organs, medical image segmentation trains a computer to see or identify the critical component of the medial image. However, one important challenge is class imbalance that occurs when a part of the image, which is of interest, for example a tumor, is far less than the rest of the image representing healthy tissue. To illustrate, a brain MRI scan could display a tumor of just 5% of the image while the 95% is made of healthy tissue [51]. This imbalance may confuse AI models that are more inclined to the majority class (healthy tissue) rather than the infected region. As a result, AI models find it difficult to detect small or rare objects like tumors. In managing these problems, experts use techniques like focal loss, which helps the AI to concentrate on critical or infected parts. Focal loss works by reducing the importance of the easy-to-classify pixels like background and more focused on the small pixels, like tumors [52].

(ii) Variability in Images: Medical images can vary significantly due to various factors such as patient anatomy, imaging equipment, and noise, which present challenge when AI models are expected to generalize. For example, MRI scans can differ in brightness, contrast, or resolution based on the machine used, scanner settings, or even patient-specific conditions like age, body, size, or health issues. Such differences can confuse AI models, thereby causing them to yield inconsistent results in segmentation of vital structures like tumors. To solve this challenge, experts use data processing to standardize image quality by reducing noise and correcting artifacts, by providing consistent input for AI models. Furthermore, domain adaptation techniques allow models to switch to new datasets, uch as scans from different hospitals, with the least changes to the entire system [53]. Jithin Saji Isaac et al. reviewed super-resolution techniques for medical image processing, emphasizing their role in enhancing image quality for accurate diagnosis. The authors have explained different techniques such as interpolation-based, deep learning-based, and reconstruction-based methods, and judged their effectiveness in providing the right resolution in all the medical data. It has been concluded that various super-resolution methods, especially based on Generative Adversarial Networks (GANs) and CNN to improve the image resolution in medical data [54].

(iii) Data Annotation: Labeling of medical images is known as data annotation. It is a challenging step in training AI models for image segmentation. It involves marking and identifying areas of interest like tumors and organs, which need a lot of time and professional training from doctors or radiologists. However, this task is complicated because different doctors may label the same image differently due to variations in interpretation, leading to inconsistency in datasets. And a large amount of data is required for training AI models, which makes manual annotation expensive and time consuming. To address these challenges, researchers have developed innovative techniques like weakly supervised learning, which allow AI models to learn incomplete or rough annotation, instead of requiring pixel-level labeling [55]. Alejandro F. Frangi et al. explored simulation and synthesis techniques in medical imaging, highlighting their significance in enhancing image quality and analysis. The authors focused on different ways of doing it such as physics-based modeling, machine learning-based synthesis, and deep learning for creating realistic images. It has been concluded that the use of these models enhances the availability of training data, supports the validation of the algorithms, and helps in clinical decision-making [56].

9.7 Emerging Trends in Medical Image Segmentation

The field of medical image segmentation is rapidly growing due to the advancements in AI, machine learning, and computational techniques. These advancements improve the accuracy and efficiency of the grayscale images. Tariq Mahmood et al. explored recent trends and future direction in active deep learning for the classification and medical image segmentation. The authors suggested that if we include AI techniques and active learning together, we have a higher chance of acquiring positive results with better accuracy of segmentation and classification, making medical imaging a more secure and time-saving process in clinical applications [57]. Miles N. Wernick et al. explored machine learning applications in medical imaging and their effect on diagnostic accuracy and automatic analysis. The research discusses some of the machine learning techniques, along with guided and self-directed learning methodologies, which facilitate pattern detection inherent in complex medical images. And this also shows that diagnosis based on images becomes better through the implementation of advanced methods such as Support Vector Machines (SVM) and neural networks. The study addresses problems with computational complexity and data quality as well as model generalization challenges [58]. Below are some of the emerging trends in image segmentation have been detailed below:

(i) Self-Supervised Learning: One of the most exciting trends in AI is self-supervised learning, which reduces the dependency on the large amount of labeled data. The self-supervised trend is the fastest trend in AI technology which learns meaningful representations or patterns from the unlabeled data without any external supervision in medical imaging, where high annotation cost and expertise are required. For example, AI models can be trained to reconstruct the corrupted data or can set the rotation of the image to get more information about the infected areas. It is especially useful in the cases in which labeled data is almost unavailable, and this approach not only reduces time and resources but also improves the quality of AI models [59].

(ii) Multimodal Image Segmentation: Medical imaging has various modalities like MRI, CT scans, X-Rays, and ultrasound, each providing different information about the patient's anatomy. Multimodal image segmentation integrates data from different modalities to produce more precise and efficient results. With the growth of deep learning technologies, AI models can now modify the weight given to each modality, based on its relevance to the task. This is particularly effective in complicated problems such as segmentation of brain tumors where the combination of functional and structural data produces better results [60].

(iii) Federated Learning: Another emerging trend is federated learning, which allows collaborative model training, while ensuring that sensitive patient information is not shared. In traditional AI development, data across various institutions or hospitals is centralized and has severe privacy concerns. Those issues are avoided in federated learning because sensitive information is stored at the institution and any number of institutions may train a local model with its data. This method is particularly applicable in medical imaging, as it enhances the overall generalizability of segmentation models to allow them to correctly recognize anatomical structures and pathologies from the various imaging protocols [61, 62].

9.8 Conclusion

Medical image segmentation is important in modern healthcare because it allows for the effective detection of anatomical structures and pathologies. The primary reason for the drastic improvement is the adoption of deep learning techniques such as CNN, U-Net-based architecture, and autoencoders, which have increased the accuracy and efficiency of diagnosis, treatment planning, and patient monitoring. Thresholding, region-based, and edge-detection techniques advanced the work, but deep learning-based solutions have solved the challenges like class imbalance, high variability of images, and data annotation.

The range of medical image segmentation applications is immense, including tumor or organ segmentation, 3D visualizations, and pathology slide analysis. These techniques are particularly useful in the field of brain imaging, cardiac imaging, lung imaging, and retinal imaging because accurate assessment of abnormalities is paramount for improving patient care. Challenges like class imbalance and high variability in imaging modalities, along with high annotation costs are barriers to overcome. Solutions for these challenges are focal loss, weakly supervised learning, and domain adaptation. Emerging trends such as self-supervised learning, multimodal imaging segmentation, and federated learning not only reduce the dependency on labeled data and enhance model generalizability but also address critical concerns like patient privacy and data security.

References

[1] Minaee S, Boykov Y, Porikli F, Plaza A, Kehtarnavaz N, Terzopoulos D. Image segmentation using deep learning: A survey. IEEE Transactions on Pattern Analysis and Machine Intelligence. 2021 Feb 17;44(7):3523–3542.

[2] Sahu S, Sarma H, Bora DJ. Image segmentation and its different techniques: An in-depth analysis. In 2018 International Conference on Research in Intelligent and Computing in Engineering (RICE) 2018 Aug 22 (pp. 1–7). IEEE.

[3] Guo Z, Li X, Huang H, Guo N, Li Q. Deep learning-based image segmentation on multimodal medical imaging. IEEE Transactions on Radiation and Plasma Medical Sciences. 2019 Jan 1;3(2):162–169.

[4] Uğurbil K. Magnetic resonance imaging at ultrahigh fields. IEEE Transactions on Biomedical Engineering. 2014 Mar 25;61(5):1364–1379.

[5] Diciotti S, Lombardo S, Falchini M, Picozzi G, Mascalchi M. Automated segmentation refinement of small lung nodules in CT scans by local shape analysis. IEEE Transactions on Biomedical Engineering. 2011 Sep 12;58(12):3418–3428.

[6] Khan W, Zaki N, Ali L. Intelligent pneumonia identification from chest x-rays: A systematic literature review. IEEE Access. 2021 Mar 30;9:51747–51771.

[7] Van Sloun RJ, Cohen R, Eldar YC. Deep learning in ultrasound imaging. Proceedings of the IEEE. 2019 Aug 21;108(1):11–29.

[8] Gupta KK, Dhanda N, Kumar U. A comparative study of medical image segmentation techniques for brain tumor detection. In 2018 4th International Conference on Computing Communication and Automation (ICCCA) 2018 Dec 14 (pp. 1–4). IEEE.

[9] Cheng Y, Li B. Image segmentation technology and its application in digital image processing. In 2021 IEEE Asia-Pacific Conference on Image Processing, Electronics and Computers (IPEC) 2021 Apr 14 (pp. 1174–1177). IEEE.

[10] Song Y, Yan H. Image segmentation techniques overview. In 2017 Asia Modelling Symposium (AMS) 2017 Dec 4 (pp. 103–107). IEEE.

[11] Sood D, Singla A. A Survey of Segmentation Techniques for Medical Images. In 2022 10th International Conference on Reliability, Infocom Technologies and Optimization (Trends and Future Directions)(ICRITO) 2022 Oct 13 (pp. 1–8). IEEE.

[12] Wang C, Yang J, Lv H. Otsu multi-threshold image segmentation algorithm based on improved particle swarm optimization. In 2019 IEEE 2nd International Conference on Information Communication and Signal Processing (ICICSP) 2019 Sep 28 (pp. 440–443). IEEE.

[13] Roy P, Dutta S, Dey N, Dey G, Chakraborty S, Ray R. Adaptive thresholding: A comparative study. In 2014 International Conference on Control, Instrumentation, Communication and Computational Technologies (ICCICCT) 2014 Jul 10 (pp. 1182–1186). IEEE.

[14] Tang J. A color image segmentation algorithm based on region growing. In 2010 2nd International Conference on Computer Engineering and Technology 2010 Apr 16 (vol. 6, pp. V6–634). IEEE.

[15] Cui X, Deng Y, Yang G, Wu S. An improved image segmentation algorithm based on the watershed transform. In 2014 IEEE 7th Joint International Information Technology and Artificial Intelligence Conference 2014 Dec 20 (pp. 428–431). IEEE.

[16] Szénási S. Medical image segmentation with split-and-merge method. In 5th IEEE International Symposium on Logistics and Industrial Informatics (LINDI 2013) 2013 Sep 5 (pp. 137–140).

[17] Gupta D, Anand RS. A hybrid edge-based segmentation approach for ultrasound medical images. Biomedical Signal Processing and Control. 2017 Jan 1;31:116–126.

[18] Gao W, Zhang X, Yang L, Liu H. An improved Sobel edge detection. In 2010 3rd International Conference on Computer Science and Information Technology 2010 Jul 9 (vol. 5, pp. 67–71). IEEE.

[19] Rong W, Li Z, Zhang W, Sun L. An improved CANNY edge detection algorithm. In 2014 IEEE International Conference on Mechatronics and Automation 2014 Aug 3 (pp. 577–582). IEEE.

[20] Dharampal VM. Methods of image edge detection: A review. Journal of Electrical and Electronic Systems. 2015;4(2):2332–2796.
[21] Bali A, Singh SN. A review on the strategies and techniques of image segmentation. In 2015 Fifth International Conference on Advanced Computing & Communication Technologies 2015 Feb 21 (pp. 113–120). IEEE.
[22] Ali NN, Kako NA, Abdi AS. Review on image segmentation methods using deep learning. In 2022 4th International Conference on Advanced Science and Engineering (ICOASE) 2022 Sep 21 (pp. 7–12). IEEE.
[23] Wang R, Lei T, Cui R, Zhang B, Meng H, Nandi AK. Medical image segmentation using deep learning: A survey. IET Image Processing. 2022 Apr;16(5):1243–1267.
[24] Yin C, Zhu Y, Fei J, He X. A deep learning approach for intrusion detection using recurrent neural networks. IEEE Access. 2017 Oct 12;5:21954–21961.
[25] Ali O, Ali H, Shah SA, Shahzad A. Implementation of a modified U-Net for medical image segmentation on edge devices. IEEE Transactions on Circuits and Systems II: Express Briefs. 2022 Jun 8;69(11):4593–4597.
[26] Thomas JM. Bio-medical image denoising using autoencoders. In 2022 s International Conference on Next Generation Intelligent Systems (ICNGIS) 2022 Jul 29 (pp. 1–6). IEEE.
[27] Ker J, Wang L, Rao J, Lim T. Deep learning applications in medical image analysis. Ieee Access. 2017 Dec 29;6:9375–9389.
[28] Liao D, Lu H, Xu X, Gao Q. Image segmentation based on deep learning features. In 2019 Eleventh International Conference on Advanced Computational Intelligence (ICACI) 2019 Jun 7 (pp. 296–301). IEEE.
[29] Abbasi H, Orouskhani M, Asgari S, Zadeh SS. Automatic brain ischemic stroke segmentation with deep learning: A review. Neuroscience Informatics. 2023 Dec 1;3(4):100145.
[30] Lassmann H. Multiple sclerosis pathology. Cold Spring Harbor perspectives in medicine. 2018 Mar 1;8(3):a028936.
[31] Gharleghi R, Chen N, Sowmya A, Beier S. Towards automated coronary artery segmentation: A systematic review. Computer Methods and Programs in Biomedicine. 2022 Oct 1;225:107015.
[32] Wollenweber T, Roentgen P, Schäfer A, Schatka I, Zwadlo C, Brunkhorst T, Berding G, Bauersachs J, Bengel FM. Characterizing the inflammatory tissue response to acute myocardial infarction by clinical multimodality noninvasive imaging. Circulation: Cardiovascular Imaging. 2014 Sep;7 (5):811–818.
[33] Li G, Cook DJ, Thabane L, Friedrich JO, Crozier TM, Muscedere J, Granton J, Mehta S, Reynolds SC, Lopes RD, Francois L. Risk factors for mortality in patients admitted to intensive care units with pneumonia. Respiratory Research. 2016 Jul 11;17(1):80.
[34] Hoffman RM, Sanchez R. Lung cancer screening. The Medical Clinics of North America. 2017 Jul;101(4):769.
[35] Alyoubi WL, Shalash WM, Abulkhair MF. Diabetic retinopathy detection through deep learning techniques: A review. Informatics in Medicine Unlocked. 2020 Jan 1;20:100377.
[36] Schuster AK, Erb C, Hoffmann EM, Dietlein T, Pfeiffer N. The diagnosis and treatment of glaucoma. Deutsches Ärzteblatt International. 2020 Mar 27;117(13):225.
[37] Sikder J, Das UK, Anwar AS. Cancer cell segmentation based on unsupervised clustering and deep learning. In International Conference on Intelligent Computing & Optimization 2020 Dec 17 (pp. 607–620). Cham: Springer International Publishing.
[38] Vu QD, Graham S, Kurc T, To MN, Shaban M, Qaiser T, Koohbanani NA, Khurram SA, Kalpathy-Cramer J, Zhao T, Gupta R. Methods for segmentation and classification of digital microscopy tissue images. Frontiers In Bioengineering and Biotechnology. 2019 Apr 2;7:53.

[39] Pan Y, Huang W, Lin Z, Zhu W, Zhou J, Wong J, Ding Z. Brain tumor grading based on neural networks and convolutional neural networks. In 2015 37th Annual International Conference of the IEEE Engineering in Medicine and Biology Society (EMBC) 2015 Aug 25 (pp. 699–702). IEEE.
[40] Wang G, Li W, Zuluaga MA, Pratt R, Patel PA, Aertsen M, Doel T, David AL, Deprest J, Ourselin S, Vercauteren T. Interactive medical image segmentation using deep learning with image-specific fine tuning. IEEE Transactions on Medical Imaging. 2018 Jan 26;37(7):1562–1573.
[41] Nemade SB, Sonavane SP. Image segmentation using convolutional neural network for image annotation. In 2019 International Conference on Communication and Electronics Systems (ICCES) 2019 Jul 17 (pp. 838–843). IEEE.
[42] Khan MZ, Gajendran MK, Lee Y, Khan MA. Deep neural architectures for medical image semantic segmentation. IEEE Access. 2021 Jun 4;9:83002–83024.
[43] Siddique N, Sidike P, Elkin C, Devabhaktuni V U-Net and its variants for medical image segmentation: Theory and applications. arXiv preprint arXiv:2011.01118. 2020 Nov 2
[44] Lin A, Chen B, Xu J, Zhang Z, Lu G, Zhang D. Ds-Transunet: Dual SWIN transformer u-net for medical image segmentation. IEEE Transactions on Instrumentation and Measurement. 2022 May 30;71:1–5.
[45] Do Hong A, Chi HD, Pham T. Medical image segmentation using deep learning and blending loss. In 2022 7th National Scientific Conference on Applying New Technology in Green Buildings (Atigb) 2022 Nov 11 (pp. 109–113). IEEE.
[46] Yu Y, Wang C, Fu Q, Kou R, Huang F, Yang B, Yang T, Gao M. Techniques and challenges of image segmentation: A review. Electronics. 2023 Mar 2;12(5):1199.
[47] Conze PH, Andrade-Miranda G, Singh VK, Jaouen V, Visvikis D. Current and emerging trends in medical image segmentation with deep learning. IEEE Transactions on Radiation and Plasma Medical Sciences. 2023 Apr 10;7(6):545–569.
[48] Azad R, Aghdam EK, Rauland A, Jia Y, Avval AH, Bozorgpour A, Karimijafarbigloo S, Cohen JP, Adeli E, Merhof D. Medical image segmentation review: The success of u-net. IEEE Transactions on Pattern Analysis and Machine Intelligence. 2024 Aug 21.
[49] Shaker A, Maaz M, Rasheed H, Khan S, Yang MH, Khan FS. UNETR++: Delving into efficient and accurate 3D medical image segmentation. IEEE Transactions on Medical Imaging. 2024 May 9;43 (9):3377–3390.
[50] Al Garea S, Das S. Image segmentation methods: Overview, challenges, and future directions. In 2024 Seventh International Women in Data Science Conference at Prince Sultan University (Wids PSU) 2024 Mar 3 (pp. 56–61). IEEE.
[51] Zhao R, Qian B, Zhang X, Li Y, Wei R, Liu Y, Pan Y. Rethinking dice loss for medical image segmentation. In 2020 IEEE International Conference on Data Mining (ICDM) 2020 Nov 17 (pp. 851–860). Ieee.
[52] Hossain MS, Betts JM, Paplinski AP. Dual focal loss to address class imbalance in semantic segmentation. Neurocomputing. 2021 Oct 28;462:69–87.
[53] Guan H, Liu M. Domain adaptation for medical image analysis: A survey. IEEE Transactions on Biomedical Engineering. 2021 Oct 4;69(3):1173–1185.
[54] Isaac JS, Kulkarni R. Super resolution techniques for medical image processing. In 2015 International Conference on Technologies for Sustainable Development (ICTSD) 2015 Feb 4 (pp. 1–6). IEEE.
[55] Peng J, Wang Y. Medical image segmentation with limited supervision: A review of deep network models. Ieee Access. 2021 Feb 26;9:36827–36851.
[56] Frangi AF, Tsaftaris SA, Prince JL. Simulation and synthesis in medical imaging. IEEE Transactions on Medical Imaging. 2018 Mar 1;37(3):673–679.
[57] Mahmood T, Rehman A, Saba T, Nadeem L, Bahaj SA. Recent advancements and future prospects in active deep learning for medical image segmentation and classification. IEEE Access. 2023 Sep 11;11:113623–113652.

[58] Erickson BJ, Korfiatis P, Akkus Z, Kline TL. Machine learning for medical imaging. Radiographics. 2017 Mar;37(2):505–515.

[59] Rani V, Kumar M, Gupta A, Sachdeva M, Mittal A, Kumar K. Self-supervised learning for medical image analysis: A comprehensive review. Evolving Systems. 2024 Aug;15(4):1607–1633.

[60] Dolz J, Gopinath K, Yuan J, Lombaert H, Desrosiers C, Ayed IB. HyperDense-Net: A hyper-densely connected CNN for multi-modal image segmentation. IEEE Transactions on Medical Imaging. 2018 Oct 30;38(5):1116–1126.

[61] Sohan MF, Basalamah A. A systematic review on federated learning in medical image analysis. IEEE Access. 2023 Mar 21;11:28628–28644.

[62] Bechar A, Medjoudj R, Elmir Y, Himeur Y, Amira A. Federated and transfer learning for cancer detection based on image analysis. Neural Computing and Applications. 2025 Feb;37(4):2239–2284.

Chetna Sharma, Astha, Manya, Ria Khurana

10 Revolutionizing Medical Image Interpretation with GANs in Healthcare

Abstract: Generative adversarial networks (GANs) are a type of AI designed to create new data, like images that look so real. The objective of a generative model is to learn the probability distribution that produced a set of training examples. This paper provides a detailed analysis of GANs, including their architecture, variants, applications, challenges, and future prospects. The study classifies the GAN applications into five broad categories: image synthesis, data augmentation, style transfer, image-to-image translation, and anomaly detection. Several sectors have been adopted using GANs including the fields of healthcare, fashion, automobiles, entertainment and media, agriculture, finance, etc. Review includes essential GAN variants including targeted synthesis using Conditional GANs (cGANs), unpaired image-to-image translation using CycleGANs, high-resolution generation in images using StyleGANs, and the Wasserstein GANs for stabilizing the training. Despite the excellent performance of GANs, it faced some major challenges, which include training instability, mode collapse, and high computational requirements. This paper discusses solutions such as improved loss functions, progressive training, and hybrid frameworks integrating GANs with other deep learning models. Ethical issues are also discussed in relation to how GAN-generated content such as deep fakes and synthetic media are misused.

Keywords: generative adversarial networks (GANs), deep learning, adversarial learning, generator network, discriminator network, conditional GANs (cGAN)

10.1 Introduction

Generative adversarial networks (GANs) are the class of machine learning (ML) models, which obtained plenty of limelight in the past years because of their ability to make and produce a duplicate copy and give rise to a high-quality similar image from probability distributions of real datasets. GANs can generate realistic data, enhance datasets, translate image-to-image, improve image resolution, and domain adaptation. This is the most interesting idea by Ian Goodfellow. GANs were first introduced in 2014 [1]. As time goes by, countless changes have been proposed: firstly there was a generator as well as a discriminator; at that time they conflicted over each other in the zero-sum game. In an ending phase of this training, the total gains for both players sum to zero, meaning one player's gain is equal to the other's loss, which ensures that the value of each player's in the action is balanced and preserved at a stable level [2]. The game reaches equilibrium in which the generator produces such realistic data

 | https://doi.org/10.1515/9783112226223-010

that the discriminator categorizes it as real with an equal probability of recognizing it as fake.

GANs embody two neural nets, the generator and the discriminator. The role of the creator is to produce a dataset as close to real whereas the discriminator's work is to recognize the difference between real data and fake ones. GANs are rooted in a minimax game, in which the generator fools the discriminator and in the meantime the discriminator maximizes its aptitude to set apart real from fake data. In 2015, GANS became one of the most exciting topics in research by improving training techniques [4]. A breakthrough in 2015 was the introduction of deep convolutional GANs (DCGANs) by Alec Radford, Luke Metz, and Soumith Chintala [4]. DCGANs improved the quality and solidity of GANs for generating such high-quality images that one was not able to tell whether it is fake or real.

Nowadays, GANs are currently being applied in several sectors, using their capability to produce high-quality synthetic data [5]. In healthcare GANs are used to enhance medical images for training, speeding up the drug discovery process [5]. In fashion GANs used in creating virtual fitting rooms as also help in generating innovative fashion designs. In other sectors [6] like entertainment and media [3], automotive [5], robotics, finance, and agriculture [7]. GANs are used to ease the task with different trends. In the industrial sector, GANs are being increasingly utilized in manufacturing [8], automobiles, aerospace, energy [9], construction, and so on. In industrial applications, GANs are adopted to improve design processes [8], maximize efficiency, generate different scenarios, and generate important findings from virtual data. Their ability to create realistic simulations makes them a powerful tool in advancing industrial technology and efficiency.

The training process involves many different steps. Firstly, initializing the generator, after that, evaluation of the discriminator, and then a feedback loop in which the discriminator provides feedback to the generator based on its classification. After the feedback loop, comes adversarial training in which the generator produces a more real dataset, while the discriminator (judge) updates itself to identify the difference between real and fake data. In the last convergence process, the creator (which is the generator) improves to such a point where its outputs become increasingly realistic, and the discriminator (judge) becomes less able to differentiate at a point between real and generated data.

Despite the success, GANs face several challenges such as training instability, while training GANs can suffer issues like cave-in mode, in which the creator produces a limited variety of outputs. GANs require significant systematic power and time to train effectively [10]. Black boxes often operate GANs, which makes it difficult to understand the generating process [11].

Different models of GANs have undergone significant changes and variations since their introduction. Original GAN was introduced by Ian Goodfellow and his colleagues. In 2015, Laplacian Pyramid GAN (LapGAN) introduced layered multi-scale image generation, although cGAN enabled controlled outputs by conditioning GANs on additional data like class labels [12]. In 2016, DCGANs and information maximizing GAN (InfoGAN) were introduced for stability and architectural improvements. In 2017, WGAN introduced the Wasserstein distance to improve training [13] and Pix2Pix was developed for paired image-to-image translation [14]. Further, every year has brought significant innovations to make GAN the most dynamic area in ML and AI.

Though GANs are highly versatile and used in countless fields there are still areas where GANs' use is limited and faces challenges. For small dataset scenarios GANs may struggle to generate high-quality outputs as GANs are typically used for large amounts of data. The use of GANs may be limited in highly regulated industries due to the need for verification processes because of the risk of unrealistic and unsafe data [15]. GANs may not be practical or deployed; therefore they might not be used in low-resource environments. GANs are not used for complex system design, data privacy, security sector, legal, and ethical concerns [15].

In this book chapter, we exhibit models and use cases of GANs, which would help in understanding diverse models, the evolution of GANs, problem-solving using different models, GANs across different industries, and limitations of GANs. Typical GAN models: CGAN, DCGAN, InfoGAN, Pix2Pix, WGAN, CycleGAN, style GAN, progressive GAN, and other GAN variants. Also, we will discuss open problems and future decisions.

10.2 GAN Architecture and Variants

This section provides an overview of GANs, the fundamental GAN models, and variants. After discussing GAN theory, we review a few of GAN variants that provide the base or building blocks for various GAN models. Numerous papers discuss GANs, and plenty of them use named GANs that are models with a particular name, which typically includes the word "GAN". GANs are categorized as generative models. The theory on which GANs are built is min-max, zero-sum game. Two neural networks – the discriminator and the generator make up GANs for this purpose. In order to mislead the discriminator, The generator learns to build fake sample distribution, but the discriminator aims to figure out and differentiate between the real and fake distribution produced by the generator. The discriminator and the generator both construct the entire structure of GAN. The generator (G) takes some random noise vector Z as input and then tries to create an image from this noise vector. The respective parameters of the generator are then updated based on the output of the discriminator. The discriminator (D) is a binary classifier that checks real and fake samples generated by G to decide which ones are real and which are spam. However, for a sample image X, the discriminator attempts to model the image being artificial or real, a likelihood or probability distribution. These probabilities are considered feedback for the G.

10.2.1 Vanilla GAN (VGAN)

Vanilla GAN (VGAN) is a basic kind of GAN, in which generator and the discriminator are basic multilayer receptors, and it adheres to the overall GAN architecture. Vanilla GAN is the simple algorithm, which implements the very simple mathematical equation on gradient descent orientation technique. The basic vanilla GAN structure takes in distorted data as an input to the generator where the discriminator is used to classify the data samples as real or fake.

10.2.2 Fully Connected GAN (FCGAN)

In his research, Goodfellow et al. [1] introduced his original GAN, which used fully connected neural networks (FCGAN) for both the generator and the discriminator. FCGAN method is generally applied on some plain image datasets such as Toronto Face Dataset, CIFAR-10 and MNIST. The fully connected neural networks do not provide ideal generalization for extra complicated images.

10.2.3 CycleGAN

CycleGAN mechanism interprets pics from one do-principal into some other. There are moreover a model – newest shifts have integrated Mocycle-GAN to translate video from one space into any other. CycleGAN fashions are educated to make sure that one can take the entire picture from one domain and translate it into a special domain or downside area belonging to the output area.

10.3 Mathematical Framework

This process includes a two-player minimax game between the generator and the discriminator. The function of this game was designed as follows:

$$\min_G \max_D E_{x\sim p_{\text{data}}(x)}[\log D(x)] + E_{z\sim p_Z(z)}[\log\log(1 - D(G(z)))] \quad (10.1)$$

in which $D(x)$ is the probability that x is real, $G(x)$ is the generated sample from noise z, $P\{\text{data}\}$ and $P(z)$ are distributions of real data and noise.

To maximize the probability that the discriminator makes a mistake we use log $(1 - D(G(z)))$. This means that the generator aimed at producing data, which would be incorrectly classified as real by the discriminator. The main objective of the discriminator is to minimize $\log D(x) + \log(1 - D(G(z)))$, i.e., improving its accuracy between real and fake data.

Discriminator training: It inputs a group of fake and real data and then processes both of samples; the output is a probability score for each sample [16]. Loss for real data measures how effectively the discriminator recognizes real samples as real and loss for fake data measures how effectively the discriminator recognizes fake samples as fake; the combination of these two losses is the total loss for the discriminator. Then discriminator's weights are revised by applying gradient descent or an equivalent efficient algorithm to reduce its loss and enhance its precision.

Generator training: It generates a bunch of fake data randomly. Then the data is forwarded to the discriminator, and after that its loss is determined based on how the discriminator responds to samples.

A lot of challenges are faced in improving the training process of both the generator and discriminator. Many advanced improvements are introduced to address many challenges such as use of IS (inception score) and FID (Frechet inception distance) to improve the quality of samples.

10.4 Typical GAN Models

10.4.1 CGAN

The advantage of GAN is that it does not require a specific data distribution for input; by applying direct sampling to input random noise Z, more realistic samples can be produced. However, one drawback is that we have to provide GAN with some limits because the generated samples are too free to be controlled and cannot concentrate on the designated samples. This concept is what brought CGAN to the GAN stage.

Mehdi Mirza and Osindero [17] devised CGAN in response to GAN's double randomization (random samples and random noise). The Y tag was added as an input to the GAN generator and discriminator. When y is added, the generator's input becomes noise and a label, and the discriminator's input becomes an actual sample that also produces a label.

In addition to creating samples with specified labels, CGAN can enhance the caliber of the samples that are produced. Then, Augustus Odena's semisupervised GAN (SGAN) transformed GAN into a semisupervised model and allowed for the evaluation of its output category labels. Auxiliary classifier GAN (ACGAN) [18], one of the two models that were received, was the idea that enhanced the quality of the samples that CGAN created.

The way that ACGAN improves on CGAN is that in order to help the generator comprehend the necessary sample structure, it incorporates tags into the actual sample in addition to using tags in the generator's input. In the discriminator's output section, two complete connection layers are employed. Using the concept of SGAN, classification results for samples are added, additionally producing findings of discriminant (fake or real).

10.4.2 DCGAN

The deep convolutional GAN (DCGAN) is among the successful and highly utilized structures for GAN networks that uses unsupervised representation learning. It is mostly made up of convolutional layers, layers that are paired together or do not have any max pooling. DCGAN uses a neural network for both the discriminator and the generator. Down sampling is achieved by using a convolutional stride, while up sampling is achieved by using a transposed convolution. In the sequence, convolutional transpose layers follow convolutional layers. For their study on deep convolutional adversarial generative networks for unsupervised representation learning, Radford et al. [19] suggested the generator's network architecture.

10.4.3 Pix2Pix

Pix2Pix is a technology that processes images by using a conditional GAN that learns to map from input to feature map. Using Pix2Pix pictures can be changed into images. The discriminator firstly is used for examining where the resultant picture is produced from the generator. The reasoning behind them, relationships between input, output, and target, is explored by the discriminator to generate an unbiased opinion on how realistic both pairs are. Weights of the discriminator are adjusted to meet the standard error of the pairs of input/output and input/target. The output of the discriminator (as well as the difference between the output and the targeted image) is used to improve repeatedly on the discriminator, which guides which images are more or less similar to the training data and guide changes made to the generator weights.

The generator is made up of two levels, which are the following: the decoder and encoder. The encoder changes the input picture to a more simplified presentation of the feature whereas the decoder evokes a regular generator in appearance but is composed of several layers called transpose-convolution that serve the opposite purposes from the layers of encoder. In place of categorizing one image as fake or real, the discriminator evaluates images in pairs (a provided visual and an unknown visual that may be a targeted visual or a created visual) and gives a label to each pair of photos as a fake or real. The definition of the loss function may be found here. The equation min G max E x,y [log $D(x, G(x))$] + E x,y [log(1–D (G (x,y)))] represents it. Pix2Pix has a disadvantage of relying on the two picture spaces that are to be pre-formatted into an X/Y image, which consists of both the closely related photographs. The images of satellite are converted from one image to other and after that changed into Google Maps using Pix2Pix.

10.4.4 WGAN and WGAN-GP

Sometimes the generator produces only a tiny subset of possible images or repeats highly similar visuals, leading to a problem known as mode collapse. To address this issue, the Wasserstein GAN (WGAN) was introduced, providing a more stable training objective based on the Wasserstein distance [25]. However, the original WGAN relies on weight clipping, which can cause optimization difficulties. To improve stability, the WGAN-GP was later proposed, replacing weight clipping with a gradient penalty for more reliable training [26].

10.4.5 CycleGAN

CycleGAN is an image translator. At the generator, it converts *X*-domain images into *Y*-domain images. It is implied by unsupervised learning that the training images lack labels. Image correlation is not required for domains. Cycle GAN is composed of *G* and *F* that are two discriminator networks as well as two generating networks. Convolutional neural networks, or DX and DY in the discriminator, are used in tracing training. To determine whether an input image is real or fraudulent $G: X \text{->} Y$ and $F: Y \text{->} X$ are used. During the process of generating images two different kinds of training observations are supplied as input throughout the training process. In one set of observations, two images are tagged "Xi, Yi," where each Xi has a matching Yi. There is no correlation between the two sets of photographs that make up the additional set of observations: one set of images from *X* and another set of images from *Y*. The generators consist of a transpose-convolutional network called a decoder that modifies the picture and a convolutional network called an encoder that minimizes the size of an image. The discriminator establishes if a picture is real or fake using cross-entropy loss. Discriminator losses are determined by the mean squared errors between the output of an image and the goal value, which can be either 0 or 1, based on which the image can be labeled as authentic or fraudulent – are what determine discriminator losses. The output of the discriminator will have 1 as the mean squared error when the picture is false and 0 if it is true [20].

10.4.6 Style GAN

The management of style of the generated images is made possible by the study "A Style-Based Generator Framework for Adversarial Networks," that makes use of Pro-GAN advancements along with adaptive instance normalization (AdaIN) and image style transfer. The Style GAN architecture is based on a mapping network that translates the input in between latent space [21]. There are roughly 18 convolutional layers overall, and the intermediate is processed by AdaIN after each layer. Compared to the original StyleGAN, Style 2 GAN offers a number of improvements, such as regularization techniques, gradually rising regularization, and normalization.

10.5 Conventional applications of GAN

Multimedia is the wide-ranging computer-based processing and handling of text, data, images, animation, sound, and other media files via a laptop that enables a user to engage with the device through a variety of senses for up-to-date information. Because the application direction of GAN primarily focuses on computer vision and verbal language processing, it is meticulously integrated with numerous multimedia generation components. The creation of GANs brings about immense scope for the development of multimedia technology.

10.5.1 Image Processing

GANs are located to be very beneficial in recognizing unique image processing-related paintings like amazing-decision pix generation, editing snap shots, generating excessive-resolution faces, manipulating facial attributes, and so forth.

10.5.2 Wonderful Decision

The conversion of low-resolution pictures into better-resolution images is known as the manner of photograph terrific resolution. The SRGAN is the first framework that may infer photo-realistic natural pix for upscaling elements of 4*x*.

10.5.3 Picture Editing

Photograph editing is described as a method of erasing or altering a portion of a picture, including hair color and smile. For example, snap shots captured during heavy rain or negative climatic situations do not possess advanced visible photos and involve guide processing for the removal of anomalies including rain drops or dust debris that have an effect on the photograph quality. Identity-CGAN authors applied GANs in overcoming the mission in unpaired image deraining problem.

10.5.4 Face Photo Era at Excessive Decision

This is any other location of image processing wherein GANs have outdone. The essential classifications into producing faces and manipulation of its attributes consist of developing whole artificial faces, the face functions or the attribute manipulation, and the face element transformation.

10.5.5 Face Thing Transformation

The objective of face aspect transformation is to change facial features such as hairstyle, color, or accessorizing with eyeglasses and lip color [22]. Authors of StarGAN have been able to translate images across multiple areas, while authors of DiscoGAN have been able to exchange hair colors. BeautyGAN can be used to convert a reference image of makeup face into a different non-makeup image while maintaining the individuality of the face. GANs can also be used for picture inpainting, which is the process of rebuilding areas of a picture that are missing.

10.5.6 Domain Transformation

That is text-to-picture, which is an input sentence to generate an image. Due to GAN's fantastic performance in photo processing, many new fashions had been created based totally on the result of this form of domain transformation.

10.5.7 GANs in Scientific and Healthcare

GANs have an extremely good application within the area of scientific image era; they may not be used most effectively for an early enhancement of prognosis but they decrease time and fee. Being short of fine, scientific picture statistics is a continual constraint, and consequently, GAN is applied as a records augmentation approach, image-to-picture translation, synthesis of artificial information, and scientific image extremely good-decision.

The magnetic resonance imaging (MRI) is the gold standard for imaging in medicine. However, this doesn't apply to patients with metal implantations because the metal that makes up the machinery may also interfere with the consequences and the patient's safety. One of the major packages of GANs in healthcare and scientific imaging is that if photograph-to-photograph translation framework, or say multimodal pictures are wanted, then it's possible to use pictures from a single domain or modality to produce images in any other domain. Like CycleGAN, MR-GAN has been applied to translate CT image segments of second-brain into second-mind MR image segments.

But in contrast to CycleGAN, that applies to unmatched photo-to-photo translation; both the unpaired and paired records are used to train MR-GAN with a combination of antagonistic loss, the dual cycle-steady loss, and the voxel-wise loss. Using the input approach of multiple-channels-multiple-landmarks colors are synthesized by MCML-GANs and retinal fundus images made up of the combination of the optic disc, vessel tree, and optic cup image. To compare their works, the authors suggested exchange architectures for the generator model and used styles that only relied on the framework that was developed in the pix2pix and CycleGAN model variant. Zhu et al. further proposed image-to-image translation models of the bathtub-GAN and the tub-sGAN for making different types of retinal images and neural images [23]. A new framework named MedGAN was devised by Armanious et al. [24] for targeting generalization of photo to picture conversion within the domains of clinical picture production by integrating an adverse framework by a new aggregate of non-adversarial declines and using CasNet, an architecture inspired by ResNets. Sandfort et al. [25] remodeled comparative CT images to non-contrast images using CycleGAN.

The synthetic images of pores and skin problems are generated by DermGAN. The version discovers how the mapping of the semantic map with prior skin state and pores is made, including their length, location, and underlying color, to a practical image that preserves the pre-designated characteristics.

Besides solving the problem of image-to-image translation, GANs are often applied in the age of synthetic clinical images. In the case of tackling the problem of the conditional retinal vascular community synthesis, an autoencoder against the opposite direction was applied by Beers et al. [26]. Costa et al. [27] have used ProGAN for enhancing the excessive resolution and fine 256 × 256 multimodal glioma and 512 × 512 retinal fundus images appropriately.

Photograph reconstruction and tremendous resolution are very crucial to obtain a high-decision picture of the area so that diagnosis may be made because it will allow very minimal results of the proscribing elements, which include the radiation required for MRI and as a result any approach in photo acquisitions might drastically affect the satisfactory images attained. The use of an effective 3D convolutional neural network layout for the architecture of generator of a multistage densely connected superb-decision network, an mDCSRN-GAN is furnished by means of Chen et al. [5], as a way to carry out extraordinary-decision for pix. The SR framework for medical pix is MedSRGAN.

10.5.8 Fashion Designing

Fashion designing may not be the initial element that pops up in our minds whilst we think of GAN; however, GANs have applications in developing designs for garments and outfits. The attribute-GAN finds a mapping from a couple of attires based totally upon garb qualities on the basis of CGAN. This model consists of two discriminators and a generator. A new edition of GAN design-AttGAN, which was to modify garment pictures mechanically primarily depending on some person-defined features, was given by Heusel et al. [28].

10.6 Applications of GAN in Security

AI is a two-edged sword, which not only demands that we should safeguard the safety of the generation implementation of AI but also adopt AI generation to create a safe environment. In this part, the introduction to packages of GAN in security is discussed.

10.6.1 Statistics Security

Information is widely spread, sharable, value-added, doable, and multifunctional, which makes it mainly essential for humans. There are meanings for the time period data protection. The first refers to protecting all styles of records assets from all forms of threats, interference, and destruction. The sport mechanism of attackers and defenders is much like GAN, which has made a few contributions to the improvement of facts safety, mainly in cryptography.

In 2016, Google proposed an encryption era primarily based on GAN, which may correctly resolve the information protection problem inside the process of data sharing. Cryptography can be both defensive and offensive, and GAN is likewise implemented to decryption generation. Sharma et al. proposed a manner of cryptography era based totally upon gadget getting to know theory to replace artificially generated cryptography regulations, which turned into named PassGAN [29]. By training on leaked password datasets, the discriminator learns real password patterns, enabling the generator to produce passwords that are closer to actual user passwords and thereby improving the effectiveness of password-guessing methods. In 2018, Akhtar and Mian [30] advanced an uncontrolled technique for cracking the code, called cipherGAN. This technique can decode the Caesar shift code or Virginia code with high constancy after unrivaled plaintext ciphertext education. Stimulated by means of CycleGAN, this model adopts unequalled plaintext and ciphertext and completes the key decoding of lengthy word level without parallel text.

10.6.2 Cyber Safety

Currently, the technology called cyber safety has attracted many researchers due to the advances in massive facts, Internet of Things (IoT), blockchain, and other warm spots. With these advances, a block of numbers threatens an increasing number of normal operation of the community, particularly, the venture Collapsar attack, disbursed denial of carrier attack, malware, and computer virus. The bizarre community conduct is not a picture or textual content, so it's far tougher to system with a GAN. In this paper, the authors take the source of the attack as the generation of the detection system of unusual community behaviors. They will apply the method of generating samples of unusual behaviors to simulate the attack and enhance the detection capability of the existing detection technology.

10.6.3 Safety in AI

Getting to know artificial intelligence improvement deep is one of the most important technologies in the field of synthetic intelligence. However, the determined vulnerabilities of deep mastering algorithms to assaults towards samples suggest that studies to improve the toughness and security of the algorithms are urgent. Wang et al. [31] summarized three major directions of deep learning set of rules in opposition to adversarial assaults, which are: the usage of changed schooling or changed enter, converting the networks, and the community of different versions appearing as the network.

Currently, the biggest percentage of this type of defense is gradient regularization/masking, but researchers proved the evasion ability of such methods. Bao et al. [32] proposed FBGAN with a purpose to seize the well-formed feature of the input and clear out the concern that isn't semantic; after pre-schooling and bidirectional mapping, it used the hostile facts de-noised classification correctly to weaken the hostile attack. The experiment showed its effectiveness.

10.7 Open Problems and Proposed Solutions

Major challenges include training instability and vanishing gradient problems by GAN. The research in GANs is speedily growing, and solutions to these problems are regularly updated.

10.7.1 Objective Function for Training GANs

Problem: The original GAN design uses a minimax game to refine the generator and the discriminator. But this function leads to many issues with non-convergence or unstable training. One of the main reasons is the generator's gradient values become very small causing the network to stop learning effectively, which can lead to the generator producing poor-quality outputs.

Proposed Solution: Researchers should explore more stable versions of the current loss to stabilize training. Introduction to address the stability by replacing the traditional loss function with the Wasserstein distance (Earth mover's distance), which reduces mode collapse and improves convergence by giving better gradients to generator could help in useful results. In big GANs, hinge loss is used to improve learning dynamics by creating better separation between real and fake datasets.

10.7.2 Mentality Behind Solving GAN Training Problems

Problem: Training GANs is not an easy task. It needs a process in which both generator and discriminator must improve continuously without one overpowering the other. Harsh training or imbalanced training can lead to poor performance. Advance training strategies can be used to manage these problems and lead to better outcomes.

Proposed Solution: Rotational updates like five updates for every generator are finer than updating both the generator and discriminator at the same time. It will help in stabilizing training. Problems like lack of paired dataset or mode collapse can be solved by trying new architectures such as CycleGAN, which adapts to unpaired image-to-image translation. Small changes can sometimes have significant effects.

10.7.3 Mode Collapse

Problem: Mode collapse is a very challenging topic in GAN training where the generator discovers how to generate limited amount of outputs, which limit the utility of GAN. With this the discriminator is being fooled by these outputs. This occurs when the generator finds few specific outputs that consistently manipulate the discriminator, which leads to ignoring other potential variations.

Proposed Solution: To address mode collapse in GAN training, many techniques can be used to enhance diversity in the generated samples. The discrimination of minibatch permits the discriminator to solve various samples simultaneously, which enables to detect if the generator is producing similar outputs, thus encouraging better variety. Training to match the statistics of real and fake dataset in some midway feature space (e.g., hidden layers) helps the generator to focus on uniqueness instead of just bluffing the discriminator.

10.7.4 Instability of Adversarial Training

Problem: Both the generator and discriminator try to outsmart each other and the balance between them is very delicate; if one of them become too strong or weak then it leads to oscillatory behavior. This makes GAN training more difficult and prone to failure if the models don't find a stable symmetry.

Proposed Solution: Use of alternative loss functions, regularization, and progressive GANs can be used to make the process stable. Progressive GANs involve starting with low-resolution images and slowly increase the resolution. Progressive GANs are used for managing complexity whereas regulation techniques are for controlling instability in the discriminator.

10.7.5 Lack of a Proper Evaluation Metric

Problem: GANs are not that easy as we don't have any simple way to calculate if the generated images are good or not in supervised learning in which we can use metrics like accuracy. There is a need of mix metrics along with human judgement and direct visual checks of images.

Proposed Solution: FID and IS are the dimensions used for image generation. Comparing both FIDs works well for evaluating GANs as it reports for both quality and diversity.

10.7.6 Vanishing Gradient

Problem: Useless feedback (gradients) to generators leads to the struggles to learn and more significant improvements as it doesn't receive enough information to understand how to create better images.

Proposed Solution: Replacing ReLU with leaky ReLU helps prevent vanishing gradients by allowing negative values to have a small gradient. Gradient penalty (WGAN-GP) ensures that discriminator's gradient remains within limits that can help in generator improvement.

10.7.7 Improving Training

Problem: Unstable loss functions, slow convergence, and difficulty balancing these factors raise the problem in the training process.

Proposed Solution: With the help of advanced techniques like spectral normalization, improvement in the generator's architecture leads to well-built pieces of training [33].

10.8 Conclusion

With the creation of news datasets like images and text complex transformations between different types of data, GANs have become key developments in artificial intelligence and machine learning. The architecture of GANs consists of a generator that creates the data and a discriminator that tries to figure out fake data. The applications of GANs are increasing rapidly, in the sectors like healthcare, media, entertainment, multiple industries etc. GANs reduce data shortages, promoting progressing fields like self-driving cars by generating virtual environments for testing and training.

The relationship between the generator and discriminator is based on game theory, in which both networks try to beat one another. Compared to older generative models, which struggled a lot with complex data, advanced ones can handle high-dimensional data. There are issues like "mode collapse," and "vanishing gradients" with which training of GANs becomes unstable. Many researchers have been working to solve these challenges by developing models like Wasserstein GANs (WGANs), which help stabilize training. Also analyzing the quality of data produced by GANs is another challenge. To judge the quality of data generated, traditional metrics are not sufficient, promoting the need for more specialized metrics like IS and FID. Areas like computer versions of GANs are being used for improving image quality or creating realistic facial images. MedGAN and seGAN GANs are used in healthcare for tasks like medical image segmentation and outlier detection. With the introduction of CGANs and DCGANs more precise data is generated. Models like Pix2Pix, CycleGAN, and StyleGAN have increased the boundaries of GANs, like converting sketches into realistic images that are used in design and fashion. However, GANs can also be misused in different ways like creating deepfakes and other misleading media. Security is also essential in GANs. GANs are likely to combine with AI technologies for better security, smarter, and more creative models. In summary, GANs have transformed generative models by giving high-quality data output. Many challenges are faced but improvements are still going on for better results. GANs have a great impact in research. The future of GANs holds great potential, but it also needs a lot of attention on security and deepfake detection.

References

[1] Goodfellow IJ, Pouget-Abadie J, Mirza M, Xu B, Warde-Farley D, Ozair S, Courville A, Bengio Y. Generative adversarial nets. Advances in Neural Information Processing Systems. 2014;27.

[2] Ahmad Z, Jaffri ZU, Chen M, Bao S. Understanding GANs: Fundamentals, variants, training challenges, applications, and open problems. Multimedia Tools and Applications. 2025 Apr;84 (12):10347–10423.

[3] Aggarwal A, Mittal M, Battineni G. Generative adversarial network: An overview of theory and applications. International Journal of Information Management Data Insights. 2021 Apr 1;1 (1):100004.

[4] Radford A, Metz L, Chintala S Unsupervised representation learning with deep convolutional generative adversarial networks. arXiv preprint arXiv:1511.06434. 2015 Nov 19.

[5] Chen Y, Yang XH, Wei Z, Heidari AA, Zheng N, Li Z, Chen H, Hu H, Zhou Q, Guan Q. Generative adversarial networks in medical image augmentation: A review. Computers in Biology and Medicine. 2022 May 1;144:105382.

[6] Wu Q, Zhu B, Yong B, Wei Y, Jiang X, Zhou R, Zhou Q. ClothGAN: Generation of fashionable Dunhuang clothes using generative adversarial networks. Connection Science. 2021 Apr 3;33 (2):341–358.

[7] Gui J, Sun Z, Wen Y, Tao D, Ye J. A review on generative adversarial networks: Algorithms, theory, and applications. IEEE Transactions on Knowledge and Data Engineering. 2021 Nov 23;35 (4):3313–3332.

[8] Haridasan PK, Jawale H. Generative AI in Manufacturing: A Review of Innovations, Challenges, and Future Prospects. JAIMLD. 2024 Sep;2(2):1418–1424.

[9] Dash A, Ye J, Wang G. A review of generative adversarial networks (GANs) and its applications in a wide variety of disciplines: From medical to remote sensing. IEEE Access. 2023 Dec 22;12:18330–18357.

[10] Salimans T, Goodfellow I, Zaremba W, Cheung V, Radford A, Chen X. Improved techniques for training GANs. Advances in Neural Information Processing Systems. 2016;29.

[11] Creswell A, White T, Dumoulin V, Arulkumaran K, Sengupta B, Bharath AA. Generative adversarial networks: An overview. IEEE Signal Processing Magazine. 2018 Jan 10;35(1):53–65.

[12] Denton EL, Chintala S, Fergus R. Deep generative image models using a Laplacian pyramid of adversarial networks. Advances in Neural Information Processing Systems. 2015;28.

[13] Arjovsky M, Chintala S, Bottou L. Wasserstein generative adversarial networks. In International Conference on Machine Learning 2017 Jul 17 (pp. 214–223). PMLR.

[14] Isola P, Zhu JY, Zhou T, Efros AA. Image-to-image translation with conditional adversarial networks. In Proceedings of the IEEE Conference on Computer Vision and Pattern Recognition 2017 (pp. 1125–1134).

[15] Benaich N, Hogarth I. State of AI Report. London: UK.[Google Scholar]; 2020.

[16] Lucic M, Kurach K, Michalski M, Gelly S, Bousquet O. Are gans created equal? a large-scale study. Advances in Neural Information Processing Systems. 2018;31.

[17] Mirza M, Osindero S Conditional generative adversarial nets. arXiv preprint arXiv:1411.1784. 2014 Nov 6.

[18] Odena A Semi-supervised learning with generative adversarial networks. arXiv preprint arXiv:1606.01583. 2016 Jun 5.

[19] Radford A, Metz L, Chintala S Unsupervised representation learning with deep convolutional generative adversarial networks. arXiv preprint arXiv:1511.06434. 2015 Nov 19.

[20] Mao X, Li Q, Xie H, Lau RY, Wang Z, Paul Smolley S. Least squares generative adversarial networks. In Proceedings of the IEEE International Conference on Computer Vision 2017 (pp. 2794–2802).

[21] Karras T, Aila T, Laine S, Lehtinen J Progressive growing of gans for improved quality, stability, and variation. arXiv preprint arXiv:1710.10196. 2017 Oct 27.

[22] Tolosana R, Vera-Rodriguez R, Fierrez J, Morales A, Ortega-Garcia J. Deepfakes and beyond: A survey of face manipulation and fake detection. Information Fusion. 2020 Dec 1;64:131–148.

[23] Zhu JY, Park T, Isola P, Efros AA. Unpaired image-to-image translation using cycle-consistent adversarial networks. In Proceedings of the IEEE International Conference on Computer Vision 2017 (pp. 2223–2232).

[24] Armanious K, Jiang C, Fischer M, Küstner T, Hepp T, Nikolaou K, Gatidis S, Yang B. MedGAN: Medical image translation using GANs. Computerized Medical Imaging and Graphics. 2020 Jan 1;79:101684.

[25] Sandfort V, Yan K, Pickhardt PJ, Summers RM. Data augmentation using generative adversarial networks (CycleGAN) to improve generalizability in CT segmentation tasks. Scientific Reports. 2019 Nov 15;9(1):16884.

[26] Beers A, Brown J, Chang K, Campbell JP, Ostmo S, Chiang MF, Kalpathy-Cramer J High-resolution medical image synthesis using progressively grown generative adversarial networks. arXiv preprint arXiv:1805.03144. 2018 May 8.

[27] Costa P, Galdran A, Meyer MI, Niemeijer M, Abràmoff M, Mendonça AM, Campilho A. End-to-end adversarial retinal image synthesis. IEEE Transactions on Medical Imaging. 2017 Oct 2;37(3):781–791.

[28] Heusel M, Ramsauer H, Unterthiner T, Nessler B, Hochreiter S. Gans trained by a two time-scale update rule converge to a local Nash equilibrium. Advances in Neural Information Processing Systems. 2017;30.

[29] Sharma P, Kumar M, Sharma HK. A generalized novel image forgery detection method using generative adversarial network. Multimedia Tools and Applications. 2024 May;83(18):53549–53580.

[30] Akhtar N, Mian A. Threat of adversarial attacks on deep learning in computer vision: A survey. IEEE Access. 2018 Feb 19;6:14410–14430.

[31] Wang Z, She Q, Ward TE. Generative adversarial networks in computer vision: A survey and taxonomy. ACM Computing Surveys (CSUR). 2021 Feb 9;54(2):1–38.

[32] Bao J, Chen D, Wen F, Li H, Hua G. Towards open-set identity preserving face synthesis. In Proceedings of the IEEE Conference on Computer Vision and Pattern Recognition 2018 (pp. 6713–6722).

[33] Miyato T, Kataoka T, Koyama M, Yoshida Y Spectral normalization for generative adversarial networks. arXiv preprint arXiv:1802.05957. 2018 Feb 16.

Gagandeep Kaur, Ashima Kukkar, Ramanjot Kaur

11 AI-Driven Healthcare: Case Studies in Intelligent Systems and Applications

Abstract: Artificial intelligence (AI) is increasingly transforming healthcare by offering refined diagnostic accuracy, streamlining workflow systems and improving patient outcomes. The AI algorithms have the ability to extract patterns, detect abnormalities, and create predictive insights out of huge medical datasets at a faster pace and more accurately as compared to the traditional methodologies. Machine learning models trained on medical records, imaging, and even genomic data can assist the clinical workforce in early-stage diagnosis, eliminating misdiagnosis and creating treatment plans for each individual. This chapter provides some real-time case studies where AI is used for predictive analytics, medical imaging, personalized care, and clinical decision support. It also traces the issues pertaining to privacy of the data, interpretability of the models, and their interfacing with existing systems. By drawing attention to AI for early detection, automation, and optimization of resources, the chapter demonstrates how truly transformative AI can be in forging next-generation smart, efficient, and patient-centric healthcare solutions.

Keywords: NLP, healthcare, machine learning

11.1 Introduction

Artificial intelligence (AI) is now a part of everyday life. Modern healthcare AI started its development in the past decades by using rule-based systems such as MYCIN, an early expert system developed in the 1970s to diagnose bacterial infections, before evolving to the current neural network technologies. The rapid growth of AI in healthcare is due to the high availability of medical data in the form of medical history. AI is based on technologies such as deep learning and machine learning, which require large amounts of data, and help in managing and analyzing information in hospitals and clinics. The development of deep learning and neural networks also played a major role in the impact of AI in healthcare. AI has grown as a healthcare technology through multiple decades to cover medical diagnostics and prognostics, as well as individualized therapeutic approaches and personalized treatment plans tailored to a patient's specific condition. The processing capabilities of natural language processing (NLP) within chatbots help patients complete symptom assessments, leading to faster medical care accessibility and shorter wait periods. Artificial intelligence facilitates

 | https://doi.org/10.1515/9783112226223-011

both diagnostic tasks and organizes administrative operations, as well as automates insurance claims and enhances hospital operational efficiency. With advances like zero-shot learning, an AI method that enables a model to make predictions on new, unseen tasks without explicit training, artificial intelligence could be a potential game changer in more effective and efficient healthcare in the future.

11.2 Evolution of AI in Healthcare

The term "artificial intelligence" was first described in the 1950s by John McCarthy at a conference at Dartmouth College. However, its application in medicine began in the 1940s with the development of neural networks, marking the beginning of AI-powered medical advancements. In the 1950s, Alan Turing invented the Turing Test, giving a benchmark against which machine intelligence would be measured. By the 1960s, AI had gone on to work on creating the first chatbot in the world, an early example of NLP that would go on to find applications in AI-powered medical assistants.

In the 1980s, AI healthcare, enhanced with expert systems, was applied to assist with medical decision-making, for example, MYCIN, a computer-assisted consultation system to assist physicians in diagnosing and selecting therapy for bacterial infections [1]. The 1990s saw the advent of computer vision being an integral part of AI-driven medical imaging that assisted in diagnostic processes like radiology and pathology. The 2000s were milestone years with Google DeepMind Health and IBM Watson using AI for medical research, data analysis, and disease prediction. The cognitive power of IBM Watson allowed it to process data for patients, both medical history and laboratory test results, to give insights and suggest personalized treatment plans [2].

In the 2020s, AI played an important role in healthcare, particularly in the development of the COVID-19 vaccine, whereby machine learning models accelerated drug discovery and pandemic prevention. The present uses of AI in healthcare are managing medical information, early disease diagnosis, diagnostics, medical assistance, and decision-making. The development of AI in healthcare is illustrated in Figure 11.1.

Litjens et al. [3] conducted a thorough review of deep learning in medical imaging and how they can revolutionize radiology and pathology. Esteva et al. [4] further demonstrated the capability of deep learning algorithms to offer dermatologist-level accuracy in the classification of skin cancer, documenting the revolutionary nature of AI in the diagnosis of healthcare.

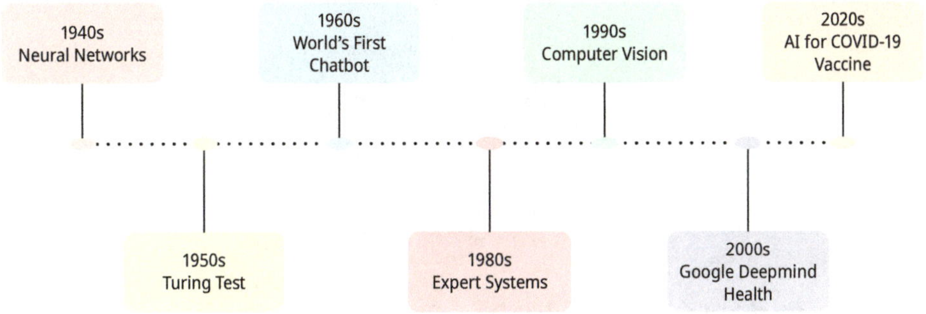

Figure 11.1: Timeline of the evolution of AI.

11.3 Importance of Intelligent Systems in Modern Medicine

AI is helping us by providing various applications in medicine and healthcare that help with surgery, medical imaging, diagnosis, clinical decision-making, data collection, administrative tasks, and getting real-time data. AI is benefiting healthcare organizations by implementing cognitive technology to unwind huge amount of medical records and helping in providing any power diagnosis. For example, Nuance is a service provider that uses AI and machine learning to predict customer intent and take the right action to proactively meet their needs. With the help of neural networks and deep learning models, AI is revolutionizing the image diagnosis field in medicine. One major application of AI in medical diagnosis is MRI scans. AI has taken over the complex analysis of MRI scans and made it easier and less time consuming. One of the examples of this application is VoxelMorph, which takes a few seconds to perform MRI scans [5]. AI has played a very important role in the early detection of medical conditions such as heart attacks. There are many AI-based wearable health trackers that are being developed to monitor the health of a person and display warnings. Prashamsh Takkalapally et al. conducted a study combining AI and wearable sensor technology to monitor patient health parameters in real-time and identify anomalies promptly [6]. Use of surgical robots can minimize human errors and increase efficiency of surgeons. One such surgical robot is Da Vinci that allows surgeons to implement with better flexibility and control than any other conventional approach. Global healthcare companies use chatbots such as Sensely to provide users with medical help. Sensely uses the bot to allow patients to talk about symptoms and receive appropriate self-treatment advice, and to communicate using Chat mode and Voice mode [7]. The applications of AI in healthcare are summarized in Figure 11.2.

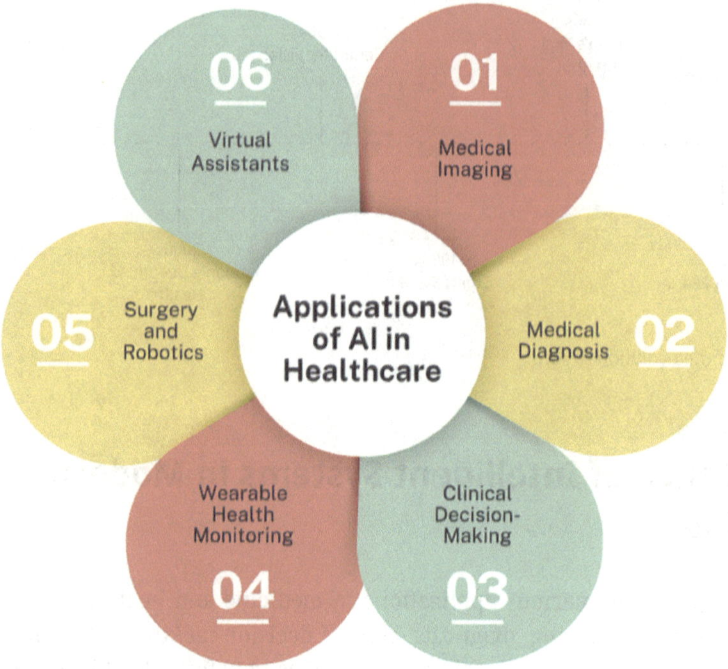

Figure 11.2: Applications of AI in healthcare.

11.4 Fundamentals of AI in Healthcare

AI is improving efficiency, accuracy, and decision-making in medical applications. The fundamental technologies of AI in healthcare include machine learning, deep learning, and NLP, as illustrated in Figure 11.3 These technologies help AI to analyze medical data, enhance medical imaging, and support healthcare administration. With the ability to process and analyze vast amounts of complex healthcare data, AI supports doctors, clinicians, and researchers in delivering faster and more precise diagnoses, personalized treatments, and better patient care. These AI technologies also enhance healthcare administration by automating routine tasks such as patient record management, appointment scheduling, and billing processes. This reduces the workload on medical staff and ensures greater accuracy and efficiency in hospital operations. Overall, the integration of these AI technologies is playing a crucial role in building a smarter, more responsive, and patient-centered healthcare system.

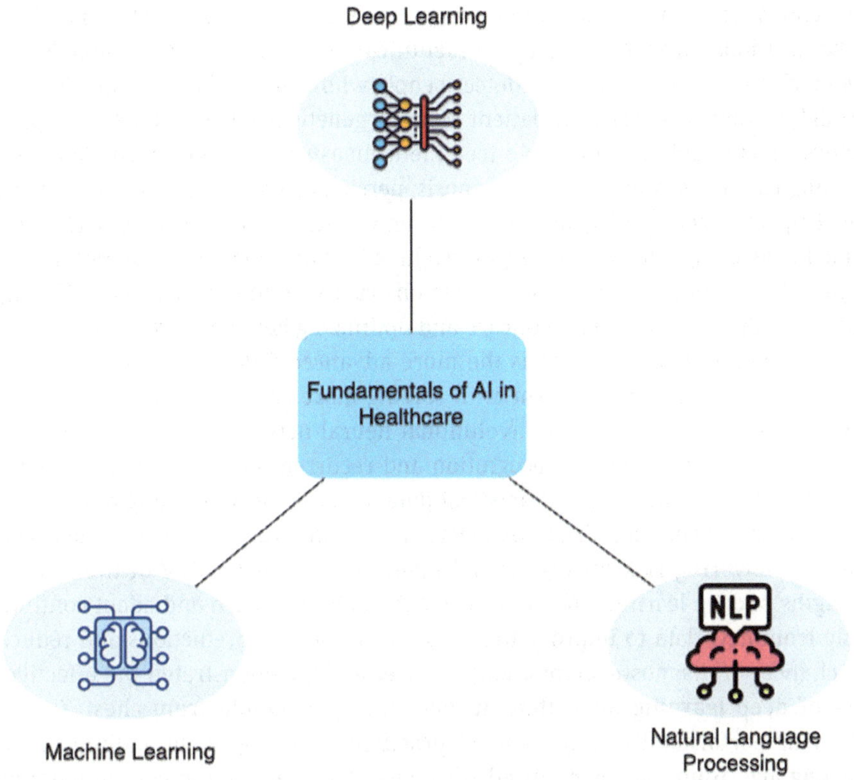

Figure 11.3: Fundamentals of AI in healthcare.

11.4.1 Overview of Machine Learning, Deep Learning, and NLP in Healthcare

The adoption of AI-based technologies in healthcare is transforming the manner in which medical data is analyzed and utilized. Machine learning, deep learning, and NLP are key contributors to the enhancement of diagnostic accuracy, forecasting of disease progression, and in automating administrative processes. By empowering these technologies, healthcare systems can deliver more effective, individualized, and evidence-based patient care.

- **Machine Learning:** Machine learning is the core of many futuristic technology advancements in our world. Machine learning is the subfield of AI that focuses on the design of the system that can make predictions based on the data. The healthcare sector utilizes machine learning to enhance medical diagnosis, while adapting individual patient therapies, and finds medicine at faster rates and achieves fewer mistakes by doctors. This technology supports disease outbreak predictions while managing patient workflows for hospitals and helps clinicians make decisions through data pattern identification. Healthcare professionals utilize machine learning models to detect people who will develop chronic diseases through combined analysis of patient records, genetic profiles, and demographic factors, thus enabling preventive treatment measures to begin early. Machine learning can be used to improve diagnosis, personalize the treatment of patients, speed up the process of drug discovery, and reduce human errors. Israa Mohamed et al. demonstrated the role of machine learning and data analytics in predicting the readmission rates of chronic obstructive pulmonary disease (COPD) patients, enhancing the quality of care and optimizing healthcare resources [8].

- **Deep Learning:** Deep learning is the more advanced field of machine learning that uses artificial neural networks in solving more complex problems that require high-dimensional data. Convolutional neural networks are used in radiology and pathology for image recognition and recurrent neural networks are applied in patient monitoring and medical data analysis. Deep learning models are also applied in genomics, drug discovery, and robotic surgeries, increasing precision and lowering risk during crucial medical procedures. One of the unique strengths of deep learning in healthcare is the ability to learn and adapt continuously from new data to improve the accuracy of medical predictions and reduce the chances of diagnostic errors. Rajpurkar et al. [9] demonstrated the effectiveness of deep learning algorithms in detecting pneumonia from chest X-rays, achieving performance comparable to practicing radiologists. In addition, deep learning algorithms are being fused with wearable health equipment to monitor vital signs, predict emergencies, and raise alerts to medical teams in real-time, enabling timely medical attention and boosting patient safety.
- **NLP:** A machine learning technique called NLP helps computers understand human language. It helps in taking out important information from clinical notes and research articles. In healthcare, it also provides medical guidance and administrative support. NLP is most essential in converting unstructured medical data to structured formats, enabling fast information retrieval and assisting with clinical documentation, prescription analysis, and patient feedback assessment. It helps summarize lengthy medical records to make sure that healthcare providers can always access the most important information about a patient in the least amount of time possible, hence saving on consultation time and improving care efficiency. Remya Mavila et al. [10] introduced iCare, an AI-powered virtual assistant designed to assist individuals with mental illnesses, through conversational interfaces. It supports multilingual patient communication, thus making it possible for caregivers to overcome language barriers and provide competent care to various populations. NLP is also revolutionizing the field of medical research by automatically scanning and analyzing thousands of scientific publications and helping researchers keep abreast of new medical discoveries and the results of clinical trials.

11.4.2 Key AI techniques for Medical Applications

AI has revolutionized many medical applications, making them more efficient and accurate. The traditional methods of healthcare often revolve around manual analysis and expert opinions, whereas AI-driven methods use the benefits of machine learning, deep learning, and NLP to strengthen their diagnostic capabilities, monitoring of patients, and in hospital management. Table 11.1 is a comparative overview of traditional healthcare methods and those based on AI technologies, giving information on which technology prevails in which domain, and the other.

- **Computer Vision:** Computer vision empowers machines to recognize and process images. Computer vision algorithms use pattern recognition. It helps in medical imaging for disease detection, for example, tumor detection using MRIs, X-rays, and CT scans. This technique also helps to analyze microscopic images of tissues, which helps in diagnosing diseases, such as cancer, more accurately. These tools benefit medical professionals by performing specific tissue division within CT images and microscopic slide evaluation, to assist in both early identification of cancer and treatment design focused on individual patient needs. According to Mahadevkar et al. [11] emerging approaches such as zero-shot learning enable rare disease detection, while self-supervised learning optimizes model performance when facing limited labeled data, indicating the substantial transformative capacity of AI-driven computer vision in medical applications.
- **Predictive Analytics:** Early predictions can help in preventing the progression of diseases. It allows healthcare businesses to take quick actions in time and avoid the effects of epidemics like coronavirus. Suicide prevention is one of the best use cases of predictive analytics in healthcare. Patient's behavior patterns and electronic health records (EHRs) can be integrated with predictive analytics tools for suicide risk detection. Mohammed et al. [12] investigated how AI-driven predictive analytics are changing the game in the healthcare sector, with the power to predict disease outcomes from clinical data via machine learning and various datasets. The authors emphasize that AI is able to manage unstructured data and is able to couple real-time wearable metrics into the model for proactive care approaches.
- **Robotics and Automation:** Automation and intelligent machines are transforming medical procedures, making them more precise, efficient, and safer for patients. One of the important applications of robotics in healthcare is in surgery. Many robotic surgical systems have been introduced to enable surgeons perform minimally invasive surgeries with enhanced precision. It also plays a crucial role in rehabilitation therapies for patients recovering from injuries or surgeries. Okamura et al. [13] explained robotic surgical systems, such as telerobotic platforms, enabling surgeons to perform minimally invasive surgeries with enhanced accuracy, leading to lesser recovery period and risk. On the other hand, these systems are important in rehab therapies where robotic exoskeletons or adaptive tools

are used to enable patients to recover from injuries or surgeries through tailored high-intensity training.

Healthcare Administration: With the help of AI tools, hospitals can improve efficiency, reduce costs, and provide better patient care. Virtual assistants can help with administrative tasks like appointment scheduling or updating medical records. By automating insurance claim processes, we can speed up the reimbursement process for patients. Gebreab et al. [14] proposed a large language model (LLM)-based multi-agent framework for automating administrative duties such as medical record retrieval and other tasks associated with billing and appointment scheduling within the field of healthcare. The system employs coordinated LLM agents that parse instructions, decompose complex tasks, and execute workflows across databases and web-based EMR platforms. Tested on real-world scenarios, the solution has been shown to be effective in streamlining administrative processes and reducing the workload of clinicians, while enhancing efficiency and security in data processing.

Table 11.1: Comparison of traditional and AI-based technologies.

References	Category	Traditional methods	AI-based technologies
Okamura et al. [13]	**Surgical assistance**	Conventional robotic-assisted surgery with human control	AI-enhanced robotic surgeries, with improved precision and automated assistance
Geert Litjens et al. [3]	**Medical diagnosis**	Manual analysis of patient symptoms; radiology interpretation by human experts	AI-powered imaging analysis; deep learning models for disease detection (e.g., pneumonia and cancer)
Hwang et al. [7]	**Patient consultation (chatbots)**	Patients call or visit doctors for health advice	NLP-powered chatbots assess symptoms, provide preliminary diagnoses, and guide patients
Mahadevkar et al. [11]	**Medical imaging**	Radiologists manually analyze X-rays, CT scans, and MRIs	AI-driven image recognition (e.g., convolutional neural network s for tumor detection, automated histopathology analysis)
Himi et al. [23]	**Wearable health monitoring**	Patients self-monitor or visit clinics for regular checkups	AI-integrated smartwatches track vitals, predict anomalies, and send alerts in real time
Javed et al. [15]	**Cardiovascular disease prediction**	Risk assessment based on patient history and clinical examination	Machine learning models predict cardiovascular disease from ECG, EHR, and wearable device data
Gebreab et al. [14]	**Hospital administration**	Manual record keeping, appointment scheduling, and insurance processing	AI-driven automation (RPA, blockchain-based EHR management, virtual assistants)

Table 11.1 (continued)

References	Category	Traditional methods	AI-based technologies
Mohammed et al. [12]	**Predictive analytics in healthcare**	Retrospective analysis of past cases and epidemiological studies	AI models predict disease outbreaks, patient deterioration, and personalized treatment plans
Alabri et al. [17]	**Cancer diagnosis**	Pathologists manually examine tissue samples	AI-assisted digital pathology with deep learning models to detect cancer at early stages

11.5 Case Studies in AI-Driven Healthcare Applications

AI is revolutionizing the way diseases are diagnosed, tracked, and treated. For example, AI is enabling physicians to detect heart issues at an early stage, aiding in cancer diagnosis with greater accuracy, and even driving chatbots that can converse with patients and respond to their queries in natural language. Meanwhile, wearable technologies such as fitness trackers can now track our health in real-time, warning us of any problems before they increase. Aside from patient care, AI is also utilized to streamline hospital and clinic administrative functions. By automating routine tasks such as scheduling, record management, and insurance claims processing, AI reduces costs, saves time, and takes some of the burden off healthcare professionals. These instances illustrate how versatile and capable AI can be in healthcare. This shift is helping create a healthcare system that works better for everyone – patients and doctors alike.

11.5.1 AI for Early Detection of Cardiovascular Diseases

Early diagnosis is important due to the significant implications for management, especially with the evolution of effective medical therapy. Cardiovascular diseases are one of the leading causes of death around the world. Early identification of people at risk due to life-threatening events such as heart attacks, strokes, or heart failure would help cut mortality. Health workers can create ways to help prevent it and plan for personalized treatment of their patients, which would enhance monitoring of patients' health, thus saving millions of lives. If heart diseases are diagnosed in an early stage, health workers can create ways to help prevent it and plan for personalized treatment of their patients that would enhance monitoring of patients' health, thus saving millions of lives. Methods based on AI like machine learning and deep learning can analyze large and complex datasets, including medical imaging, electrocardiogram (ECG) signals, echocardiograms, blood pressure trends, and EHRs, to identify patterns and risk factors. Machine learning models such as logistic regression can predict the likelihood of cardiovascular diseases, based on various patient parameters. The probability of CVD occurrence can be estimated using the following equation:

$$P(Y=1|X) = \frac{1}{1+e^{-\left(\beta_0+\beta_1X_1+\beta_2X_2+\ldots+\beta_nX_n\right)}} \tag{11.1}$$

This helps healthcare providers identify high-risk patients and implement targeted prevention strategies.

Haseeb Javed et al. emphasized the importance of AI in predictive cardiology and supported it with studies conducted by Abubaker and Babayigit [13, 14] in their ECG-based CVD detection using machine learning and deep learning methods. AI-based tools are also incorporated into wearable health devices and remote monitoring systems for continuous cardiac surveillance, alerting patients well in advance of a possible serious event. These improvements not only enhance diagnostic accuracy but also help make cardiovascular care more proactive, personalized, and available, making people healthier, and safer in the end.

11.5.2 AI-Assisted Detection of Cancer in Histopathology

AI is playing an important role in the field of digital pathology. It is driving real diagnosis improvements in cancer. Given the increasing burden of cancer worldwide, there is a pressing need for diagnostic tools that are faster, more accurate, and cost-effective. AI is coming in to fill this void by assisting the pathologists to analyze difficult tissue samples with minimum errors. For example, in prostate cancer, there are no symptoms, unlike other cancer types, making it important to identify the disease at the initial stage and reduce death rates. Early diagnosis not only enhances the chances of survival but will also prepare patients for light and less-aggressive treatments, thus maintaining their quality of life. There are several AI models that improve the accuracy of the diagnosis. These models help deal with huge amounts of histopathological images and extract minute details of the structure of tissues that are often overlooked by the human eye, ensuring that no subtle hints of malignancy are missed. Indeed, Yahya Khalid Alabri et al. [15] examined an AI model that aids in prostate cancer diagnosis and identifies cancer in its earliest form for healthcare professionals. The model integrates machine learning and deep learning techniques for the analysis of medical imaging and patient data for future optimal predictions. This combination not only enhances diagnostic accuracy but also helps predict disease progression and patient outcomes, which is crucial for long-term treatment planning. Through the process of automated feature extraction and classification, cancers could be identified with greater accuracy and efficiency. Moreover, AI algorithms are continuously learning from new datasets and adapt their performance in time, allowing them to act as trustworthy partners in clinical practice. Such systems, powered by AI, have greatly reduced the timescales for diagnostics and chances of human error, enabling speedier decisions and quicker treatment commencement. As digital pathology evolves more and more, integrated, yet key, technologies for AI are bound to become essential components in the daily evaluation of pathology, which can assure growth in the standards needed for scalable and superior evaluations for the rapidly growing demands of global healthcare.

11.5.3 NLP-Based Chatbots for Patient Consultation

Chatbot is one of the biggest applications of NLP. Due to the long waiting times and limited healthcare resources, patients often struggle with accessing medical advice. Such a lack of timely access to healthcare professionals may lead to delayed diagnosis and treatment, with the attendant worsening of patient outcomes. NLP-based chatbots assess symptoms using pre-trained databases to recommend medical advice to patients. Chatbots utilize natural conversational language to make it easier for patients to describe their cases so that the system may analyze their inputs and provide personalized responses. Such chatbots are available 24/7, so a patient can get immediate help, no matter the hour or location, thereby especially benefitting rural areas with limited medical facilities. Such an NLP-based chatbot named MedBot for Diabetes Prediction is studied in a research conducted by Sreekar et al. [18], with the aim of predicting diabetes and offering a portable and accessible solution to patients living with diabetes. MedBot is capable of evaluating specific responses given by the patient regarding common symptoms, family history, and lifestyle factors, and estimating the patient's risk of developing diabetes and then recommending consultation with a health professional. Besides symptom assessment, NLP-powered chatbots include medication reminders, patient progress monitoring, mental health programs, and answering questions on common health concerns, thus reducing the burden on healthcare professionals. With the latest upgrades in AI, these chatbots keep on improving their abilities to interpret complex medical terminology and detect emotions expressed by patients, upgrading the quality of virtual consultations and thus enhancing the inclusiveness and efficiency of healthcare.

11.5.4 Automating Administrative Processes with AI

The process of admitting and discharging patients is a paper-based, intensive process. To admit or discharge a patient, a healthcare provider has to go through lengthy processes like verifying patient insurance availability, checking doctor availability, manage EHRs, and follow up with patients for test procedures and reports after discharge. Very high workloads for healthcare staff, leading to larger operational costs and delays, could also negatively impact patient care and satisfaction. By using AI-driven systems, healthcare organizations can streamline repetitive workflow, diminish administrative burden, and enable medical staff to concentrate more on patient-based tasks.

This entire process can be visualized in Figure 11.4, illustrating how AI tools like RPA (robotic process automation), blockchain, and wearables automate each step, from patient registration to post-discharge monitoring. RPA automates patient data entry and insurance claim processing, minimizing errors. Through RPA, tasks once taking hours now take, literally, minutes, with less paperwork and live updates throughout the systems. Alaa Haddad et al. [19] evaluate the use of blockchain technology and AI in EHR management. It guarantees greater security and transparency in the storage and access of sensitive patient data, with just the authorized personnel being able to access it. AI further analyzes this data for better decision-making. Chinmay Mistry et al. introduced MedBlock, an AI and blockchain-based system for improving the management of EHRs during the pandemic [27, 28]. MedBlock had showcased how combining AI with blockchain could handle large volumes of patient records, while maintaining privacy, preventing data tampering, and allowing seamless sharing of medical information among healthcare providers. To further ease the administrative load and speed up or improve communication, AI chatbots and virtual feature referees are also used to schedule appointments, send reminders, or respond to patients' inquiries. Hence, as the health sector moves further toward digitization, AI will be crucial in automating administration work to speed up, save, and secure patient services, thus creating a more efficient healthcare ecosystem.

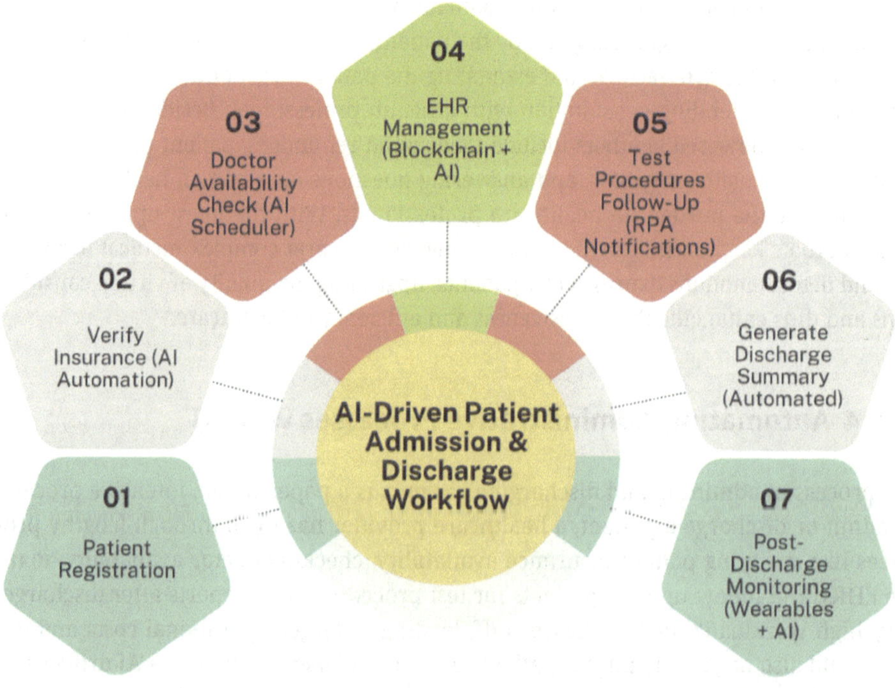

Figure 11.4: AI-driven patient admission and discharge workflow.

11.5.5 Wearable AI for Real-Time Health Monitoring

Wearable devices such as SmartWatches, fitness trackers and biometric sensors have become essential tools for monitoring real-time health metrics. These gadgets, combined with AI, are revolutionizing healthcare by providing continuous tracking of vital signs like heart rate, sleep patterns, and physical activity. This integration of AI in wearable technology is shaping the future of connected health, improving early disease detection and enhancing personalized wellness programs. Wearable devices were initially made for fitness enthusiasts. However, the new developments in AI and sensor technology have turned such devices into powerful health monitoring systems. Himi et al. presents a smartwatch-based application framework named MedAi that uses ML algorithms to predict diseases, including diabetes, hypertension, and cardiovascular disorders [21]. The authors focus on physiological data gathered from the smartwatch, including heart rate, blood pressure, and activity, to apply machine learning techniques. One of the key parameters monitored is heart rate variability (HRV), which provides insights into cardiac health and stress levels. The HRV is computed using the following equation:

$$\mathrm{HRVF} = 1 - \frac{C_{\mathrm{max1}} + C_{\mathrm{max2}}}{N_{<50\mathrm{ms}}} \tag{11.2}$$

where

- C_{max1} is the first maximum consecutive count of adjacent R–R intervals
- C_{max2} is the second maximum consecutive count of adjacent R–R intervals
- $N_{((<50\ \mathrm{ms}))}$ is the total number of R–R interval differences less than 50 ms

MedAi crosses the bridge from wearable technology to proactive healthcare by enabling real-time alerts and personalized health recommendations. Similarly, Skibinska et al. have given a contemporary insight into the potential of smartwatches and ML techniques that can provide real-time clues of COVID-19 patients [22, 28, 29]. They further highlighted the use of wearable technology as a cost-efficient and relatively noninvasive avenue for presenting economic reasoning, in favor of allowing continuous monitoring of health parameters throughout the pandemic. With the help of real-time data acquisition and analysis, their framework supports many interventions that could prevent overload in healthcare systems. The following case studies illustrate AI's transformative role in healthcare. Table 11.2 summarizes their objectives, techniques, and outcomes.

Table 11.2: Summary of key case studies in AI-driven healthcare.

References	Case study	Objective	AI technique used	Outcome/Impact
Abubaker et al. [16]	Early detection of cardiovascular diseases	Detect heart diseases using ECG and EHR data	Machine learning and deep learning	Improved accuracy in CVD detection
Haddad et al. [21]	Automating administrative processes	Streamline patient data and insurance claims	Robotic process automation (RPA) and blockchain	Reduced errors and optimized workflows
Himi et al. [23]	Wearable AI for health monitoring	Monitor real-time health metrics	Machine learning and sensor technology	Early disease detection and personalized insights
Alabri et al. [17]	AI-assisted cancer diagnosis	Identify cancer in histopathology slides	Deep learning	Comparable accuracy to expert pathologists
Sreekar et al. [20]	NLP-based chatbots for diabetes prediction	Provide medical advice and predict diabetes	Natural language processing (NLP)	Faster and reliable disease prediction

11.6 Challenges and Ethical Considerations

While AI promises great potential in transforming healthcare, it also triggers critical challenges and ethical concerns that will have to be heeded. One challenge centers on privacy and security of data. Since healthcare data can be very sensitive, in many instances, the implementation of an AI system will require an access path to the massive datasets of EHRs. The question then is how to use this information to achieve the best outcome while, at the same time, protecting patient confidentiality. Another challenge is algorithmic bias, which directly influences the equity of healthcare outcomes. If the training datasets are not equally diverse or representative, the AI model will perform less adequately for some people and, thus aggravate existing disparities.

Transparency and explainability are other challenges. Most AI systems, especially those based on deep learning, are considered to behave like "black boxes," leaving clinicians with great uncertainty as to how the decision was made. Andreas S. Panayides et al. highlighted this concern in an analysis of medical imaging informatics that standardized methods and interpretable AI tools need to be developed in order to inspire trust among healthcare providers [23].

Furthermore, ethical dilemmas prove to be troublesome in cases where AI systems influence patient care decisions. A comprehensive literature review conducted by Sudeep Pasricha [24], covering issues around transparency, bias, privacy, safety, liability, fairness, and patient autonomy, identified these considerations as important for steering the advancements of AI in healthcare onto an innovative yet safe ethical road.

11.7 Future of AI in Healthcare

AI has the ability to revolutionize the area of healthcare by enhancing patient care, improving work in operations, and facilitating preventive solutions that ultimately enhance health outcomes. Hospitals and physicians who use AI in their workflow are most likely to experience enhancements in patient outcomes, operational efficiency, and economic savings. For instance, AI systems diagnose diseases quicker and more accurately than human professionals, help in robot surgeries for greater accuracy, and tailor medicine according to individual genetic histories and health records. Such diagnosis could help cancer, cardiovascular diseases, and neurological diseases to be diagnosed early enough to save millions of lives. As Nasr et al. [25] say, the use of AI in health care is stimulating a move toward "smart health care," whereby continuous monitoring, predictive analytics, and data-based decision-making are facilitated by powerful technologies. Research highlights how AI-based diagnostic tools have begun to offer improved advantage, particularly in instances like deep learning algorithms for radiology imaging that excel in the detection of cancer at its early stages as well as in the prediction of heart disease threats, in comparison with traditional techniques. These results support research on the emerging AI in healthcare, which determines the significance of tackling ethics and cooperation among technologists, clinicians, and policymakers to ensure the transition to its full capability. Overcoming these challenges will enable AI to offer healthcare solutions that are more efficient, equitable, and patient-centric [26].

11.8 Results and their Significance

This chapter reveals the ways in which AI technologies like deep learning and NLP significantly enhance the speed and accuracy of diagnosis and enable earlier and more precise detection of disease. It also depicts how AI streamlines clinical processes and also administrative work through automation, reducing the workload on healthcare personnel and improving patient access to timely treatment. The integration of AI into personalized treatment planning and continuous health monitoring is a breakthrough in itself, allowing for more patient-centric and predictive care. The chapter notably highlights such on-going issues as data privacy, algorithmic bias, and the need for open and ethical AI use. In regard to the three challenges that were raised, regulatory relations require the combined effort and balancing of technologists, clinicians, and policymakers. This work provides a comprehensive and balanced overall view, considering emerging issues, while also acknowledging new developments by synthesizing the recent series of case studies and advances in technologies. Overall, this work provided evidence for how AI can be used to enhance the delivery of care, both efficiently and effectively, and ensuring equity in delivery of care, while calling for careful and considered engagement with the integration into existing care delivery structures.

11.9 Novel Contributions

This chapter contributes several novel advances to the field of AI-driven healthcare. It provides a cohesive summary of not only the different types of AI technologies like machine learning, deep learning, NLP, and robotic process automation and their applications in everything from diagnostics, predictive analytics, personalized treatment, and even administrative automation in clinical settings, but it also brings together meaningful examples that demonstrate real-life case studies and measurable outcomes to bridge the gap that exists between research and practice evidence. In providing an overview that balances technological advancement with operational issues such as data privacy, ethics, and integrating systems, this chapter builds an overall view that moves beyond established AI literature in healthcare. This provides a view that not only illustrates the capacity for AI to disrupt the sector, but it includes actionable pathways for responsible and effective utilization, distinguishing this from other available literature that focused on just the technical or standalone aspects of healthcare. In the end, the focus on practical application, integration across disciplines, and duty to ethics of this chapter sets it apart from previous publications that either focus on theoretical models, without application, or application, in isolation of anything else, to take the discussion further along the developmental trajectory of AI in healthcare toward a higher maturity and greater contact [32–34].

11.10 Future Work

Moving forward, future AI-enabled healthcare research must attend to a few main areas of priority, to have the most impact and tackle current challenges. First, we must work to improve the interpretable and transparent nature of AI algorithms, so clinicians and patients can develop more trust in them, allowing for more interpretability and acceptance of AI-enhanced decisions. Next, privacy-preserving methods such as federated learning and secure multi-party computation allow individuals and organizations train their AI together on sensitive health-related data, without revealing patient information or infringing patient confidentiality. Third, in order to eliminate algorithmic bias, we must create more inclusive and representative datasets that are realistic of populations around the globe to ensure that AI systems yield equitable healthcare choices. Fourth, emerging methods in AI, such as zero-shot and few-shot learning, promise to improve AI's ability to generalize from limited data, which is vital to dealing with rare diseases, or new medical conditions. Also, using AI applications for healthcare, including rehabilitation, mental health, and chronic disease management, instead of exclusively diagnostics, may enhance healthcare access and personalization. Finally, we should study the additive nature of AI in actions with other existing healthcare infrastructures like EHRs and telemedicine platforms so as to optimize workflows and patient engagement.

11.11 Conclusion

AI is revolutionizing the healthcare system by ensuring better patient care, efficient operations, and timely disease detection. As per the study, AI evolution takes us along with its important real-world applications in cancer diagnosis, predictive cardiology, NLP-enabled chatbots, and health monitoring through wearables. These developments assist with better accuracy of diagnosis, personalized treatment, and addressing shortages resulting from inefficiencies. However, the considerations of data privacy, bias, transparency, and ethics should be addressed with caution. Scientists, clinicians, and policymakers must act in unison to build fair and equitable AI. The future of AI in healthcare is full of promise. Through continuous monitoring, prediction analytics, and evidence-based decisions, AI supports intelligent and patient-centric care. This might be realized with sustained innovation and integration of AI in a responsible and ethical manner into healthcare systems.

References

[1] Shortliffe EH. A rule-based computer program for advising physicians regarding antimicrobial therapy selection. In Proceedings of the 1974 Annual ACM conference-Volume 2 1974 Jan 1 (pp. 739–739).

[2] Pinnaka M, Zaidi S, Viswanathan V. Integrating advanced IBM cloud-based AI/machine learning platform to develop predictive models for medical applications. In 2024 IEEE Integrated STEM Education Conference (ISEC) 2024 Mar 9 (pp. 1–2). IEEE.

[3] Litjens G, Kooi T, Bejnordi BE, Setio AA, Ciompi F, Ghafoorian M, Van Der Laak JA, Van Ginneken B, Sánchez CI. A survey on deep learning in medical image analysis. Medical Image Analysis. 2017 Dec 1;42:60–88.

[4] Esteva A, Robicquet A, Ramsundar B, Kuleshov V, DePristo M, Chou K, Cui C, Corrado G, Thrun S, Dean J. A guide to deep learning in healthcare. Nature Medicine. 2019 Jan;25(1):24–29.

[5] Balakrishnan G, Zhao A, Sabuncu MR, Guttag J, Dalca AV. Voxelmorph: A learning framework for deformable medical image registration. IEEE Transactions on Medical Imaging. 2019 Feb 4;38 (8):1788–1800.

[6] Takkalapally P, Kasula BY, Patil SC, Logeshwaran J. Examining the Benefits of AI in Wearable Sensor-based Healthcare Solutions. In 2024 15th International Conference on Computing Communication and Networking Technologies (ICCCNT) 2024 Jun 24 (pp. 1–6). IEEE.
[7] Hwang TH, Lee J, Hyun SM, Lee K. Implementation of interactive healthcare advisor model using chatbot and visualization. In 2020 International Conference on Information and Communication Technology Convergence (ICTC) 2020 Oct 21 (pp. 452–455). IEEE.
[8] Mohamed I, Fouda MM, Hosny KM. Machine learning algorithms for COPD patients' readmission prediction: A data analytics approach. IEEE Access. 2022 Feb 1;10:15279–15287.
[9] Rajpurkar P, Irvin J, Ball RL, Zhu K, Yang B, Mehta H, Duan T, Ding D, Bagul A, Langlotz CP, Patel BN. Deep learning for chest radiograph diagnosis: A retrospective comparison of the CheXNeXt algorithm to practicing radiologists. PLoS Medicine. 2018 Nov 20;15(11):e1002686.
[10] Mavila R, Jaiswal S, Naswa R, Yuwen W, Erdly B, Si D. iCare - An AI-Powered Virtual Assistant for Mental Health. In 2024 IEEE 12th International Conference on Healthcare Informatics (ICHI) 2024 Jun 3 (pp. 466–471). IEEE.
[11] Mahadevkar SV, Khemani B, Patil S, Kotecha K, Vora DR, Abraham A, Gabralla LA. A review on machine learning styles in computer vision – Techniques and future directions. IEEE Access. 2022 Sep 26;10:107293–107329.
[12] Mohammed MA, Mohammed VA, Ganesh A, Kiruthiga T. Investigating the Advantages of AI-driven Predictive Analysis in Healthcare. In 2024 IEEE 3rd World Conference on Applied Intelligence and Computing (AIC) 2024 Jul 27 (pp. 167–172). IEEE.
[13] Okamura AM, Matarić MJ, Christensen HI. Medical and health-care robotics. IEEE Robotics & Automation Magazine. 2010 Sep 9;17(3):26–37.
[14] Gebreab SA, Salah K, Jayaraman R, ur Rehman MH, Ellaham S. LLM-based framework for administrative task automation in healthcare. In 2024 12th International Symposium on Digital Forensics and Security (ISDFS) 2024 Apr 29 (pp. 1–7). IEEE.
[15] Javed H, Muqeet HA, Danesh A, Rehman AU, Javed T, Bermak A. Impact of AI and dynamic ensemble techniques in enhancing healthcare services: Opportunities and ethical challenges. IEEE Access. 2024 Aug 14.
[16] Abubaker MB, Babayiğit B. Detection of cardiovascular diseases in ECG images using machine learning and deep learning methods. IEEE Transactions on Artificial Intelligence. 2022 Mar 15;4(2):373–382.
[17] Alabri YK, Dhanalakshmi V, Bino D. AI-Assisted Model to Classify Prostate Cancer Diagnosis for Early Intervention. In 2024 15th International Conference on Computing Communication and Networking Technologies (ICCCNT) 2024 Jun 24 (pp. 1–5). IEEE.
[18] Kaur S, Singla J, Nkenyereye L, Jha S, Prashar D, Joshi GP, El-Sappagh S, Islam MS, Islam SR. Medical diagnostic systems using artificial intelligence (ai) algorithms: Principles and perspectives. IEEE Access. 2020 Dec 3;8:228049–228069.
[19] Campanella G, Hanna MG, Geneslaw L, Miraflor A, Werneck Krauss Silva V, Busam KJ, Brogi E, Reuter VE, Klimstra DS, Fuchs TJ. Clinical-grade computational pathology using weakly supervised deep learning on whole slide images. Nature Medicine. 2019 Aug;25(8):1301–1309.
[20] Sreekar K, Aswin V, Narayanan V, Thangavel SK, Somasundaram K, Shanmugam SK. MedBot – an NLP based ChatBot for Diabetes Prediction. In 2024, 15th International Conference on Computing Communication and Networking Technologies (ICCCNT) 2024 Jun 24 (pp. 1–11). IEEE.
[21] Haddad A, Habaebi MH, Islam MR, Hasbullah NF, Zabidi SA. Systematic review on AI-blockchain-based e-healthcare records management systems. IEEE Access. 2022 Aug 26;10:94583–94615.
[22] Mistry C, Thakker U, Gupta R, Obaidat MS, Tanwar S, Kumar N, Rodrigues JJ. MedBlock: An AI-enabled and blockchain-driven medical healthcare system for COVID-19. In ICC 2021-IEEE International Conference on Communications 2021 Jun 14 (pp. 1–6). IEEE.

[23] Himi ST, Monalisa NT, Whaiduzzaman MD, Barros A, Uddin MS. MedAi: A smartwatch-based application framework for the prediction of common diseases using machine learning. IEEE Access. 2023 Jan 11;11:12342–12359.

[24] Skibinska J, Burget R, Channa A, Popescu N, Koucheryavy Y. COVID-19 diagnosis at early stage based on smartwatches and machine learning techniques. IEEE Access. 2021 Aug 18;9:119476–119491.

[25] Panayides AS, Amini A, Filipovic ND, Sharma A, Tsaftaris SA, Young A, Foran D, Do N, Golemati S, Kurc T, Huang K. AI in medical imaging informatics: Current challenges and future directions. IEEE Journal of Biomedical and Health Informatics. 2020 May 29;24(7):1837–1857.

[26] Pasricha S. AI ethics in smart healthcare. IEEE Consumer Electronics Magazine. 2022 Nov 7;12 (4):12–20.

[27] Nasr M, Islam MM, Shehata S, Karray F, Quintana Y. Smart healthcare in the age of AI: Recent advances, challenges, and future prospects. IEEE Access. 2021 Oct 8;9:145248–145270.

[28] Kaur G, Goyal S, Kaur H. Brief review of various machine learning algorithms. Proceedings of the International Conference on Innovative Computing & Communication (ICICC). 2021 December.

[29] Bindal R, Sarangi PK, Kaur G, Dhiman G (2019). An approach for automatic recognition system for Indian vehicles numbers using k-nearest neighbours and decision tree classifier.

[30] Kukkar A, Mohana R, Nayyar A, Kim J, Kang BG, Chilamkurti N. A novel deep-learning-based bug severity classification technique using convolutional neural networks and random forest with boosting. Sensors. 2019;19(13):2964.

[31] Wang H, Hao L, Sharma A, Kukkar A. Automatic control of computer application data processing system based on artificial intelligence. Journal of Intelligent Systems. 2022;31(1):177–192.

[32] Reddy K, Reddy MA, Kaur V, Kaur G. Career guidance system using ensemble learning. Akhila and Kaur, Veerpal and Kaur, Gagandeep, Career Guidance System Using Ensemble Learning July 14, 2022 2022.

[33] Kaur V, Kumar P, Kukkar A, Kaur G, Kaur A. A Review on Facial Anti-spoofing Techniques. In International Conference on Mobile Radio Communications & 5G Networks 2023, August (pp. 323–335). Singapore: Springer Nature Singapore.

[34] Kukkar A, Kaur G. AEC: A novel adaptive ensemble classifier with LIME and SHAP-Based interpretability for fake news detection. Expert Systems with Applications. 2025;281:127751.

Sartaj Singh, Jasleen Kaur Paintal, Namarta Vij

12 Balancing Innovation and Integrity: Challenges in Deep Learning for Medical Imaging

Abstract: Deep learning (DL) has significantly advanced the role of medical imaging, improving diagnostic performance, automating disease detection, and streamlining workflows. However, while there are many advantages to deploying DL in clinical practice, implementing this technology will pose a number of serious ethical challenges. This chapter examines the often-adversarial relationship between innovation and ethical accountability in the deployment of artificial intelligence within medical imaging. Three central ethical challenges emerge: the challenge of obtaining an appropriately large-scale data-sharing agreement in light of strict data privacy challenges; the risk of algorithmic bias when training data is not representative; and the lack of interpretability of complex DL models in practice. As algorithms develop rapidly, they often outpace existing regulatory transparency, raising concerns surrounding validation, accountability, and patient safety. To explore the tension between technical feasibility and ethical accountability within the medical imaging space, this chapter provides a technical and ethical framework for evaluating DL technologies applied to radiology, oncology, ophthalmology, and pandemic responses [1]. Real-world case studies across radiology, oncology, and public health examine how ethical frameworks affect impreciseness in performance, logistical constraints, and fairness. The chapter evaluates interpretability tools (Grad-CAM and SHAP) and fairness metrics (demographic parity and equalized odds), among other ethical tools and frameworks. Ultimately, the chapter presents knowledge and best practices in developing more contextually equitable responses and responsibly leveraging DL in medical imaging settings for patient benefit. This also highlights the evolution of federated learning for secure and collaborative model training without sacrificing patient privacy. This speaks to the three-part technical-ethical-regulatory framework proposed above. With fairness, interpretability, and accountability made key factors within DL pipelines in medical imaging, this area would instill, rather than undermine, equity and trust in patient care through innovation.

Keywords: deep learning, medical imaging, interpretation, data bias, fairness metrics, federated learning, patient privacy, ethical AI, regulatory compliance

 | https://doi.org/10.1515/9783112226223-012

12.1 Introduction

In the last 10 years, deep learning (DL) has transformed medical imaging as we know it; it has altered the way clinicians review, evaluate, and act on diagnostic data. The traditional method of interpreting diagnostic imaging has often subjugated itself to the manual extraction of features and heuristics from an expert, whereas data-driven clinical algorithms can discern complex, nonlinear representations in large volumes of imaging data via data mining and machine learning techniques. The application of convolutional neural network (CNN), recurrent neural network (RNN), and generative adversarial network (GAN) models has prompted innovative research behind diagnostic accuracy, segmentation of imaging studies, and enhancement and manipulation of imaging across CT, MRI, PET, and X-ray imaging systems. However, with the implementation of DL into a clinical environment, inherent ethical, regulatory, and operational challenges can arise. The collision of innovation and integrity in medical artificial intelligence (AI) raises questions about trust, fairness, accountability, and patient privacy. AI models that are deployed are positioned to provide operational efficiencies and benefits on the medical end, and the "black-box" complexity of these algorithms, alongside the need for large-scale annotations to train these models and potential biases in algorithms for implementation, is an area of adaptation for equitable adoption. In this, healthcare organizations must continuously consider and navigate tight regulations of privacy laws such as Health Insurance Portability and Accountability Act (HIPAA) and General Data Protection Regulation (GDPR) into the challenges of sharing data between regions and validating AI models on limited patient data sets in an ethical manner.

This chapter intends to present a balanced assessment of DL's role as an impetus for innovation alongside an impetus for ethical contemplation within medical imaging. It methodically examines and discusses both technical and ethical features that impede the application of DL models in real-world clinical contexts. The chapter has four central aims:

- To review recent developments in DL architectures for disease detection, segmentation, and prognosis prediction.
- To delineate central challenges, including data privacy, bias, limited interpretability, and regulatory uncertainty, that complicate clinical translation.
- To assess ethical frameworks and governance models to support the ethical use of AI in healthcare.
- To provide case studies and comparative examples of successful and unsuccessful uses of DL in imaging workflows.

By blending these technical and ethical aspects of medical imaging into one chapter, this chapter seeks to highlight that the development of responsible AI is not just a regulatory obligation but a responsibility of medicine. The next sections outline the evolution of DL in imaging, explain federated learning and interpretability, and summarize the successful real-world implementations that demonstrate innovation and integrity can coexist in AI-enabled healthcare.

12.2 Overview of Deep Learning in Medical Imaging

Deep learning (DL) is a subsection of AI that allows computers to automatically learn sophisticated representations from data through multiple levels of abstraction. DL has become a vital tool in medical imaging, serving multiple tasks, from detecting diseases and segmentation to planning treatment and predicting prognosis. Unlike pre-DL image analysis techniques that depended on feature engineering by hand, with DL, an architecture learns discriminative patterns autonomously from raw image data designs. This is especially useful for handling large, high-dimensional imaging datasets.

One of the biggest advantages of machine learning in medical imaging is that it can often learn to be end-to-end, such as in a direct mapping from input image (i.e., MRI, CT, ultrasound, and PET scan) to output for diagnostic purposes. This end-to-end learning ability has spurred advances in radiology, oncology, ophthalmology, cardiology, and neurology. For example, automated systems can identify and indicate specific pulmonary nodules, segment tumors, indicate and grade retinal lesions, and classify cardiac anomalies, all with comparable ability to, or sometimes superior to, professional human experts. Common DL architectures in medical imaging include:

- CNNs: These are the foundational architectures for almost all imaging applications. CNNs may extract spatial hierarchies of features such as edges, textures, and shapes to identify anomalies like tumors, lesions, or fractures in computed tomography, magnetic resonance imaging, and X-ray studies.
- For instance, CNN-based models have reached over 94% accuracy in classification of lung nodules in CT scans, which increases the early diagnosis of cancer.
- RNN: These are especially useful at analyzing sequential or temporal imaging data, such as echocardiograms or MRIs that are based on time series, as they are able to capture changes in anatomical structures over time. This temporal analysis may provide prognostic prediction and modeling disease progression.
- GAN: These are used extensively for data augmentationimage reconstruction, and noise reduction, in this case, generating synthetic medical images similar to real datasets, in order to overcome class imbalance and improve generalization in machine learning model training. GANs can also be used to convert low-resolution scans into high-quality images and enhance visualizing in clinical diagnostic studies.
- The successful application or combination of these architectures is not solely dependent on computational capacity but also requires curated datasets and validation processes that yield data-driven and reproducible results. Figure 12.1 shows a schematic of how these architectures are applied across different imaging modalities.

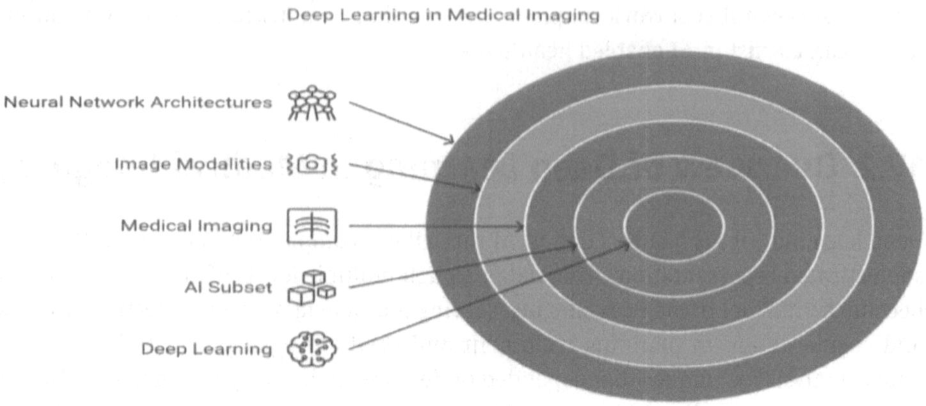

Figure 12.1: Application of these architectures in different imaging tasks.

The development of DL in medical imaging continues to develop rapidly with hybrid models that incorporate the best of multiple architectures (e.g., CNN-RNN hybrids for longitudinal imaging analysis or transformer-based models to capture global context). The next section presents a chronologic account of this evolution, demonstrating the evolution of early rule-based systems into contemporary intelligent data-driven diagnostic pipelines.

12.3 Evolution of Deep Learning in Medical Imaging

The use of computational intelligence in medical imaging has changed quite a bit over the last 60 years – from simple, rule-based expert systems to the current advanced DL architectures [2]. This development signifies the increasing synergy with computational capacity, the availability of data, and algorithmic developments.

12.3.1 Early Years: Rule-Based Systems to Machine Learning

In the 1960s and 1970s, medical image analysis depended upon rule-based computer-aided diagnosis systems with explicit logic to a priori identify abnormal patterns based on handcrafted features, including shape, intensity, and textures from radiography and computed tomography. Despite being groundbreaking, the systems did not generalize well with changes in modality or patient populations. Each rule for diagnosis was manually crafted and adjusted to fit a particular case, which limited scalability.

In the 1990s, traditional machine learning became accepted when support vector machines, *k*-nearest neighbors, and random forest algorithms automated a portion of the decision-making process; still, those algorithms relied on manual feature extraction, which continued to limit adaptations and accuracy of the model with complex medical data.

12.3.2 The Deep Learning Revolution

The change really began around the early 2010s, and it progressed through the confirmation of the value of deep CNNs following the ImageNet competition in 2012, as shown in Figure 12.2. AlexNet blew past all previous models for the task of object classification. Shortly afterward, the community began to seek out similar work in medical imaging using the AlexNet architecture. CNNs trained from pixel data circumvented the typical performance-gaps for features that captured the complexity of classification in prior models through the normal need for feature engineering.

Once CNNs were deployed, it became evident very shortly afterward that their performance for imaging modalities for tumor detection, organ segmentation, and disease classification daily outperformed the existing methods. For instance, U-Net (2015) uniquely exploited the encoder-decoder architecture to segment biomedical images while particularly accounting for spatial detail while at the same time, using context, which changed the field of biomedical image segmentation as work with U-Net led and continues to lead segmentation advances across multiple fields, including radiology, pathology, and ophthalmology.

12.3.3 Modern Advancements: Beyond CNNs

In recent years, the field has progressed not just in CNNs but also in vision transformers (ViTs), graph neural networks, and self-supervised learning approaches. These networks typically utilize large datasets and unsupervised learning objectives to achieve a significant level of DL and contextual understanding. For example, ViTs utilize a global attention mechanism to learn about relationships among areas within an image, regardless of distance, and typically produce gains in performance that can be translated to medical tasks such as tumor classification and organ segmentation.

Transfer learning has also started to become a significant development in the field. Transfer learning enables the use of pretrained AI models to fine-tune those models to specific medical tasks when there is a small medical dataset to be used. Further, in addition to pretrained settings, institutions can collaborate and contribute their own data to a shared model. By blending pretrained models and approaches in the medical field with methods developed in the nonmedical field, they can contribute to federate learning and privacy-preserving computation so multiple institutions can share improvements or help promote development for shared AI efforts while maintaining patient privacy and confidentiality.

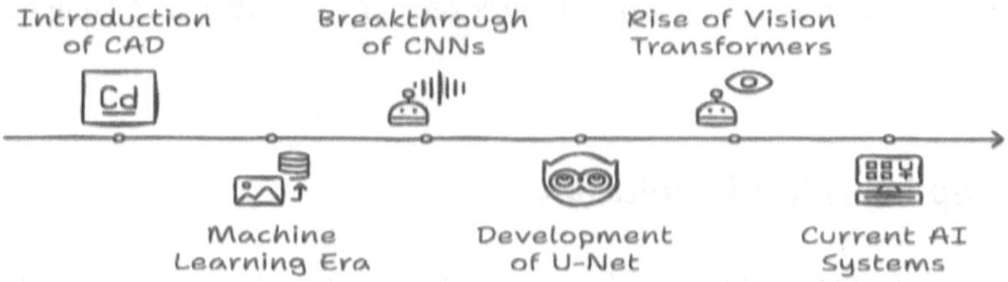

Figure 12.2: Timeline highlighting key milestones in the evolution of deep learning in medical imaging.

The swift development of DL within medical imaging represents not only a technological transformation but also a fundamental paradigm shift in the realm of clinical diagnostics. AI-based systems are evolving in their role from solely decision-support tools into integral partners with clinicians in delivering high-quality care as part of clinical workflows. In the next section, we will explore how these technologies are being employed in medical imaging across different areas of care as a demonstration of their real-world clinical implications.

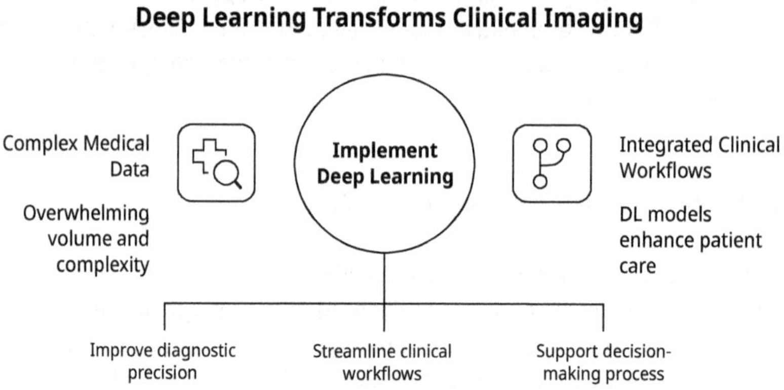

Figure 12.3: Broad clinical spectrum where DL has been successfully deployed.

12.4 Clinical Applications

DL has moved beyond being simply a research idea to being a pillar of clinical imaging applications. From radiology, oncology, ophthalmology, cardiology, and neurology, DL-based modalities are now employed in clinical workflows for diagnosis, prognostication, and treatment planning, as shown in Figure 12.3. These new systems help improve accuracy and efficiency, and process a developing array and volume of medical information clinically.

12.4.1 Diagnosis

DL systems facilitate the detection of diseases such as lung cancer and diabetic retinopathy (DR). For example, a CNN trained on more than 40,000 lung CT scans was able to report an AUC of 94.4%, which is better than human radiologists. In Figure 12.4, you will see a heatmap produced by this model, which identifies the suspicious nodules.

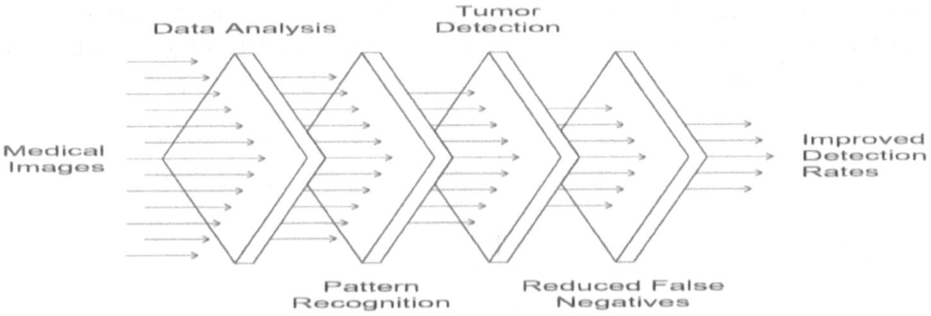

Figure 12.4: CNN-generated heatmap.

12.4.2 Prognosis

AI models have the ability to forecast disease progression [3]. For example, in the case of Alzheimer's disease, MRI-based DL models correctly predicted cognitive decline years prior with an accuracy of 92%. Figure 12.5 shows an example of disease trajectory mapping generated using AI-based models.

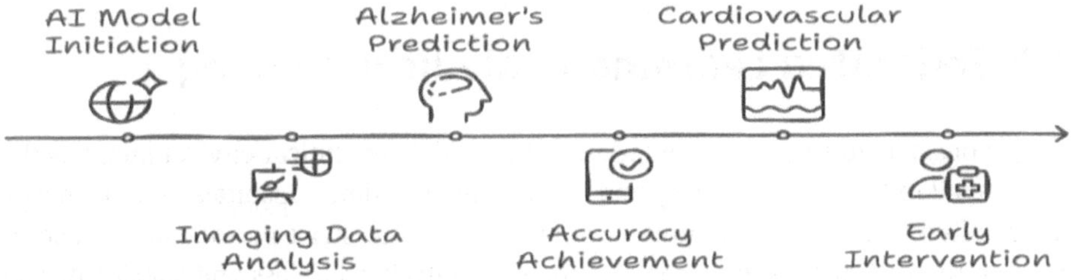

Figure 12.5: Example of an MRI-based deep learning model predicting Alzheimer's progression.

12.4.3 Treatment Planning

DL likewise promotes treatment improvement and surgical planning and provides assistance for physicians to deliver accurate, individualized interventions. In the field of radiation oncology, DL models have been developed to assist in the automated contouring of tumors and organs at risk, enhancing accuracy in the radiation dose delivered and potentially decreasing radiation to normal tissues. Similarly, for neurosurgery and orthopedics, three-dimensional CNNs are generating enhanced anatomical reconstructions to prompt surgeons to visualize complicated anatomy prior to surgery.

Additionally, there is an increasing use of reinforcement learning models to recommend the optimal radiation dosing schedule or optimal procedural steps during robot-assisted surgeries. The incorporation of DL-driven methods offers not only greater accuracy during procedures but also decreases patient exposure. Figure 12.6 shows an AI-driven re-construction model supporting neurosurgical planning.

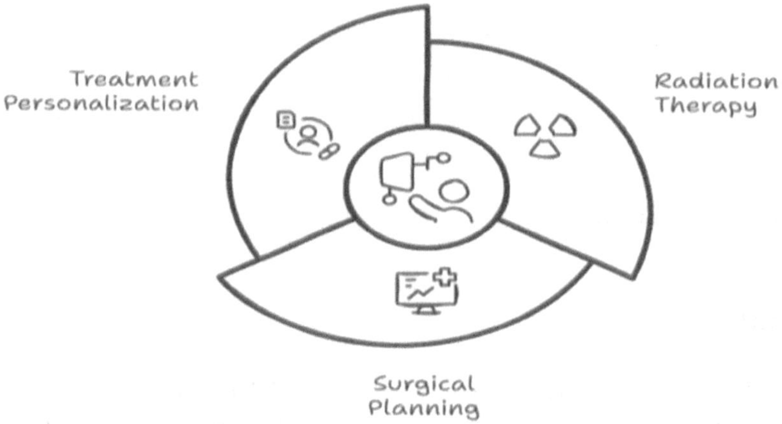

Figure 12.6: AI-driven three-dimensional reconstruction model supporting neurosurgical planning for improved accuracy and safety.

12.5 Federated Learning in Medical Imaging

A major obstacle to the implementation of DL models in clinical environments is the limited availability of high-quality and diverse medical data. Hospitals and diagnostic centers often collect vast amounts of imaging data, but strict privacy laws, such as HIPAA in the USA and GDPR in the EU, prevent centralized access and sharing of sensitive patient data. These laws create data silos that hinder the generalizability and fairness of AI models.

Federated learning (FL) has emerged as a game-changing approach to this problem. FL allows many medical institutions to collaboratively train AI models without the use of raw patient data. In FL, each medical institution will train a local model using its own data and simply share model parameters (gradients or weights) to a central server. A central aggregator collects the model parameters from each local model and combines them to produce a global model that captures all local knowledge without any patient data ever leaving the local site.

12.5.1 How Federated Learning Works

In a classic FL setting, we have the following process:

1. A central server initializes a global model.
2. The hospitals that are involved, train their local copies of the model independently, separately, and in their own privacy protected datasets.
3. The hospitals then upload the models' parameters (not data) back to the central server.
4. The central server then combines the updates via federated averaging and shares a new and improved global model back to the hospitals.
5. The centralized averaging process is repeated in a loop until convergence occurs.

This decentralized training technique minimizes risks to patient data privacy, reduces compliance issues, and provides the ability to collaboratively improve models across institutions from similar but geographically independent locations (Figure 12.7).

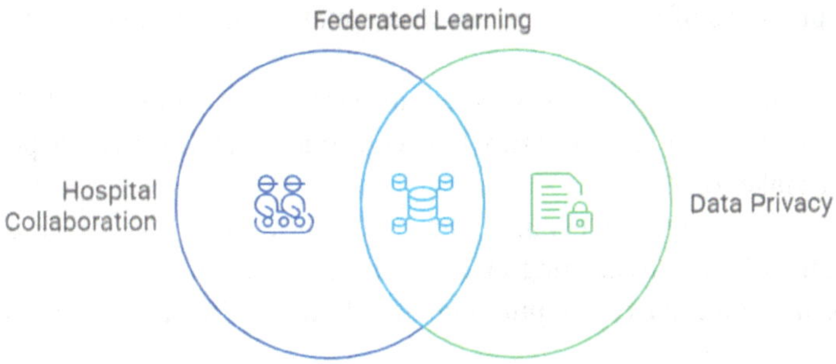

Figure 12.7: Federated learning workflow in medical imaging.

12.5.2 Advantages of Federated Learning in Healthcare

- Preservation of Data Privacy: Patient safety and patient data privacy concerns dictate that sensitive patient data never leaves a secure server or a hospital set-

ting if it is being used for training AI/ML algorithms that are subject to privacy compliance and institutional ethics.
- Broader Generalizability of the Model: The data used to develop the model parameters is coming from different, heterogeneous datasets across hospitals, which gives the model global robustness across different demographics or geographic locations.
- Collaboration Without Data Sharing: Institutions can partake in research and innovation without the legal and logistical complications associated with data sharing and data transfer.
- Rare Disease Improvement: Research can help pool datasets that are distributed and used in rare diseases, and can help create models that may otherwise not be feasible to create or impossible due to limited datasets.

The EXAM project (2021) is an excellent example of FL and the power of multi-institutional collaboration, wherein hospitals ($n = 20$) spanning five continents used FL to create a model for the detection of COVID-19 from chest X-rays. The global model improved generalizability by 16% over any single-site model, while data remained at the data site.

12.5.3 Challenges and Future Directions

Even though federated learning has a lot of usefulness, it also comes with its own set of limitations:
- Data Heterogeneity: Institutions may differ in data resolution, acquisition protocols, and labeling practices of medical images, which may inhibit model convergence.
- Communication Overhead: Regular assessments of parameters would result in considerable overhead on the bandwidth when working collaboratively, especially with multiple stakeholders.
- Security: Regularly updating models may uncover knowledge via reverse-engineering-style attacks, and preexisting methods of protocols, such as differential privacy and secure multi-party computation, are already being developed for FL security.

While it uses existing FL technologies, future FL medical imaging studies are being explored for personalized FL for the model refinement to local data distributions, and blockchain FL will provide transparency and proof of contribution.

12.6 Bias and Fairness in Deep Learning Models

DL has demonstrated great potential to automate medical image analysis, yet algorithmic bias remains a critical concern that can result in inequitable or unjust outcomes for different patient sub-populations. In a field like healthcare, where decisions are made about human beings, even minor algorithmic bias can be judged unethical and/or unclinically appropriate. Thus, improving fairness, transparency, and accountability in AI-based medical imaging systems is important for creating public trust and improving equitable access to healthcare.

12.6.1 Sources of Bias in Medical Imaging Datasets

Most biases in DL models stem from non-representative and biased training datasets. Medical imaging datasets are often biased toward demographics, geographical locations, or clinical conditions. A model trained primarily with data from Western hospitals may perform poorly when applied clinically to patients in Asian or African countries, for instance, due to different human anatomy, imaging protocols, or disease prevalence. Some of the common bias sources include:

- Sampling Bias: Certain groups (women, children, ethnic minorities, etc.) are often not adequately represented in datasets.
- Labeling Bias: Insufficient or inconsistent annotating by radiologists can become part of the training process.
- Acquisition Bias: Systematic bias can arise from differences in imaging hardware, resolution, or scanning parameter protocols from institution to institution.
- Socioeconomic Bias: Models can indirectly learn from biases in the data pattern that are associated with access to care, thereby reinforcing the existing health disparities.

While these biases can influence model performance, they can also exacerbate inequality; for example, misdiagnosis can occur at a much higher rate with underrepresented groups.

12.6.2 Fairness Metrics and Evaluation Approaches

Following are the fairness metrics, and their details are also presented in Table 12.1:

- To identify and reduce bias, researchers apply numerous fairness metrics that assess model performance across demographic subgroups. Many of the metrics that apply are those described by the following labels of bias:
- Demographic Parity: The model should be statistically independent of sensitive characteristics (e.g., race and gender) in its predictions.

- Equalized Odds: True and false positive rates should be similar across demographic groups.
- Predictive Parity: Predictive value (precision) should be fairly consistent between groups.
- Counterfactual Fairness: In a synthetic counterfactual scenario, changing a sensitive characteristic (e.g., gender) should not change the model's prediction.

Table 12.1: Key fairness metrics used in medical imaging AI.

Metric	Description	Goal
Demographic parity	Predictions independent of sensitive attributes	Equal representation
Equalized odds	Equal true/false positive rates across groups	Uniform model accuracy
Predictive parity	Consistent precision across demographics	Reliable predictions
Counterfactual fairness	Outcome invariant to sensitive attributes	Bias-free reasoning

12.6.3 Bias Mitigation Strategies

Several approaches have been introduced to enhance fairness in medical AI systems and mitigate bias built into the model development process (Figure 12.8):

- Preprocessing Approaches: Prior to model development, the dataset can be modified to ensure that under-represented groups are managed through oversampling, data augmentation, or data set weighting adjustments so that patients from different groups have equitable representation.
- In-processing Approaches: During model development, adversarial debiasing is an example of a technique of the model development process that embeds fairness rules in the model, so it is not reliant upon the sensitive attribute such as race or gender regarding prediction.
- Post-processing Approaches: After the model is derived, its thresholds for a patient's response can be modified to mitigate inequitable outcomes between groups.
- Transfer and Federated Learning: Model training, especially on data from numerous hospitals or regions, can help bridge bias, especially regarding a specific institution-based patient population, as the model is not reliant on only one patient population when developing its learning capabilities.

These approaches allow for a combination of many approaches to create equity, reliability, and generalizability across several healthcare settings.

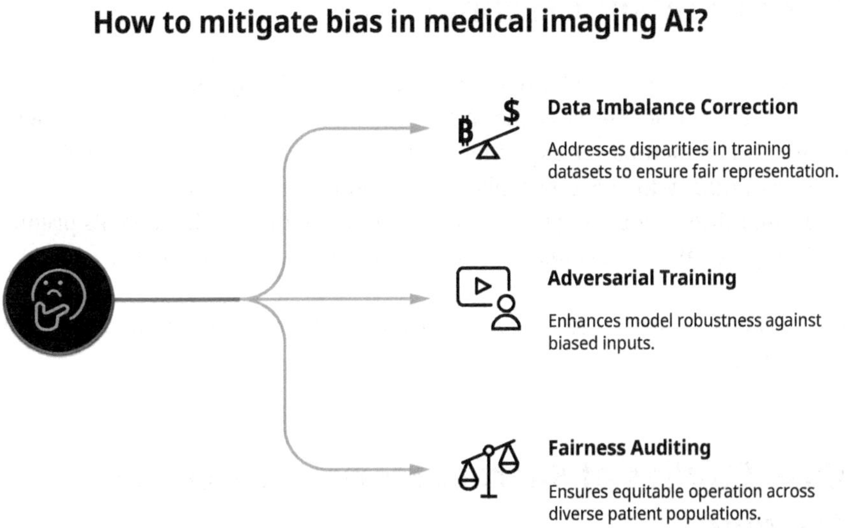

Figure 12.8: Illustration of bias sources and mitigation strategies in medical imaging AI.

12.6.4 Ethical and Regulatory Considerations

Regulatory agencies, such as the U.S. Food and Drug Administration (FDA) and the European Commission, are increasingly embedding fairness assessment into AI approval processes. Developers will have to provide evidence that their models provide equitable performance across groups by age, gender, and ethnicity prior to use in clinical settings. In addition, developing governance frameworks – like the Ethical AI in Health (EAIH) guidelines – are encouraging transparency through audits of bias, systems to enhance model interpretability, and inclusive data curation.

12.6.5 The Path Forward

In order to reach fairness in DL for medical imaging, both a technical approach and an ethical approach are necessary. Generating large-scale, demographically diverse datasets and agreeing on standardized fairness benchmarks are two critical steps. Working closely with clinicians, data scientists, and ethicists guarantees that AI systems reinforce equity and do not undermine it in the care of patients.

12.7 Interpretability Techniques

Explainable AI methods enhance trust and transparency in DL systems:
- Grad-CAM: It identifies parts of the input image that contributed to the model's prediction. It is a method for explainable AI that provides visual evidence for the model's reasoning, which is especially valuable in radiology [7, 8].
- SHAP: It quantifies the contributions of each input feature to the model's prediction, while providing both global and local interpretations of the model.

These methods allow clinicians to validate model decisions, as well as understand and identify sources of error.

12.8 Case Studies of Success Stories in Medical Imaging

12.8.1 Case Study 1: CT Imaging Augmented with AI for Diagnostics of Lung Cancer

Lung cancer is one of the leading causes of cancer-related morbidity and mortality worldwide, and interventions provide improved patient outcomes with early detection. Imaging evaluations through conventional radiology have subjective human interpretation factors in the diagnostic process to account for. CNNs are a specific type of DL model, have shown promise to objectively differentiate between cancerous nodules on CT imaging.

As an example, a multidisciplinary group of investigators from Stanford University studied a DL model using over 40,000 lung CT scans, improving the area under the curve (AUC) value of 94.4%, higher than the average radiologist's lung cancer diagnosis AUC value. Once studies were instrumented and developed, the targeted hospitals began integrating the AI application as background into practice as a tool within radiologist workflows. This implementation resulted in increased detection rates of lung cancer earlier in the disease progression, along with a decrease in false positives. This case study highlights the potential of audiologists in providing patient therapies using AI-based imaging, and positive outcomes these interventions can bring to clinical decision-making and diagnostic accuracy.

12.8.2 Case Study 2: Screening for Diabetic Retinopathy Using AI-Powered Fundus Imaging

Diabetic retinopathy (DR), a leading cause of vision loss, affects millions worldwide. Traditional screening of retinal images is slow and needs trained specialists [12, 13]. AI-based tools are now being used to speed up and automate detection. For example, DeepMind trained a DL model on over 100,000 retinal fundus images. It achieved 98.2% accuracy in identifying referable DR, outperforming general ophthalmologists. This AI system makes large-scale screening in clinical practice possible, reduces the screening workload for providers, and finds patients who need early treatment but might otherwise be missed [14]. This is one way AI is improving access to and efficiency of medical diagnostics.

12.8.3 Case Study 3: Detecting Breast Cancer Using AI-Assisted Mammography

The diagnosis and treatment of breast cancer are still highly relevant in public health, and the time of diagnosis directly affects the efficacy of these treatments. Mammography screening, while extremely relevant, can produce significant numbers of false positives and false negatives that ultimately lead to unnecessary biopsies and potentially missed detections. AI-assisted mammography systems that are powered by DL have been demonstrated to be effective in improving accuracy of early detection.

As part of an MIT study [15, 16], an AI model that was trained on more than 60,000 mammograms demonstrated 93% accuracy for early-stage detection of breast cancer, producing a better outcome than traditional methods. This AI model was assimilated into existing radiology workflows and used by radiologists as a second opinion. This assimilation of an AI model ultimately benefited the patient and increased diagnosing accuracy while decreasing radiologists' workload.

12.8.4 Case Study 4: Deep Learning-Based Automated Stroke Detection in MRI

Stroke is a major cause of death and disability, and the management and treatment of stroke depend on an accurate and timely diagnosis. While often MRI and CT scans are used to visualize the stroke, there can be challenges to timely diagnoses when used, especially in a time-sensitive inpatient care environment. A study conducted by researchers from Harvard Medical School developed a DL-based model that has further assessed the capacity for accuracy in identifying ischemic strokes in MRI scans. The accuracy level was close to 96% [17]. The DL-based model can help identify and treat these patients in minutes instead of an hour. The DL-based approach can quickly characterize ischemic stroke while quickly processing multiple imaging sequences (during the source imaging process) as well as real-time predictions. There are many supporter accounts, often anecdotal, in the hospitals from those who utilized the AI application about improved and faster treatment and response times in healthcare professionals working with acute stroke patients in emergency situations.

12.8.5 Case Study 5: Detecting COVID-19 with AI-Enhanced Chest X-Rays

The COVID-19 pandemic highlighted the critical need for rapid and precise diagnostic tools [18, 19]. Due to the extensive time and resource burden associated with RT-PCR testing for COVID-19, AI-based analysis of chest X-rays has augmented pneumonia diagnosis, specifically for COVID-19. Researchers at Johns Hopkins University developed an AI model which identified COVID-19-associated lung abnormalities by analyzing thousands of chest X-ray images. The model achieved 91% accuracy in distinguishing COVID-19 pneumonia from other pneumonia-related infections [20]. This AI-supported diagnostic tool was built and deployed in hospitals around the world as a means to triage patients faster and optimize resource allocation.

12.8.6 Comparative Summary of Case Studies

The previous case studies (Sections 12.8.1–12.8.5) illustrate how DL can make a difference in diverse fields of medical imaging, spanning oncology, ophthalmology, neurology, and infectious disease detection. Each of the case studies has its own specific advantages, along with practical and ethical challenges that must be resolved prior to large-scale implementation in clinical practice.

To illustrate these various elements summarized and organized, Table 12.2 parallels the five published, real-world implementations of DL that were discussed previ-

Table 12.2: Comparative summary of major deep learning applications in medical imaging.

Case study/ domain	Dataset size/ source	Reported accuracy/ AUC	Regulatory challenges	Ethical concerns	Key takeaways
Lung cancer detection (Stanford University)	>40,000 CT scans	94.4% AUC	FDA approval and model validation	False positives and overreliance on AI	Demonstrated DL superiority in nodule detection; improved early diagnosis

Table 12.2 (continued)

Case study/ domain	Dataset size/ source	Reported accuracy/ AUC	Regulatory challenges	Ethical concerns	Key takeaways
Diabetic retinopathy screening (DeepMind)	>100,000 retinal fundus images	98.2% accuracy	Clinical validation required	Data consent and equitable access	Enabled large-scale DR screening, and reduced ophthalmologist workload
Breast cancer detection (MIT study)	>60,000 mammograms	93% accuracy	Data generalizability	Explainability and AI dependency	Early cancer detection improved, and validated the value of AI as a diagnostic aid
Stroke detection (Harvard Medical School)	Multi-hospital MRI dataset	96% accuracy	Emergency use validation	Real-time liability and timeliness	Accelerated stroke diagnosis from hours to minutes, and enhanced outcomes
COVID-19 detection (Johns Hopkins University)	>10,000 chest X-rays	91% accuracy	Rapid deployment compliance	Data bias due to pandemic urgency	Enabled fast triage and diagnosis during outbreaks, and supported crisis response

ously. The specific areas of comparison in the table are diagnostic performance (accuracy or AUC), regulatory potential, and ethical issues.

In all examples, we showcase how diagnostic acuity and clinical decision-making have been accelerated by the use of DL, frequently surpassing human-level performance in some contexts. Yet, performance alone is insufficient to justify clinical readiness. Regulatory approval requires an assessment of performance and certain interpretability, reproducibility, and bias metrics.

12.8.7 Cross-Domain Observations

1. Diagnostic Ability: Every model displayed a high predictive performance (91%–98%), indicating that DL could capture complex information from images that could not be obtained by human visual imaging.
2. Data Disparity and Scale: The increased scale and data disparity have resulted in better generalizability of multi-center training, showcasing the benefit and function of ML and federated learning environments.

3. Barriers to Regulatory Approval: The majority of systems evaluated have at least some sort of clinical validation and regulatory approval barrier, typically around data provenance, explainability, and patient consent.
4. Barriers to Ethical Approval: Reported ethical barriers were often related to data privacy, explainability of algorithms, and access to AI-related technologies, specifically in resource-poor environments.
5. Integration: AI tools provide good prediction in controlled trial testing, but integration into a workflow of normal clinical practice requires interoperability, clinician training, and monitoring of the clinical model.

12.9 Limitations

Primarily, this chapter is literature-based and not framed in real-time experimentation. That is not to say the examples provided are not interesting, but the examples mostly derive from high-resource clinical settings, which can exclude underrepresented populations. The technical methods of SHAP and Grad-CAM are presented as concepts without evaluation or performance comparisons. This chapter also discusses technical methods within the framework of evolving statutory and work processes, and therefore, some information may require an update in a future reading.

12.10 Conclusion

DL has changed the medical imaging field. It has moved the field from manual interpretation to an algorithm-driven, data-centric science. Complex neural architecture (CNN, RNN, GAN, and transformers) ride on automated disease detection, image segmentation, and predictive diagnostics that reach and in many cases surpass human accuracy. The pace of technology transfer represents a shift in healthcare toward machine intelligence that seeks to partner with human intelligence to improve accuracy, diagnosis, and efficiency. Nonetheless, the incorporation of DL into medical imaging is not just a technical issue; it is also an ethical and regulatory issue. The introduction of these models provides clinical stakeholders a different lens on health information, but it also raises difficult concerns about data privacy, algorithmic bias, interpretability, and accountability. The "black-box" design of many DL systems limits our capacity to validate how a machine arrived at its predictions, while data-driven bias can exacerbate health inequities if there are no checks and balances. The complexity of the duality of innovation vs. accountability, efficiency vs. fairness, and automation vs. accountability reflects the current debate in the context of AI-enabled healthcare. A key lesson from this chapter is that technical excellence must be accompanied by ethical oversight. To create a trustworthy AI system, researchers need to consider:

- Ensuring an appropriate level of diversity and fairness in datasets to prevent biased predictions.
- Using explainable/understandable mechanisms, such as Grad-CAM and SHAP, to clarify the basis for model decisions.
- Promoting the use of federated learning frameworks which can foster data-sharing projects while protecting patients' privacy.
- Building a comprehensive, regulated pathway for model verification, monitoring, and post-deployment auditing.

Interdisciplinary collaboration also plays a key role. The responsible implementation of AI-supported healthcare will depend on independent data scientists, engineers, clinicians, ethicists, policymakers, and lawyers collaboratively develop responsible innovation ecosystems. The only way to ensure trust in the healthcare community is by ensuring that AI-supported clinical care enhances trust between patients, clinicians, and healthcare organizations rather than eroding the trust altogether.

With the continued progression of DL, our attention now must move on from just achieving peak accuracy to promoting ethical adequacy – ensuring all predictions are transparent, accountable, and take into account the needs of the patient. When fairness, interpretability, and privacy are purposefully designed from step one, DL can become not only a means of innovation but also a vehicle for equity and trust in healthcare for the twenty-first century.

To improve interpretable, transparent, and trustworthy practices, future work should work to improve and establish the practice of using DL algorithms. Important areas of future work include:

- Advancing Federated Learning: Expand FL structures to enable models to develop across multiple health organizations while keeping data secure and private (e.g., hospitals, medical clinics, and family health teams).
- Trustworthy Explainable AI Models: Building interpretable AI systems via explainable AI (XAI) will provide assurance for both doctors and patients in understanding and trusting AI-based recommendations and decisions.
- Reducing Bias with Diverse Datasets: Reducing and eliminating bias from AI-based diagnosis will require developing and de-risking datasets that represent diverse populations in order to provide the most comprehensive-based recommendations for patients.
- Smooth Inclusion of AI into Clinical Workflows: There is mounting emphasis on co-designing with health professionals and AI developers for AI systems to be integrated in workflow, called "Smoothly incorporating AI into Clinical Workflows."
- Healthcare AI Regulatory Standardization: Establishing internationally accepted guidelines for ethical AI use, AI validation, and compliance systems to direct the responsible deployment of AI.

Addressing these issues can enable DL in medical imaging to drive the development of more transparent, equitable, and efficient AI-enabled healthcare solutions. Much of a responsible AI-powered future for medical imaging will be a result of partnering with researchers, policymakers, healthcare providers, and regulatory agencies.

References

[1] Esteva AK, Novoa RA, Ko J, Swetter SM, Blau HM, Thrun S. Deep learning-enabled medical computer vision. Nature Biomedical Engineering. 2021;5(6):541–553.

[2] Lundervold S, Lundervold A. An overview of deep learning in medical imaging. Frontiers in Medicine. 2019;6:14.

[3] Topol EJ. High-performance medicine: The convergence of human and artificial intelligence. Nature Medicine. 2019;25(1):44–56.

[4] Rajpurkar P, Irvin J, Zhu K, Yang B, Mehta H, Duan T, Ng A. CheXNet: Radiologist-level pneumonia detection on chest X-rays with deep learning. PLoS ONE. 2020;15(3):e0230301.

[5] Wong HH, Schuster J, Yala T. Ethical considerations in AI-based medical imaging. Artificial Intelligence in Medicine. 2021;113:102024.

[6] Ardila D, Kiraly B, Bharadwaj JP, Choi R, Chang AP. End-to-end lung cancer screening with three-dimensional deep learning on low-dose CT scans. Nature Medicine. 2019;25(6):954–961.

[7] Ronneberger PF, Brox T. U-Net: Convolutional networks for biomedical image segmentation. Medical Image Analysis. 2020;65:101776.

[8] Litjens G, Kooi T, Ehteshami Bejnordi B, Setio AA, Ciompi F. A survey on deep learning in medical image analysis. Medical Image Analysis. 2019;42:60–88.

[9] Kundu S, Chen J, Zhao Y. AI in medical imaging: Current trends and future directions. Journal of Medical Imaging. 2022;9(1):012002.

[10] McKinney SM, Sieniek M, Godbole V, Godwin J. International evaluation of an AI system for breast cancer screening. Nature. 2020;577(7788):89–94.

[11] Gulshan V, Peng L, Coram M, Stumpe K. Development and validation of a deep learning algorithm for detection of diabetic retinopathy in retinal fundus photographs. Jama. 2019;316(22):2402–2410.

[12] Zhang Y, Yang J, Wang B. Deep learning in medical imaging: General overview and future directions. Annual Review of Biomedical Engineering. 2021;23:221–251.

[13] Kim DW, Jang SJ, Kim JW, Lee TK. Deep learning-based diagnosis in medical imaging: A survey. Journal of Imaging Science. 2020;64(4):345–357.

[14] Ravi D, Wong C, Deligianni H. Deep learning for health informatics. IEEE Journal of Biomedical and Health Informatics. 2019;21(1):4–21.

[15] Chen J, Wu X, Li Y. Federated learning in medical imaging: Applications, challenges, and future directions. Journal of Medical Internet Research. 2022;24(3):e28331.

[16] Wang H, Lin R, Zhang T. A review of deep learning applications in medical image analysis. Health Informatics Journal. 2021,27(3):14604582211007277.

[17] Kaushal PS, Gupta S. Explainable AI in medical imaging: Challenges and opportunities. Medical Image Analysis. 2022;78:102424.

[18] Lu MT, Ikuta T, Rosenkrantz RG. Deep learning for cardiovascular imaging: A review. JACC: Cardiovascular Imaging. 2020;13(3):734–751.

[19] He K, Zhang X, Ren S, Sun J. AI-driven radiomics for personalized medicine. IEEE Transactions on Medical Imaging. 2021;40(2):371–384.

[20] Yadav SS, Jadhav SM. Deep convolutional neural network-based medical image classification for disease diagnosis. Journal of Biomedical Informatics. 2019;93:103118.

Ravneet Kaur, Chaitanya Singla, Gurpreet Singh, Manjot Kaur, Thuseethan Salavia

13 Transformers and Deep Learning in Medical Imaging: Future Directions and Impact

Abstract: Medical imaging is the core of modern diagnosis, treatment, and medication. Within the field of medical image analysis, deep learning, especially convolutional neural networks (CNNs), has achieved tremendous growth. Despite this, CNNs do not have the inherent capability of modeling long-range dependencies in medical images. In this paper, an overview of the role of transformers, a new class of deep learning models designed to handle global relationships, in transforming medical imaging is given. Additionally, the fundamentals of transformers, their strengths over CNNs, and their applications in a variety of medical imaging tasks, i.e., image classification, segmentation, and multimodal analysis, are presented. In addition, challenges and limitations, i.e., data sparsity, computational cost, and interpretability, are presented. Lastly, future directions, i.e., hybrid architecture, unsupervised learning, and integration into clinical workflow, are presented. This review emphasizes the transformers show higher diagnostic accuracy and efficiency by overcoming the limitation of CNNs in modeling long-range dependency and multimodal fusion.

Keywords: medical imaging , deep learning , transformers, convolutional neural networks (CNNs), long-range dependencies, image classification

13.1 Introduction

Convolutional neural networks (CNNs) lack inherent ability to capture long-range dependency in medical images. In this chapter, a review of the promise of transformers, a new generation of deep learning (DL) models specially tailored to manage global ***relationships***, toward transforming medical imaging is provided. In addition, the basics of transformers, their advantages over CNNs, and their applications in numerous medical imaging tasks, i.e., image classification, segmentation, and multimodal analysis, are discussed. Additionally, challenges and limitations, i.e., sparsity of data, computational cost, and interpretability, are discussed. Finally, future directions, i.e., hybrid architecture, unsupervised learning, and integration with clinical workflow, are discussed. This review identifies that the transformers exhibit higher diagnostic accuracy and efficiency by overcoming the limitation of CNNs in managing long-range dependency and multimodal fusion.

 | https://doi.org/10.1515/9783112226223-013

CNNs, a popular family of DL models, have made exemplary contributions to medical image analysis tasks, such as tumor detection, organ segmentation, and disease classification. Their hierarchical nature allows them to learn features at various levels of abstraction, making them extremely effective in automated image analysis. Though CNNs have been exceptional in feature extraction and automatic image analysis, they have some drawbacks [1]. They use local receptive fields, and therefore, capturing long-range dependencies between distant areas in an image becomes challenging. This is extremely critical in medical imaging, where it is necessary to detect anomalies of an entire organ or examine spatial interdependencies in multimodal datasets. Furthermore, CNNs need large, labeled data for training, which may become a challenge due to privacy restrictions in medical data acquisition. To overcome these restrictions, researchers have investigated architectures that are capable of modeling long-range dependencies and contextual relationships better [2]. Differently from CNNs that process sequential local features, transformers employ self-attention mechanisms for modeling global dependencies in an image [3]. This aspect allows transformers to scan spatially distant areas efficiently, and therefore, they are more suited to medical imaging tasks like tumor segmentation, multimodal image fusion, and volumetric analysis of 3D scans [4]. For instance, vision transformers (ViTs) [5] outperform traditional CNNs in medical image classification tasks and are more accurate in ChestX-Ray14 and BRATS-2021 MRI segmentation datasets. Furthermore, transformers have superior data efficiency through pretraining techniques, minimizing dependence on large, labeled datasets [7–10].

Transformers are initially developed by natural language processing and hence, excel in long-range dependencies and contextual relationships in data. For advanced medical imaging, multimodal integration, and spatial modeling, these attributes are suitable. Transformer brings revolution in medical field by enhancing accuracy and automation, though based on the strength of CNN and handles its weaknesses. This leads to the integration of transformer in clinical practices that requires a paradigm shift and hence ensures ethical artificial intelligence (AI) deployment, human oversight, and regulatory compliance. Additionally, Industry 5.0 endorses collaboration between AI-human and automation that helps the Figure 13.1 radiologist in decision-making. Therefore, it is depicted that AI-powered tools are not the replacement of medical officers, though can act as assistance; thereby minimizing the error, accurate diagnosis, minimizing error, and reducing workload while maintaining transparency and interpretability [12].

Previously, in medical imaging, the focus is on efficiency; however, Industry 5.0 emphasis on collaboration of humans with intelligent systems. This not only promotes customization and sustainability but also increases social responsibility. To serve the societal goal, this paradigm shift aims to combine technological advancement with human creativity to ensure maximum utilization of computational power of the machine. The structured representation in Figure 13.1 details the integration of DL models, particularly CNN and transformers. The first phase starts from data acquisition

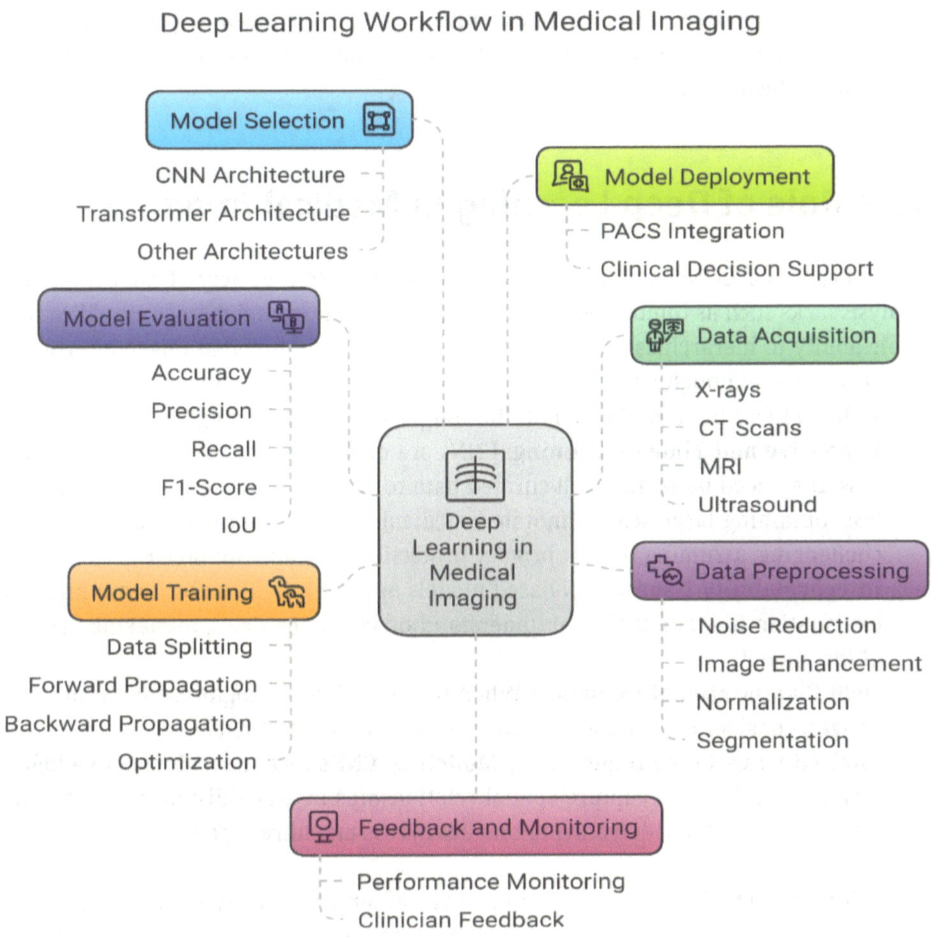

Figure 13.1: Deep learning workflow in medical imaging: integration of CNNs and transformers.

leads to preprocessing, ensuring standardized images for optimal analysis. Feature extraction is the next step, where hierarchical features can be extracted by CNN, whereas spatial and global dependencies can be extracted by transformers. Finally, the processed output is used for disease classification, segmentation, and anomaly detection [11–14].

Deep learning dramatically improves the accuracy of diagnosis, patient prognosis, and clinical effectiveness following a structured application. The combination of CNNs and transformers for medical imaging is far from trivial. This chapter investigates the changing landscape of DL in medical diagnostics – ranging from data scarcity, overfitting, and computational complexity to model interpretability and future developments, including hybrid architectures, unsupervised learning strategies, and emerging clinical workflows. By focusing on these features, this chapter emphasizes how these technologies can have a disruptive impact in terms of diagnostic accuracy and patient welfare, embodying the "sustainable" and "human-centered" efforts pursued with Industry 5.0.

13.2 Role of Deep Learning in Medical Imaging

DL, in particular CNNs, has been successfully applied to a variety of medical image analysis tasks such as tumor detection, organ segmentation, and disease classification. Their ability to hierarchically extract features supports automated image interpretation. However, although CNNs have great advantages, they also have significant limitation that affects their application in the complex medical imaging tasks [4, 5]:

- **Expensive and Time-Consuming:** CNNs are expensive and time-consuming process; they need large and well-curated data to get the accurate results. Nevertheless, obtaining large-scale annotated medical imaging data with high quality is challenging, assuming patient privacy protection and annotation cost.
- **Interpretability Issues:** CNN-based models operate as a black box, with no clear explanation for radiologists to understand or verify the decision-making process of the model.
- **High Computational Demands:** When training CNNs on high-resolution medical images, considerable GPU and memory resources are needed.
- **Limited Long-Range Dependency Modeling:** CNNs rely on local receptive fields, making it difficult to capture spatial relationships between distant regions in an image – essential for detecting anomalies across an entire organ.

To address the challenges, transformers have emerged as a promising alternative. Their self-attention mechanisms enable global feature extraction, allowing better contextual understanding and superior medical image analysis.

13.3 CNN Components and Mathematical Foundations

CNNs operate through multiple architectural components, including convolutional layers, pooling layers, and activation functions, each playing a role in hierarchical feature extraction. A comprehensive analysis of these components, their mathematical representations, and their role in medical imaging is provided in Table 13.1, offering insights into their computational efficiency and impact on classification performance [15–17].

Table 13.1: CNN architecture, operations, and implementation techniques in medical imaging.

Equation component	Description	Medical image application	Implementation techniques
Convolutional layer	Applies convolutional filters to input images to extract features.	Tumor detection: Identifying cancerous regions in CT scans or MRIs. Organ segmentation: Segmenting organs in medical images for surgical planning. Disease classification: Classifying diseases based on patterns in medical images.	TensorFlow, Keras, PyTorch, CuDNN (for GPU acceleration), transfer learning (using pretrained models), and data augmentation (to increase training data)
Equation: $(f * g)(t) = \int f(\tau)g(t - \tau)\,d\tau$	Here, f is the input image, and g is the convolutional filter. The operation integrates the product of f and the shifted version of g over all possible values of τ.		
Activation function	Introduces nonlinearity to the network, allowing it to learn complex patterns.	Enhancing feature representation: Improving the representation of relevant features in medical images Improving nonlinearity	ReLU, sigmoid, Tanh, Leaky ReLU, parametric ReLU (PReLU), and ELU (exponential linear unit)
Equation: $\mathrm{ReLU}(x) = \max(0, x)$	The ReLU function outputs the input directly if it is positive; otherwise, it outputs zero.		
Pooling layer	Reduces the spatial dimensions of the feature maps, decreasing computational complexity and controlling overfitting.	Reducing computational load: Downsampling feature maps to decrease computational requirements Controlling overfitting: Helping the model generalize better to unseen data	Max pooling, average pooling, global average pooling, and strided convolutions

Table 13.1 (continued)

Equation component	Description	Medical image application	Implementation techniques
Equation: Pool(*i*, *j*) = max(*A*[*s***i*: *s***i* + *k*, *s***j*:*s***j* + *k*])	Max pooling selects the maximum value from each region defined by a spatial neighborhood, where *s* is the stride and *k* is the kernel size. The function traverses the input matrix *A* with these parameters.		
Batch normalization	Normalizes the activations of each layer to stabilize training and accelerate convergence.	Stabilizing training: Ensuring that the training process converges more smoothly and quickly Improving generalization: Reducing the risk of overfitting by normalizing activations	Batch normalization layers in TensorFlow, Keras, or PyTorch
Equation: $x\_norm = (x - \mu) / \sqrt{(\sigma^2 + \varepsilon)}$	Here, *x* is the input, μ is the mean, σ is the standard deviation, and ε is a small constant to prevent division by zero. This equation normalizes *x* across a batch.		
Fully connected layer	Connects all neurons in the previous layer to the output layer, performing high-level reasoning and classification.	Disease classification: Making final predictions about disease categories based on extracted features Medical image reporting: Generating reports based on the analysis of medical images	Fully connected layers in TensorFlow, Keras, or PyTorch; Softmax activation (for multiclass classification)
Equation: $y = \sigma(Wx + b)$	Here, *y* is the output, *W* is the weight matrix, *x* is the input, *b* is the bias term, and σ is the activation function. This equation represents the transformation of the input through the weights and biases, followed by the activation function.		

Table 13.1 (continued)

Equation component	Description	Medical image application	Implementation techniques
Loss function	Quantifies the difference between predicted and actual values, and guiding the training process.	Training optimization: Providing a measure of error to guide the training process Evaluating model performance: Assessing how well the model performs on the given task	Cross-entropy loss (for classification), dice loss (for segmentation), mean squared error (MSE) (for regression)
Equation: CrossEntropy = $-\Sigma\, t_i \log(s_i)$	This calculates the loss as the negative sum of true labels multiplied by the log of predicted probabilities. The closer the predicted probability to the true label, the lower the loss.		

The convolutional layer processes key features from medical images, such as filters to support tumor detection, organ segmentation, and disease classification with TensorFlow or PyTorch. The use of nonlinear activation functions, such as ReLU and Sigmoid have nonlinearity to feature representation in deep networks and can help to prevent overfitting. Pooling layers, such as max pooling, contribute to the reduction of computational complexity by down-sampling feature maps, aiding in preventing overfitting. Batch normalization stabilizes training by mini batch normalizing activations, facilitating more effective convergence and generalization. The fully connected layer, which oversees high-level reasoning and classification, maps the features extracted to different disease categories using SoftMax activation. Planning minimizing a loss function like cross-entropy, or dice loss about how close the actual results are to what was predicted and use that to steer the optimization in the right direction, toward accurate medical diagnoses. These ingredients still allow CNNs to process and analyze high-dimensional medical images effectively.

13.4 Introduction to Transformers

Transformers have been a jolt to the DL industry and have provided new solutions to the problems of older neural network models, such as CNNs and RNNs. These state-of-the-art models were first explained in the seminal paper "Attention Is All You Need"

[3], published in 2017. Transformers are one such type of architecture, and they were specifically designed to handle long-range dependencies and contexts in data efficiently; hence, they are very powerful and general.

Originally developed for NLP, transformers became very popular for their ability to capture relationships across the entire sequence of data. As they were tried in various tasks, transformers moved into other domains, like computer vision, audio processing, and, more recently, medical imaging. This separates transformers from other models, as they can look beyond local structure and process data in a multimodal context, being particularly suitable for tasks that require context-aware interpretations, such as the anomaly detection in medical images or multimodality data fusion for simultaneous diagnoses [4].

The core of transformers is the self-attention mechanism, an essential component that enables the model to dynamically emphasize input elements. This puts transformers at an advantage in tasks that demand detailed and contextual reasoning [5]. In contrast to CNNs, which operate on restricted local regions (receptive fields), transformers examine interactions between all features at once, allowing them a more global view. In medical imaging, the global view plays the crucial role for more understanding of interplay and distant features in an image, including tissue structures, ligaments, anomalies, and many more [6].

Moreover, transformers are adaptive in nature that dynamically sets the patches and hence does classification as per region of interest. This scalability and hierarchical structure help them to remain pretrained on extensive datasets and further fine-tuned for the specified tasks. This is an effective approach during data scarcity for medical imaging. Considering the limitations of traditional architectures and leveraging the advantages of transformers, it is viewed as a cornerstone of current medical diagnostics that offers enhanced accuracy, and ability to handle complex scenarios.

The self-attention mechanism operates via the formula [3]:

$$\text{Attention}\,(Q,\,K,\,V) = \text{SoftMax}\,((Q \times K^\wedge\,T)/\sqrt{}(d\_k))\;V,$$

where key components Q is a query derived from input data and defined the feature to focus on, whereas K is the key that encodes information about all other features in the sequence or spatial domain. However, V value contained the contextual data tied to the key, and $\sqrt{}(d\_k)$ is the scaling factor that normalizes the dot-product to prevent excessively large output values during computation.

13.5 Detailed Workflow of Self-Attention Mechanism

In Figure 13.2, medical image processing with transformers involves three key steps. Preprocessing and tokenization begin by dividing images into smaller patches (2D square patches or 3D subregions), flattening them into tokens, and embedding these tokens into a vector space via linear projection [12]. Input representation enhances these tokens with positional embeddings to preserve spatial relationships, enabling transformers to analyzeFigure 13.2 contextual features [13]. Computing relationships leverages self-attention mechanisms, where each token (query) dynamically interacts with others (keys/values) to calculate attention weights, prioritizing diagnostically relevant regions [12]through learned spatial and semantic dependencies.

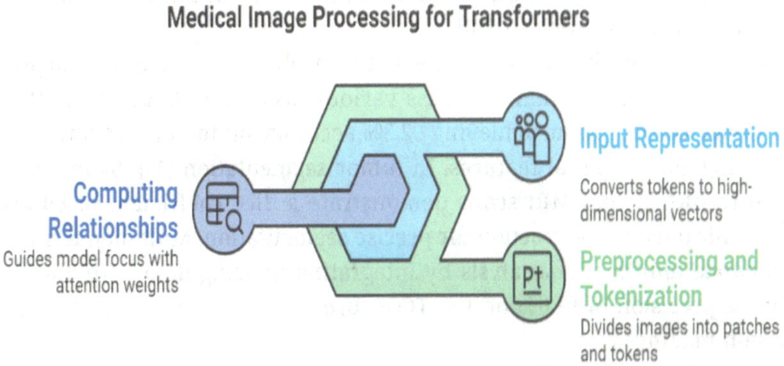

Figure 13.2: Medical image processing with transformers.

Apart from describing transformers' operation in medical imaging in general, their processing pipeline comprises several important steps that improve spatial and contextual understanding of complex anatomy. Medical images are first segmented into small patches (e.g., 16 × 16 pixels for 2D images or sub-volumes for 3D scans), which are flattened and linearly projected to a token vector separately. Tokens are positionally encoded so that spatial relationships lost upon flattening are restored, allowing the model to correctly understand anatomical location. Tokens are then passed through several layers of self-attention after embedding, where every token dynamically attends to all other tokens through dot-product attention. This allows the transformer to capture long-range dependencies critical in identifying spatially distributed patterns, e.g., tumors or tissue abnormalities. Several attention heads are used for extracting diverse spatial and semantic cues in parallel to enhance the model's ability to learn subtle variations in medical images. Final representations are then passed through task-specific heads, e.g., classifiers or segmentation decoders, to generate outputs such as disease labels, lesion maps, or structured clinical reports. In addition, the attention maps generated during inference can be visualized to highlight diagnostically informative areas, making transformer-based systems not only predictive but also interpretable and conducive to human-in-the-loop medical applications.

13.6 Comparative Advantages of Transformers over CNNS

Transformers address the challenges by leveraging self-attention mechanisms, enabling a more comprehensive analysis of spatial and contextual relationships in medical images. Their ability to process entire images holistically makes them particularly well-suited for complex medical tasks, including tumor segmentation, multimodal data integration, and 3D image analysis [16].

Table 13.2 highlights the diverse applications of transformers in medical imaging, showcasing their superior performance across various tasks. ViTs have been effectively used for image classification, achieving 92.3% accuracy on the ChestX-Ray14 [7] dataset by capturing global spatial features. In tumor segmentation [11], Swin transformers applied to BRATS-2021 MRI scans demonstrate a dice coefficient of 89.7%, leveraging hierarchical feature extraction for precise segmentation. Multimodal transformers [12] enhance multimodal analysis by integrating imaging data with clinical records, yielding a precision of 91.6% on the TCGA Breast Cancer dataset, improving diagnostic decision-making.

For 3D image analysis [13], 3D ViTs are particularly strong in the NSCLC Radiomics challenge, achieving Jaccard Index of 87.8%, allowing accurate spatial modeling of 3D medical volumes. In retinal disease diagnosis, the ViTs [14] trained by Kaggle EyePACS dataset achieve 94.2% in sensitivity, suggesting that they can effectively process high-resolution retinal scans. These findings accentuate the increasing importance of transformers in medical imaging and provide a more accurate and intrinsic spatial analysis, particularly with multimodal data integration, which is something that CNN models cannot achieve.

Table 13.2: Applications of transformers in medical imaging.

Medical imaging task	Transformer model	Dataset used	Performance metrics	Key features	References
Image classification	Vision transformers (ViTs)	ChestX-Ray14	Accuracy: 92.3%	Captures global spatial features	[7]
Tumor segmentation	Swin transformers	BRATS-2021 MRI	Dice coefficient: 89.7%	Hierarchical feature extraction	[11]
Multimodal analysis	Multimodal transformers	TCGA Breast Cancer	Precision: 91.6%	Integrates imaging with patient data	[12]
3D image analysis	3D vision transformers	NSCLC Radiomics	Jaccard index: 87.8%	Spatial modeling of volumetric data	[13]
Retinal disease detection	Vision transformers	Kaggle EyePACS	Sensitivity: 94.2%	High-resolution retinal scan analysis	[14]

The key applications of transformers are explored following to demonstrate superior performance as compared to traditional CNNs in medical imaging:

- **Global Context Modeling:** While CNNs rely on local receptive fields, transformers evaluate all input features simultaneously, capturing spatial and semantic relationships across the entire dataset.
- **Reduced Dependency on Labels:** Transformers benefit from self-supervised pretraining, requiring fewer labeled data points.
- **Multimodal Integration:** Transformers are uniquely equipped to combine imaging data with other clinical records, including genomic data and patient history.

13.7 Use Case: Tumor Segmentation Using Swin Transformers

Task: Automated segmentation of glioblastomas in 3D multiparametric MRI scans for surgical planning and treatment assessment [12]
Dataset: BraTS 2021 dataset containing 1,251 cases with four 3D MRI modalities (T1, T1ce, T2, and FLAIR)
Architecture: Hybrid Swin transformer-CNN model featuring:

The hybrid Swin transformer-CNN architecture combines the global contextual awareness of transformers with CNNs' local feature extraction capabilities through several innovative components: Enhanced local self-attention (ELSA) transformer blocks optimize local feature enhancement by refining attention patterns within shifted windows, while channel squeeze and spatial excitation modules dynamically prioritize informative channels and spatial regions to reduce redundancy. The 3D Swin UNETR backbone processes volumetric medical data efficiently using shifted window self-attention, capturing long-range dependencies across MRI slices without quadratic computational costs. This architecture achieves parameter efficiency (4.47 M parameters) via the IMS^{2}Trans implementation, which integrates lightweight multiscale interactions through depth-wise separable convolutions and bottleneck layers. The design addresses key challenges in medical imaging, balancing computational efficiency with state-of-the-art segmentation performance across modalities like MRI and CT.

Table 13.3 presents performance metrics comparing the Swin transformer model against traditional CNN baselines in this segmentation task. For comparison, the Swin transformer yields a dice coefficient of 0.897 versus 0.847 achieved by CNN models, a 5% boost in the quality of segmentation. In spatial accuracy, the 95% Hausdorff distance (HD) is much smaller with Swin, for which it is 8.90 mm compared to 12.34 mm for CNNs (a 28% drop). The Jaccard Index also exhibits the benefit of the Swin transformer at 0.86 versus 0.81 for the baseline CNN method (+6% gain). Further, we highlight that the Swin transformer achieves quicker inference time in 18 s per scan, which is 24 s per scan in CNNs, leading to a 25% efficiency boost in processing. Altogether, these measures highlight the efficacy of the hybrid Swin transformer-CNN model to promote automated glioblastoma segmentation in medical imaging.

Table 13.3: Performance metrics (Swin transformer vs. CNN baselines).

Metric	Swin transformer	Traditional CNN	Improvement
Dice coefficient	0.897	0.847	+5%
95% Hausdorff distance	8.90 mm	12.34 mm	28%
Jaccard index	0.86	0.81	+6%
Inference speed	18 s/scan	24 s/scan	+25%

This dual-axis bar chart in Figure 13.3 presents a comparative analysis of the performance metrics between the Swin transformer and traditional CNN models. The metrics under evaluation include dice coefficient, 95% Hausdorff distance, Jaccard Index, and inferenceFigure 13.3 speed, which are critical measures for assessing model efficiency in computational tasks such as image segmentation.

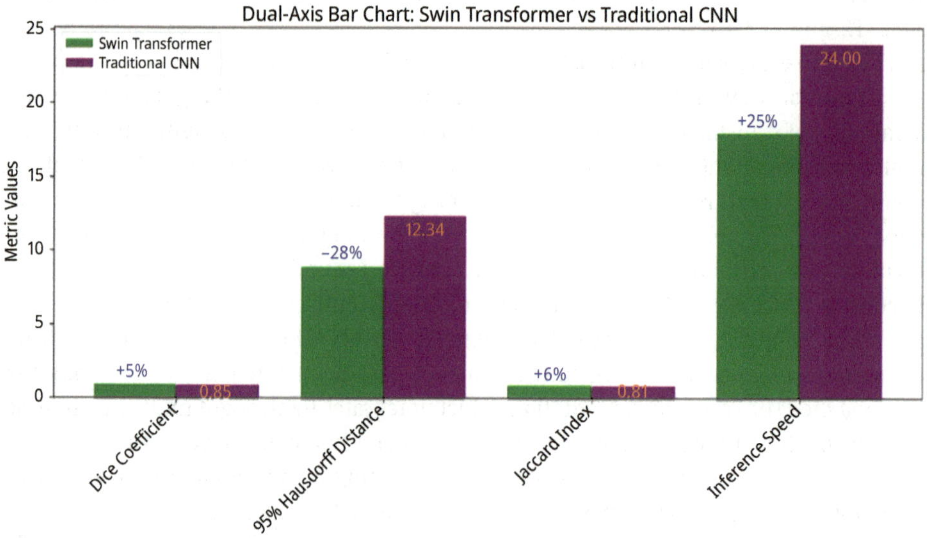

Figure 13.3: Performance comparison between Swin transformer and traditional CNN for tumor segmentation.

In Figure 13.3, comparison of Swin transformer and traditional CNN is done. The green bar defines Swin transformer, whereas the purple bar defines traditional CNN. It has been depicted from the graph that Swin transformer outperforms for dice coefficient by 5% and achieve the value 0.897, whereas CNN shows 0.847, indicating high accuracy in pixel-wise classification. The Hausdorff distance metric is defined for reduction of error; therefore, lower the value better the performance. In this case, Swin transformer achieved 8.90 mm Hausdorff distance as compared to CNN, i.e., 12.34 mm reflecting 28% substantial reduction in boundary deviation errors. However, the Jaccard Index shows an increase in 6% with Swin transformer that signifies superior spatial overlap accuracy. For inference speed, the scan in Swin transformer happens for 18 s and thus outperforms CNN 24 s by the margin of 25%, and thus significantly increases in computational efficiency. These results underline the efficacy of the Swin transformer in generating enhanced accuracy, reliability, and efficiency across multiple performance dimensions.

Key Advantages:

1. Superior handling of long-range dependencies in 3D volumes
2. Robust performance with missing MRI modalities
3. Efficient computation through window-based self-attention
4. State-of-the-art results on BraTS benchmarks

13.8 Challenges and Limitations

Despite the transformative potential of DL in medical imaging, several challenges hinder its widespread adoption and effectiveness [18–20]:

1. **Data Scarcity and Quality**: DL models require large, high-quality, and well-annotated datasets. However, medical image data is often scarce due to privacy concerns, ethical issues, and the need for expert annotations. Variability in imaging protocols across different institutions can introduce inconsistencies in datasets, leading to biased models.
2. **Computational Complexity:** Training deep models, particularly transformer models, demands high computational costs, which, in most cases, necessitate high-end GPUs or cloud computing. Real-time processing of high-resolution medical images is still an issue due to model complexity and inference time.
3. **Explainability and Interpretability:** DL models, i.e., neural networks, are "black boxes," and clinicians are unable to comprehend their predictions. It can reduce confidence and hinder clinical use if model decision-making is not explainable.
4. **Integration into Clinical Workflow:** Hospital infrastructures and clinical workflows of today may not be adequate for the introduction of AI-based systems. Ensuring regulatory compliance, standardization, and interoperability of AI systems and medical imaging devices is a massive challenge.
5. **Ethical and Security Issues:** Patient confidentiality and regulations like HIPAA and GDPR are top-of-mind concerns in healthcare uses of AI. AI-based diagnosis and treatment planning have legal and ethical concerns like liability in case of wrong predictions. Security of DL models against adversarial attacks remains a persistent issue.

Despite these difficulties, new DL methods are constantly working to overcome these limitations. In contrast, hybrid models of CNN and transformers enable more powerful feature extraction, and federated learning provides better privacy for AI training. Furthermore, self-supervised learning is also aiming to minimize the reliance on huge amounts of labeled data. These developments, as well as human-centric AI guidelines provided by Industry 5.0, are envisioned to drive DL in medical imaging for a future that is more accurate, efficient, and ethical for AI applications in healthcare.

Several countermeasures have been found effective to tackle the problem of data limitation in medical imaging. Transfer learning enables models to use knowledge gained in large sets (e.g., ImageNet or publicly available medical history) and apply it to small domain-specific sets by fine-tuning; performance is significantly enhanced with less labeled training data. Data augmentation includes rotation, scaling, flipping, elastic deformation, etc., artificially expanding the size and variety of the dataset, and helping the model generalize better. The information feedback system is designed to enhance the data without releasing precise patient data. Federated learning enables multiple institutions to get together to train the model without exposing the raw patient data, using encryption.

In addition, self-supervised learning has received tremendous popularity because it allows models to learn effective representations from unlabeled data by optimizing pretext tasks to reduce the need for human-annotated datasets. Such methods are particularly useful in healthcare, where annotation is laborious and subject to severe ethical and regulatory limitations.

With advancing technology, effective deployment of AI/ML in healthcare also depends on multidisciplinary collaborations. Data scientists bring model development and optimization skills, clinicians bring domain knowledge and clinical utility validation, and government agencies bring ethics compliance, patient confidentiality, and standards of safety. Such coordination will facilitate development of explainable, transparent, and clinically accurate AI systems that underpin real-world healthcare processes. Such multidisciplinary strategies then become the key bridge for translating results into practice, addressing legal considerations, and establishing end-user trust. With increasing dependence on AI for diagnosis and decision-making, such multidisciplinary endeavors will form the foundation to providing accountable and patient-centered healthcare innovations.

13.9 Conclusion and Future Work

DL, especially transformers, has transformed medical imaging by improving image classification, segmentation, and multimodal analysis. Despite CNNs being the foundation of medical image analysis, their inadequacy in addressing long-range dependencies has led to a shift to transformer-based models. Transformers bypass these shortcomings by employing self-attention mechanisms to enable proper contextual understanding and enhanced diagnostic accuracy. However, several challenges, like the unavailability of good data, computational complexity, interpretability, and issues of integration, have to be addressed in order for these models to be deployed widely in hospitals. Additionally, ensuring the ethical use of AI, data privacy protection, and model explainability will be crucial so that DL technologies continue to develop in medical diagnosis.

Virtually overcoming these issues allows AI-powered medical imaging technologies to enhance the detection of diseases at an early stage, improve patient care, and contribute to a more effective healthcare system. The combination of DL in medical imaging and Industry 5.0 innovation is of immense potential to transform diagnostic capabilities and make healthcare services more accessible and accurate. The future of DL in medical imaging will have to overcome present limitations while expanding its potential through a range of new approaches. Hybrid models will be the focus, merging CNNs and transformers to take advantage of each's strength, as well as investigating graph neural networks and attention mechanisms to better understand space. In addition, the shift to unsupervised and self-supervised learning will decrease dependence on large, labeled sets of data, enabling models to learn from unlabeled medical images and decreasing dependence on expert-labeled data. This will enable more stable models to learn to adapt to diverse imaging conditions.

Along with this, federated learning is expected to be a key method of privacy-focused AI that enables collaborative model training across institutions without jeopardizing confidential patient data. The method is going to enhance model generalization on heterogeneous data sets and resolve privacy-related challenges simultaneously. The use of real-time and edge AI will optimize DL models for mobile and edge device deployment, thus enabling timely diagnosis in resource-constrained environments. The increase in regulatory and ethical considerations will make the development of standard frameworks and the use of explainable AI methods enables regulatory compliance as well as instills confidence among medical professionals. Lastly, the integration of multimodal data from genomic sources, electronic health records, and wearable devices will enable overall patient diagnostics and tailored treatment plans, while enhancing security and resilience against adversarial attacks will enable dependability in clinical environments. Overall, these technologies align with Industry 5.0 frameworks that focus on human-focused AI that promotes cooperation among AI systems and medical professionals.

References

[1] Suzuki K. Overview of deep learning in medical imaging. Radiological Physics and Technology. 2017 Sep;10(3):257–273.

[2] Smith NB, Webb A. Introduction to Medical Imaging: Physics, Engineering and Clinical Applications 2010 Nov 18. Cambridge university press.

[3] Shamshad F, Khan S, Zamir SW, Khan MH, Hayat M, Khan FS, Fu H. Transformers in medical imaging: A survey. Medical Image Analysis. 2023 Aug 1;88:102802.

[4] Pu Q, Xi Z, Yin S, Zhao Z, Zhao L. Advantages of transformer and its application for medical image segmentation: A survey. BioMedical Engineering Online. 2024 Feb 3;23(1):14.

[5] Kim JW, Khan AU, Banerjee I. Systematic review of hybrid vision transformer architectures for radiological image analysis. Journal of Imaging Informatics in Medicine. 2025 Jan;27:1–5.

[6] Wang Y, Deng Y, Zheng Y, Chattopadhyay P, Wang L. Vision transformers for image classification: A comparative survey. Technologies. 2025 Jan 12;13(1):32.

[7] Hatamizadeh A, Nath V, Tang Y, Yang D, Roth HR, Xu D. Swin unetr: Swin transformers for semantic segmentation of brain tumors in mri images. In International MICCAI Brainlesion Workshop 2021 Sep 27 (pp. 272–284). Cham: Springer International Publishing.

[8] Mustapha B, Zhou Y, Shan C, Xiao Z. Enhanced pneumonia detection in chest X-rays using hybrid convolutional and vision transformer networks. Current Medical Imaging. 2025 Jan; 9: e15734056326685.

[9] Rifaldi D, Famuji TS, Wijaya SA, Abougarair AJ, Chotikunnan P, Ma'arif A. Machine Learning 5.0 In-depth Analysis Trends in Classification. Scientific Journal of Computer Science. 2025 Feb 9;1(1):1–5.

[10] Tang Y, Yuan Y, Tao F, Tang M. Cross-modal Augmented Transformer for Automated Medical Report Generation. IEEE Journal of Translational Engineering in Health and Medicine. 2025 Jan 29.

[11] Fu X, Lin R, Du W, Tavares A, Liang Y. Explainable hybrid transformer for multi-classification of lung disease using chest X-rays. Scientific Reports. 2025 Feb 24;15(1):6650.

[12] Nakach FZ, Idri A, Goceri E. A comprehensive investigation of multimodal deep learning fusion strategies for breast cancer classification. Artificial Intelligence Review. 2024 Oct 12;57(12):327.

[13] Zhang X, Zhang G, Qiu X, Yin J, Tan W, Yin X, Yang H, Liao L, Wang H, Zhang Y. Radiomics under 2D regions, 3D regions, and peritumoral regions reveal tumor heterogeneity in non-small cell lung cancer: A multicenter study. La Radiologia Medica. 2023 Sep;128(9):1079–1092.

[14] Balaha HM, Hassan AE, Ahmed RA, Balaha MH. Advancing eye disease detection: A comprehensive study on computer-aided diagnosis with vision transformers and SHAP explainability techniques. Biocybernetics and Biomedical Engineering. 2025 Jan 1;45(1):23–33.

[15] Priyadharsini S, Saravanakumar S. Mathematical Foundations of CNN in Soft Computing. In InSoft Computing (pp. 303–324). CRC Press.

[16] Kuo CC. Understanding convolutional neural networks with a mathematical model. Journal of Visual Communication and Image Representation. 2016 Nov 1;41:406–413.

[17] Bhatt D, Patel C, Talsania H, Patel J, Vaghela R, Pandya S, Modi K, Ghayvat H. CNN variants for computer vision: History, architecture, application, challenges and future scope. Electronics. 2021 Oct 11;10(20):2470.

[18] Dhar T, Dey N, Borra S, Sherratt RS. Challenges of deep learning in medical image analysis – Improving explainability and trust. IEEE Transactions on Technology and Society. 2023 Jan 4;4 (1):68–75.

[19] Razzak MI, Naz S, Zaib A. Deep learning for medical image processing: Overview, challenges and the future. Classification in BioApps: Automation of Decision Making. 2017 Nov;14:323–350.

[20] Zhou SK, Greenspan H, Davatzikos C, Duncan JS, Van Ginneken B, Madabhushi A, Prince JL, Rueckert D, Summers RM. A review of deep learning in medical imaging: Imaging traits, technology trends, case studies with progress highlights, and future promises. Proceedings of the IEEE. 2021 Feb 26;109(5):820–838.

Kanika Sharma, Parul Khurana, and Balraj Kumar

14 Cognitive Imaging: AI-Driven Transformation in Medical Diagnostics

Abstract: Medical imaging is one of the foundations of modern healthcare, as it allows clinicians to see the internal anatomy and physiology without the need of any invasive procedures. X-ray, computed tomography, magnetic resonance imaging, ultrasound, positron emission tomography, and single-photon emission computed tomography are some examples of modalities that have developed over the last century out of primitive imaging systems up to advanced multimodal platforms that have the capability of producing high-fidelity three-dimensional information. Meanwhile, the explosive growth in the quantity of volumetric image data and the sophistication of diagnostic processes has put more and more demands on radiologists and imaging specialists. Meanwhile, paradigms of artificial intelligence (AI), and in fact deep learning, have developed to the point of being able to handle complex image patterns, assist with segmentation, classification, and anomaly-detection tasks, and, in certain applications, rival human performance levels in tasks like diabetic retinopathy recognition and tumor segmentation. Although this has been achieved, there are still some major hurdles in the form of data heterogeneity, model generalizability, transparency, regulatory compliance, and clinical adoption. This chapter is an integrative and systematic account of the ways in which DL methods are being integrated into medical imaging processes, including underlying principles, modality-specific use-cases, performance indicators, clinical validation issues, and regulatory concerns. Besides that, this chapter lists technical and practical pitfalls, including but not limited to model bias, lack of interpretability, data shortage, and cross-site variability, and describes the new directions, like federated learning, multimodal AI systems, imaging-genomics integration, and real-time cognitive imaging assistance. This work is designed to fill this gap that currently exists between algorithmic innovation and clinical use with conceptual, methodological, and operational perspectives that researchers, imaging professionals, and healthcare stakeholders will require to realize the vision of genuinely intelligent, data-driven diagnostic ecosystems.

Keywords: artificial intelligence, deep learning, medical imaging, diagnostic imaging, imaging workflows, cognitive imaging, radiomics, federated learning, interpretability, clinical deployment

 | https://doi.org/10.1515/9783112226223-014

14.1 Introduction

Medical imaging is one of the best things about modern healthcare right now. It helps doctors find, watch, and treat a wide range of diseases. Imaging technologies have come a long way in the last hundred years. They have gone from simple machines to X-ray machines to today's advanced, multimodal platforms that show internal anatomy and physiology in three dimensions. Computed tomography (CT), magnetic resonance imaging (MRI), ultrasound, nuclear medicine (positron emission tomography (PET) and single-photon emission computed tomography (SPECT)), and digital radiography have all made it possible to see organs, tissues, and diseases that were never possible to see before. This imaging equipment is currently used within clinical workflows that aid physicians in locating issues at an early stage, like fractures, tumors, vascular issues, and neurological disorders. Medical imaging is not only a means of diagnosing symptoms, but it can also assist in the surgery planning process, guiding the minimally invasive operations, and monitoring the progress of the disease increasing and the effectiveness of the treatment with time. The amount of image data that is made has grown by a huge amount because of better image equipment and easier access. The proliferation of this brings both opportunity and challenge as how many more data can provide deeper insights but also adds a lot of stresses in radiologists and specialist responsible for interpreting these images that are fast and entitled [1].

14.1.1 Artificial Intelligence (AI) and Deep Learning (DL)

AI is taking over the business world, and healthcare is still one of its most promising fields. Deep learning (DL) is a part of machine learning (ML) that works very well when it comes to processing complex data patterns, like images or signals [2]. The DL mechanism, which is based on an artificial neural network (ANN), mimics the structure of the human brain to process information in layers. Convolutional neural networks (CNNs) and other, more advanced architectures have become the mainstay of automated image analysis in medical imaging. DL models can automatically learn important features from raw data, which is better than traditional image processing methods like image classification, segmentation, and anomaly detection [3].

For that, DL can find diabetic retinopathy (DR) in fundus photographs, pulmonary nodules in CT scans, and brain tumors in MRIs with the same level of accuracy as a person. Enhance diagnostic accuracy to ensure reproducible and consistent results essential for clinical decision-making. DL systems can also keep running all the time and on a large scale, which is a solution to the growing shortage of radiologists and imaging experts in many parts of the world. This integration into current imaging systems allows for real-time analysis, faster reporting, and helps with triage by clearly identifying high-risk findings that need to be reviewed right away. The use of DL in medical imaging marks the beginning of a new era of smart diagnostics. It is not only a technological innovation, but it also brings the real improvement to healthcare outcomes. DL may automate the process of complex imaging data assessment and this facilitates the work of doctors and lowers the probability of an error, particularly in regions with an intense population or with limited resources. Another great thing about DL is that it can find diseases early. A DL algorithm trained on large annotated datasets can find small pathological changes that manual reviews might miss. This is particularly relevant given that the administration of such diagnostic tests can occur in the initial phases of cancers, neurodegenerative diseases, and cardiovascular conditions, where earlier intervention significantly enhances prognosis and survival rates [4]. It also enables the application of quantitative imaging biomarkers and radiomics that are diagnostic methods that identify and characterize a large number of characteristics in images and the efficacy of interventions. This is because they can assist you to determine what you or the other person can do to ensure that treatment is effective to all people rather than only the few. DL replaces time-consuming tasks in the workplace, such as the measurement of lesions, segmentation of organs, and report production. It does not only reduce the time required to arrive at a diagnosis but also allows the radiologists to practice more complex cases that require expert judgment [5].

14.1.2 Deep learning integration in medical imaging

However, the integration is hard because you have to think about data quality, how well the model works in different situations, ethical issues, and following the rules. It is very important to make DL systems clear, easy to understand, and clinically validated in order to build trust and keep patients safe [6].

14.1.3 Scope of Chapter

This chapter aims to give a profound interpretation of the revolutionary role of DL in medical imaging and how it has now been used to transform the overall health care system. Its principal objectives are as follows:

- It defines the conceptual framework of DL and outlines its differences with other paradigm of AI. It discusses valuable breakthroughs and demonstrates the way DL has been applied to real life, including X-ray, CT, MRI, and ultrasound. Currently, changes in research directions, data, and challenges are used as the benchmark of DL expansion in medical imaging.
- Determining what is good and bad about the utilization of DL in clinical processes concerning precision, effectiveness, and bad judgment matters.

- It presents problems that arise due to both technical and practical challenges, such as data heterogeneity, model bias, interpretability, and rules. Future directions (the role of federated learning (FL), multimodal AI systems, and the convergence of imaging with genomics and electronic health record (EHR)) are identified.

Through this structured conversation, this chapter seeks to give readers a deep understanding of how DL is moving medical imaging to fundamentally different places and what's ahead for smarter, data-driven healthcare.

Medical imaging is essential for modern healthcare because it allows us to see the body's internal structures and functions without having to cut into it. It is a key part of diagnosing, planning treatment, and keeping an eye on many diseases. This section provides in-depth details about all major imaging modalities, including their clinical applications and associated challenges [4].

14.2 Major Imaging Modalities

There are many different types of medical imaging, and they all have different physical forms and uses for the images they create to help doctors figure out how the body works and what its anatomy is. The main types are described below and also represented in Figure 14.1:

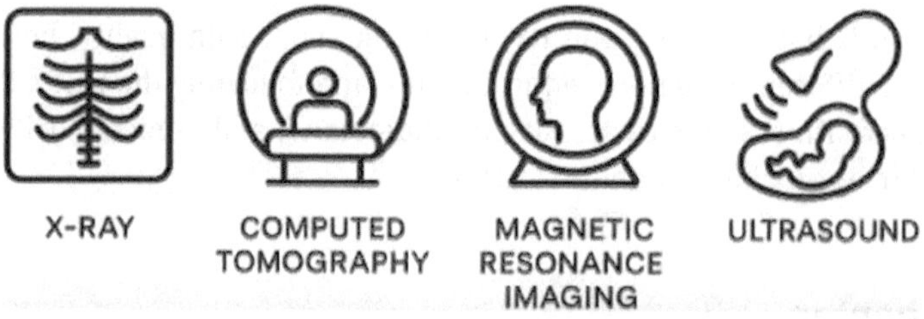

Figure 14.1: Depicts the major imaging modalities.

a. X-ray (Radiography): X-ray imaging is one of the oldest and most common ways to take pictures. It uses ionizing radiation to make 2D pictures of solid things, like bones and some organs. Absorption of X-rays: The principle says that X-rays pass through the body and that different tissues respond to them in different ways. Soft tissue absorbs less, while dense materials, like bone, absorb more and look white. Mammography, chest infection, dental assessment, and bone fracture are all common names for this test [7].
b. Computed Tomography (CT): CT scanning, or computerized tomography, is a modernized version of a special kind of X-ray that makes cross-sectional pictures of

the inside of the body by taking slices through it. Technique: To get the detailed 3D image, you need to combine the multiple X-ray images taken from different angles and process them through a computer. What they are used for: finding tumors, injuries inside the body, vascular diseases, and guiding biopsies and surgeries.

c. Magnetic Resonance Imaging (MRI): MRI uses strong magnetic fields and radiofrequency pulses to make high-resolution pictures of soft tissues. This principle makes hydrogen atoms line up with the magnetic field. When the field is turned off, the atoms send out signals that are used to make the images. Common uses: brain and spinal cord problems, joint injuries, soft tissue tumors, and heart imaging [8].
d. Sonography (Ultrasound): Ultrasound imaging uses high-frequency sound waves to make real-time pictures of organs and structures. This is the main idea: Sound waves go through the body and bounce off of tissues. Echoes that are detected and used make up images. Uses: monitoring pregnancy, checking the health of abdominal organs, echocardiography (heart imaging), and helping to place needles.
e. Positron Emission Tomography (PET): PET is a nuclear medicine method for looking at how the body uses energy. To follow the principle, a radioactive tracer is put into the body. When it decays, it makes positrons that hit electrons, which makes gamma rays that the scanner can pick up. Common uses: Finding cancer, checking brain disorders, and checking how well the heart works [9].

14.3 Deep Learning

An important application of DL is in medical imaging, where moving images can be interpreted better and analyzed automatically. DL is a complex branch of ML, which is part of AI that approximates the human brain's process of hierarchical learning. The given section discusses the theory of DL, examines the major architectures, such as CNNs, emphasizes the models applied in the healthcare field, and examines the notion of the pretrained-model and transfer learning in medicine [10].

- **ML in Medical Imaging:** ML allows computers to understand how to respond to information without the assistance of code because it uses data to identify trends and make plans. Unlike rule-based systems, ML algorithms are dynamic and are enhanced upon exposure of greater data. Among the tasks with which ML has been employed in medical imaging are:
  a) Classification of images (such as differentiating between benign and malignant tumors),
  b) segmentation (such as defining organs or lesions), and
  c) detection of anomalies, such as spotting odd growths or structures.
- **Deep Learning:** DL is a subset of ML based on ANNs. DL involves multilayered complex networks as a modification or extension of traditional ML. Such net-

works are particularly applicable when analyzing high-dimensional image data, where the networks can learn complex and hierarchical data representations [11]. Characteristics of DL entail:

- **Automatic Feature Extraction:** DL models avoid the need of human feature engineering by automatically extracting important patterns on unprocessed images. (a) Scalability: The more data is available and the more computational power is utilized, the greater DL models can become. (b) End-to-End Learning: These models can translate inputs into outputs with minimal indications and guidance of people. (c) CNNs represent a special type of deep neural network that is intended to work with image data. The local features recognition and spatial hierarchies' representation in images are very well suited to CNNs since they are based on the human visual system. Table 14.1 provides a complete summary of various fundamental concepts of DL that can be used in healthcare.

Table 14.1: Summary of deep learning concepts.

Topic	Description	Applications/examples
Machine learning (ML)	Subset of AI; learns patterns from data without explicit programming	Tumor classification, anomaly detection, and segmentation
Deep learning (DL)	Advanced ML using neural networks with multiple layers for complex data analysis	High-dimensional image analysis (e.g., MRI and CT)
Key features	– Automatic feature extraction – Scalability – End-to-end learning	Enables direct image-to-diagnosis models
Convolutional neural networks (CNNs)	DL models designed for image data are inspired by the visual cortex	Mammography, CT scan analysis, and lung X-ray classification
Components of CNNs	– Convolutional layers – Pooling layers – Fully connected layers – Activation functions	Extract edges, shapes, textures; reduce size, and compute load
Why CNNs?	Effective at learning spatial hierarchies and subtle features	Tumor detection, organ segmentation, and disease classification
Deep learning architectures	Different architectures for classification, segmentation, and detection	Each model has specific use cases
Pretrained models	VGG, ResNet, Inception; Model Zoos; self-supervised models	Retina diagnosis, COVID-19 detection, and tumor segmentation

14.4 Applications of Deep Learning in Medical Imaging

DL has profoundly transformed medical imaging by automating complex tasks, enhancing image quality, and facilitating faster and more accurate diagnosis. The capabilities of deep neural networks, particularly CNNs, have led to breakthroughs across the imaging pipeline – from disease detection and segmentation to prognosis and workflow optimization. Figure 14.2 discussed the various applications of DL in medical imaging. This section explores the practical applications of DL in medical imaging, demonstrating its pivotal role in reshaping clinical diagnostics and decision.

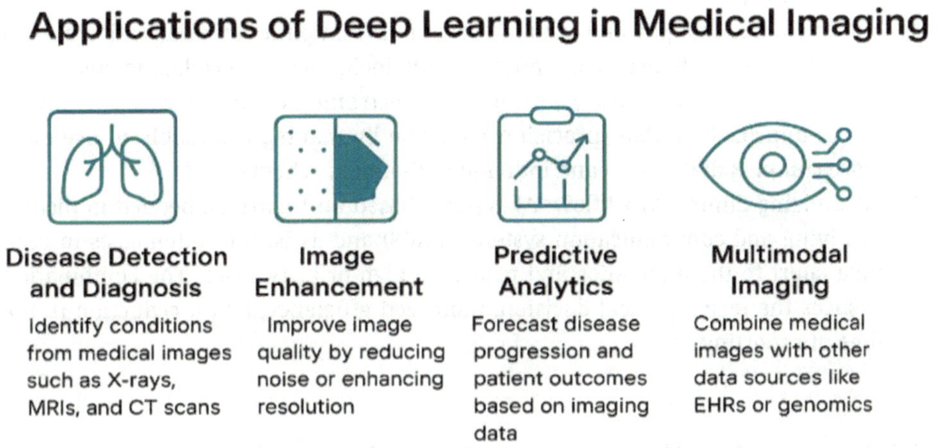

Figure 14.2: Applications of deep learning in medical imaging.

(a) Automated Detection and Diagnosis of Diseases: CNNs are especially forms of DL commonly employed to detect abnormalities in medical imaging, such as MRIs, CT scans, and X-ray images. The precision through which these models are able to identify all sorts of conditions, like pneumonia, cancer, fractures, and retinal disorders, is similar to or even better than that of a human expert. In a clinical environment characterized by a high burden of patients, automated detection reduces the chances of over-surveillance and decreases the time it takes to make a diagnosis [12].

(b) Advanced Neural Networks: ANNs can enhance medical images, and this is because they can make noise reduction, resolve improvements, and also fix artifacts to polish up the images. This becomes specifically useful in case of fast scans in MRI and low-dose CT, where high-quality imaging is not easily acquired due to speed or safety reasons attributed to the patient. Better pictures will permit a better interpretation and a better knowledgeable clinical decision [13].

(c) Anatomical Structures: Segmentation and Quantification: The next component is segmentation and quantification of organs, tumors, and lesions of complex imaging data that requires DL. Automated segmentation is used to measure the size of tumor, monitor the progression of disease, and plan treatment (medical procedures such as surgical resection and radiation therapy). Taking illustration of the DL-based segmentation, it reduces drastically the duration radiologists require to establish tumor boundaries in brain imaging [14].
(d) Prognosis Estimation and Predictive Analytics: DL models can analyze imaging data and combine it with historical medical records to predict patient outcomes, interaction with treatment, and disease progression. Such forecasting tools are important in guiding specific treatments and to optimally allocate resources within a medical setting.
(e) Multiple-Mode Imaging and Data Integration: DL simplifies the integration of data to offer an in-depth picture of patient health, including various data including imaging, genomics, laboratory testing, and electronic health records. This multimodal approach enables precision medicine by making it possible to use even more complex diagnostics and individual treatment schemes [15].
(f) Maximizing Clinical Workflow: AI systems based on DL are embedded in picture archiving and communication systems (PACS) and assist the radiologists in real-time, alert to the anomalies and prioritize high-priority cases. The combination ensures the faster clinical decision, improved efficiency, and a reduction in the time of reporting.

14.5 Case Studies and Real-World Deployments

While DL has demonstrated exceptional performance in laboratory environments, its true impact is being realized in real-world clinical settings. The deployment of DL solutions in hospitals, diagnostic centers, and commercial healthcare platforms has not only improved diagnostic accuracy but also enhanced workflow efficiency, patient outcomes, and healthcare accessibility. This chapter explores practical success stories from hospitals and research labs, commercial DL-powered tools, and how these innovations are being effectively integrated into clinical workflows [16]:

– *Hospital and Research Lab Success Stories*: Numerous success stories from top hospitals and research institutions demonstrate how DL has the potential to revolutionize real-world medical imaging environments. A breakthrough in pneumonia detection was achieved at Stanford University with the creation of Chex Net, a 121-layer CNN. The Chex Net, after training on ChestX-ray14 dataset, was capable of detecting pneumonia and other thoracic diseases as well as it was the case with practice radiologists, or even better. It had a successful track record and demonstrated that DL could assist radiologists in high-pressure environments such as emergency rooms and sparked a surge in similar developments in the field of thoracic imaging and however led to AI tools having FDA approval [17].

- Harvard Medical School and Massachusetts General Hospital (MGH) have used DL in brain tumor segmentation using DL based on the 3D U-nets architectures. After being introduced into clinical practice and being trained using the BraTS set, the time it took to manually segment the tumors shrank to less than 2 min as opposed to 30 to 40 min. Radiotherapy treatment planning has been one of the main areas that has actively benefited because of this significant time saving in the neuro-oncology care [18].

  DL was used by Aravind Eye Hospital in India, in collaboration with Google AI, to screen fundus images for DR. Their AI model produced diagnostic results comparable to those of skilled ophthalmologists, allowing for efficient eye screening even in remote locations without access to specialists. Since then, the system has been implemented in numerous clinics, screening hundreds of thousands of patients and making a substantial contribution to the decrease of avoidable blindness in underprivileged areas.

  Researchers at the Mayo Clinic created DL-based classifiers to use multiphasic CT images to differentiate between benign liver lesions and hepatocellular carcinoma. The model demonstrated the usefulness of DL in supporting complex hepatological diagnoses by lowering the number of needless liver biopsies while simultaneously increasing diagnostic confidence. Together, these illustrations show how DL can improve patient outcomes, expedite clinical procedures, and improve diagnostic capabilities in a variety of medical fields [19].
- *Deep Learning-Powered Commercial Tools and Platforms*: DL technologies' increasing maturity has sparked the creation of FDA-approved and CE-marked commercial medical imaging tools, allowing for extensive clinical use. For example, Aidoc focuses on radiology case triage in real time. Aidoc's tools, which are directly integrated into PACS, are capable of identifying serious conditions like spinal fractures, pulmonary embolisms, and intracranial hemorrhages. By prioritizing urgent cases, these real-time alerts assist radiologists in making better decisions in emergency situations and cutting down on turnaround times.
- Preventative screening and population health are the main priorities of Zebra Medical Vision. Numerous applications, such as vertebral fracture detection, breast cancer risk assessment, and coronary artery calcium scoring, are supported by its AI1 platform. Zebra's tools, which have been implemented in radiology networks in the United States, Israel, and the UK, have been crucial for early disease detection and effective resource allocation in extensive screening initiatives.
- Another well-known player is PathAI, which specializes in digital pathology and image analysis for histopathology. Its DL models support clinical trials, help detect biomarkers, and grade cancer. Major pharmaceutical firms like Roche and

Bristol-Myers Squibb collaborate with PathAI to speed up drug development and lower diagnostic error rates in busy pathology labs.

- AI solutions for mammograms and chest X-rays are available from Lunit INSIGHT. Its models, which are used in Saudi Arabia, Brazil, and South Korea, identify more than ten anomalies, including pneumonia and tuberculosis. These tools improve the efficacy and consistency of screening, which is especially helpful in areas with a lack of radiology expertise. With products like qXR for chest X-ray interpretation and qER for head CT triage, Qure.ai focuses on accessible AI for emerging markets. Qure.ai's tools are widely used in public health initiatives in Africa, Indonesia, and India. They give frontline healthcare workers instant diagnostic support, allowing for prompt interventions in low-resource environments. When taken as a whole, these commercial platforms demonstrate the clinical and financial feasibility of DL-based imaging technologies in both established and emerging healthcare systems [20].
- *Clinical Workflow Integration*: DL models must be seamlessly integrated into current healthcare workflows in order to reach their full clinical potential. The integration of AI tools into Radiology Information Systems (RIS) and PACS has been a crucial tactic. This integration enables AI-generated annotations to show up directly within the radiologist's interface, real-time identification of important findings, and automated retrieval and analysis of medical images. For instance, Aidoc's alerts are integrated into the PACS viewer, enabling prompt action without interfering with workflow.
- To make DL inference easier, hospitals are also implementing cloud-based and edge computing models. While edge computing allows for real-time analysis in remote or bandwidth-constrained environments, cloud platforms such as AWS Health Lake and Google Cloud Healthcare provide scalable computational power for high-throughput environments. These deployment choices improve diagnostic speed and offer resource allocation flexibility.
- Collaboration between clinicians and AI is essential for successful adoption. By highlighting areas of interest, producing probability scores, and providing visual explanations like heatmaps, DL tools enhance human judgment rather than replace it. In order to guarantee that AI outputs are interpreted within the entirety of the clinical context, clinicians maintain the final interpretive authority.
- Clinical trials, reader studies, and retrospective analyses are used to rigorously validate DL models in order to guarantee safety and regulatory compliance. Clinical deployment necessitates regulatory approvals, such as the FDA's 510(k) clearance in the United States and the CE marking in Europe. Additionally, tools need to abide by regulations like General Data Protection Regulation (GDPR) and Health Insurance Portability and Accountability Act (HIPAA) that protect health data.
- Lastly, trust-building and user training are essential. Clinicians must be informed about the features, restrictions, and advantages of DL tools for implementation to

be successful. In order to promote acceptance, interpretability strategies, such as saliency maps and open documentation, are essential. As these components come together, DL is playing a crucial role in clinical imaging, facilitating more precise diagnosis, increasing workflow effectiveness, and eventually enhancing patient outcomes.

14.6 Challenges and Limitations

Although there is massive potential of DL in medical imaging, its implementation is not devoid of massive challenges. Clinical setting provides complex and high-stake criteria, which even the most advanced algorithms struggle to solve. This section presents the key hurdles that still need to be seemingly conquered in order to ensure that DL is implemented throughout the medical field in a sustainable and ethical way [16]:

- Data Privacy and Regulatory Concerns (HIPAA, GDPR): This integration of DL into medical imaging presents serious concerns associated with issues of data privacy and regulatory compliance, as the data is patient health data, which is complex in nature. In the United States, the use and disclosure of the Protected Health Information (PHI) are strictly overseen by the HIPAA. Any medical images to be trained on an AI model should either be de-identified so that all identifying patient data has been removed or an overt patient consent should be received. In the European Union, under the GDPR, informed consent in the use of data is required, patients are given the right to access their data and data portability, and the right to explanation is established in the case of decision-making about them being done by automated systems. These regulatory systems play an essential role in protecting the rights of patients but also delay the process of innovation, as they require strict adherence to the regulations. In addition, the incompatibility of global privacy regulations hinders the creation and implementation of AI models internationally, thus making the process of cross-border collaborations and scalable solutions even more challenging.
- Requirement of Large-Scale, Annotated Data: DL-based systems perform best when given a large volume of high-quality data, yet the medical field is lacking in large, annotated datasets. Medical image annotation involves experts who identify the medical images, which means that both labeling and annotation would be a time-consuming and costly process as it involves people with professional training and experience like radiologist or pathologist, etc. Such a bottleneck is especially acute in rare diseases, whose underrepresentation in the datasets leads to their imbalance and subsequent inability of the models to generalize. Furthermore, institutional silos, where hospitals and medical centers store data in isolated systems with limited sharing capabilities, exacerbate the problem by re-

stricting collaboration and dataset diversity. Without access to large-scale, representative, and well-annotated datasets, it becomes challenging to train robust AI models. This lack of data diversity can lead to overfitting, reduce diagnostic accuracy across different populations, and limit the models' applicability in real-world scenarios.

- Model Generalizability and Interpretability: Another critical challenge in applying DL to medical imaging is ensuring that models are both generalizable and interpretable. Many models are trained on data from a single institution or scanner type, which may not reflect the variability present in other healthcare settings. For example, a chest X-ray model trained using images from Hospital A may perform poorly on images from Hospital B due to differences in imaging protocols, equipment quality, or patient demographics. Such limitations restrict the scalability and reliability of AI tools. Additionally, most DL models, especially CNNs, function as "black boxes," offering little insight into how they arrive at their conclusions. This lack of transparency makes it difficult for clinicians to trust AI outputs and poses significant barriers to regulatory approval, which increasingly demands interpretability to ensure patient safety and accountability.
- Ethical Concerns and Bias in Algorithms: Ethical concerns and algorithmic biases are pressing issues in the deployment of AI systems in healthcare. Models trained on datasets that are not diverse or representative can inadvertently perpetuate or even amplify existing disparities in healthcare outcomes. For instance, an algorithm trained primarily on data from one ethnic group, age group, or gender may yield biased predictions for underrepresented populations. This can result in misdiagnosis, inappropriate treatments, and delayed care, potentially causing harm and raising legal and ethical questions. Addressing these challenges requires a proactive approach – ensuring inclusive and balanced datasets, conducting fairness audits, and maintaining transparency in model development and evaluation. Without systematic efforts to identify and mitigate bias, AI in medicine risks entrenching inequalities rather than alleviating them.
- Human-in-the-Loop Approaches: Despite the push toward full automation in AI, human-in-the-loop (HITL) approaches are emerging as a pragmatic and safer solution for clinical environments. In these systems, human experts – such as clinicians and radiologists – interact with the AI outputs, validating, refining, or overriding them before any final decisions are made. This collaborative approach ensures accountability, maintains clinical oversight, and strengthens user trust by keeping healthcare professionals actively engaged in the diagnostic process. For example, radiologists might use AI tools to segment tumors in medical scans but ultimately confirm or adjust the results based on their expertise. Similarly, pathologists may use AI-generated diagnostic suggestions as decision support tools. HITL systems blend the speed and efficiency of AI with the nuanced judgment of medical professionals, enhancing safety, accuracy, and clinician accep-

tance. As a result, HITL is increasingly viewed as a viable path toward responsible AI adoption in healthcare.

Table 14.2: Key findings of AI-driven transformation in medical diagnostics.

Key area	Findings	Impact
Disease detection and diagnosis	Deep learning models outperform traditional systems in detecting pneumonia, tumors, DR, etc.	Enhances diagnostic accuracy, reduces time to diagnosis, and improves early detection rates.
Image quality enhancement	AI improves resolution, noise reduction, and artifact removal in imaging (MRI, CT, etc.).	Enables better clinical interpretation, particularly in low-dose or fast-acquisition scenarios.
Predictive analytics	AI combines imaging with clinical data to predict disease progression and treatment response.	Enables proactive care and personalized treatment planning.
Workflow optimization	Tools like Aidoc, Qure.ai, and PathAI automate triage, annotation, and reporting.	Increases efficiency, reduces radiologists' workload, and prioritizes critical cases.
Clinical deployment	Integration with PACS/RIS systems and cloud/edge deployments ensures seamless workflow use.	Supports real-time decision-making and large-scale adoption.
Ethical and legal compliance	Adoption depends on addressing HIPAA, GDPR, bias, and interpretability concerns.	Essential for trust, regulatory approval, and global scalability of AI solutions.
Collaborative AI-human models	Human-in-the-loop systems combine AI insights with expert validation.	Maintains accountability and enhances clinicians' trust in AI-assisted decisions.
Global access to healthcare	AI enables diagnostics in remote and under-resourced areas via portable devices and cloud AI.	Improves healthcare accessibility and equity, especially in rural or low-income regions.
Multimodal learning integration	Combines imaging with EHR, genomics, and clinical notes	Supports precision medicine and deeper phenotyping.
Explainability and trust	Use of saliency maps, Grad-CAM, and attention mechanisms	Enhances transparency and is crucial for clinical adoption and legal accountability.

Table 14.2 discusses the key findings of AI-driven transformation in medical diagnostics. It provided various key finding in this area. It also discussed the important findings and impact of it on various factors.

14.7 Future Directions and Innovations

The future of DL in medical imaging lies in collaborative, interpretable, and personalized AI systems. Emerging innovations aim to address current limitations and enhance the precision, trust, and reach of medical AI. Below are the possible area which demands future attention:

Privacy-Preserving AI and FL: Centralized datasets are necessary for training traditional DL systems, which frequently raise serious issues with patient privacy, data security, and regulatory compliance. By allowing model training across several institutions without requiring the sharing of raw patient data, FL provides a revolutionary way to get around these restrictions. Every participating institution in FL trains a local model on its own dataset independently, sharing only the learned model weights or gradients with a central server for aggregation. In addition to protecting sensitive health information's privacy and ownership, this decentralized approach complies with ethical and legal requirements like GDPR and HIPAA. Additionally, FL encourages cooperation between institutions that are spread out geographically, which makes it easier to create reliable, broadly applicable AI models that have been trained on a variety of patient populations. Examples from the real-world include the Clara FL platform from NVIDIA, which facilitates safe cross-site training for medical imaging applications, and innovative projects run by King's College London and the Mayo Clinic that show the viability and efficiency of FL in clinical settings:

- Fusion of Multimodal Data: Another new area of medical AI is multimodal data fusion, which combines medical imaging with other patient data sources, including genomic profiles, laboratory test results, clinical notes, and EHRs, to provide a more complete picture of a patient's health. DL models can create context-aware diagnostics that take into account a wider range of patient variables by integrating various data modalities. This combination facilitates more precise treatment outcome prediction, improves disease stratification, and supports individualized healthcare delivery. For example, combining MRI data with genetic markers has shown promise in the early detection and subtyping of Alzheimer's disease, and predictive models that evaluate CT scans in addition to lab results can predict how well cancer patients will respond to treatment. These multimodal methods are fundamental to deep phenotyping and precision medicine, giving doctors more comprehensive, data-driven insights for customized treatment regimens.
- Medical Diagnostics with Explainable AI (XAI): Explainability is essential for gaining the trust of clinicians, obtaining regulatory approval, and guaranteeing the safe and responsible use of AI in high-stakes industries like healthcare. The goal of XAI in medical diagnostics is to increase the transparency and interpretability of DL models' decision-making processes. Methods like attention mechanisms, gradient-weighted class activation mapping (Grad-CAM), and saliency maps make it easier to see which areas of an image affected the model's prediction. Clinicians can comprehend, validate, and even challenge AI-generated conclusions by using

these visual cues in conjunction with diagnostic outputs. Additionally, clinical workflows and reports can incorporate both local (case-specific) and global (model-wide) explanations, increasing their applicability in real-world situations. In addition to bridging the gap between clinical practice and black-box algorithms, XAI integration increases user accountability, trust, and engagement in AI-assisted diagnostics.

- Real-time monitoring and integration with wearable technology: Beyond conventional clinical settings, new opportunities for ongoing, real-time health monitoring have been made possible by the combination of DL and wearable technology. Wearable AI-powered devices, including photoplethysmography sensors, electrocardiogram patches, and portable ultrasound devices, can now produce physiological or imaging-like data streams that can be immediately analyzed using on-device or cloud-based DL models. Even in remote or underserved areas, these models can provide point-of-care diagnostics, monitor pulmonary conditions such as lung congestion in heart failure patients, and identify cardiac arrhythmias. These real-time features are transforming home-based care and telemedicine by enabling proactive chronic condition diagnosis and management, and lowering readmission rates to hospitals. These systems have enormous potential to expand healthcare access and encourage ongoing, individualized care as they get more precise, small, and reasonably priced.
- Predictive analytics and personalized medicine: The trend toward predictive healthcare and personalized medicine is centered on DL. Large-scale patient data, such as genomic data, medical imaging, and old EHR records, can be analyzed by DL models to find patterns that allow for personalized treatment planning and early disease detection. Long before symptoms appear, DL-powered predictive analytics can identify at-risk individuals, enabling prompt intervention and a better prognosis. AI can, for example, tailor cancer treatment by analyzing tumor characteristics and patient-specific biological markers, or it can predict the onset of diseases like diabetes or cancer by analyzing subtle changes in imaging data over time. In addition to improving treatment effectiveness, this change from a reactive to a proactive model of care also reduces side effects and maximizes resource use. A major step toward truly individualized medicine will be taken as the field develops, with DL playing a key role in providing targeted, data-driven therapies that are customized to each patient's particular profile.

14.8 Conclusion

Medical imaging has undergone a significant transformation thanks to DL, which has ushered in a new era of accuracy, effectiveness, and accessibility in healthcare. Its impact extends across various fields, including radiology, pathology, ophthalmology, cardiology, and more, from automating disease detection and diagnosis to enhancing image quality and enabling predictive diagnostics. There are already noticeable advantages to the practical application of DL models in actual clinical settings, including improved patient outcomes, streamlined workflows, and fewer diagnostic errors. However, stakeholders from all areas of the ecosystem must actively support the responsible development and application of AI in healthcare if its full potential is to be realized. The emphasis for researchers needs to change to developing models that are not only accurate but also fair, interpretable, and generalizable across a range of imaging modalities and populations. Building trust in AI-driven diagnostics will require addressing data biases and guaranteeing transparency in model decisions. Conversely, clinicians are essential to the adoption, ethical integration, and validation of AI tools in clinical practice. Their knowledge of patient contexts and domain expertise are essential in directing the proper application of these technologies, making sure that they support human judgment rather than take its place. The job of developers is to create interoperable, scalable, and secure solutions that fit in with the current healthcare systems. Usability, data privacy, and regulatory compliance must be given top priority in their efforts. The development and application of AI in medical imaging will ultimately depend on a multidisciplinary approach involving clinicians, data scientists, ethicists, and policymakers. Developing reliable systems will require rigorous data governance frameworks, ethical considerations, and ongoing validation in real-world scenarios. As technology develops, this cooperative, human-centered approach will be crucial to realizing AI's full potential and guaranteeing that its advantages are shared fairly and in line with the main objective of enhancing global health outcomes.

References

[1] França RP, Monteiro AC, Arthur R, Iano Y. An overview of deep learning in big data, image, and signal processing in the modern digital age. Trends in Deep Learning Methodologies. 2021 Jan;1:63–87.
[2] Cardoso MJ, Li W, Brown R, Ma N, Kerfoot E, Wang Y, Murrey B, Myronenko A, Zhao C, Yang D, Nath V. Monai: An open-source framework for deep learning in healthcare. arXiv Preprint arXiv:2211.02701. 2022 Nov 4.
[3] Nguyen HV, Huang SX, Xue Y. Data Augmentation, Labelling, and Imperfections. Springer Cham, 2022.
[4] Anderson NG, Butler AP. Clinical applications of spectral molecular imaging: Potential and challenges. Contrast Media & Molecular Imaging. 2014 Jan;9(1):3–12.
[5] Cai Y, Ji S, Crimi A, Menze B, Maier O, Reyes M, Winzeck S, Handels H Brainlesion: glioma, multiple sclerosis, stroke and traumatic brain injuries.
[6] Kebaili A, Lapuyade-Lahorgue J, Ruan S. Deep learning approaches for data augmentation in medical imaging: A review. Journal of Imaging. 2023 Apr 13;9(4):81.

[7] Nath R, Kazemi A, Callahan S, Stoddard MF, Amini AA. 4Dflow-VP-Net: A deep convolutional neural network for noninvasive estimation of relative pressures in stenotic flows from 4D flow MRI. Magnetic Resonance in Medicine. 2023 Nov;90(5):2175–2189.

[8] Panayides AS, Amini A, Filipovic ND, Sharma A, Tsaftaris SA, Young A, Foran D, Do N, Golemati S, Kurc T, Huang K. AI in medical imaging informatics: Current challenges and future directions. IEEE Journal of Biomedical and Health Informatics. 2020 May 29;24(7):1837–1857.

[9] Veasey BP, Broadhead J, Dahle M, Seow A, Amini AA. Lung nodule malignancy prediction from longitudinal CT scans with Siamese convolutional attention networks. IEEE Open Journal of Engineering in Medicine and Biology. 2020 Sep 11;1:257–264.

[10] Farhangi MM, Petrick N, Sahiner B, Frigui H, Amini AA, Pezeshk A. Recurrent attention network for false positive reduction in the detection of pulmonary nodules in thoracic CT scans. Medical Physics. 2020 May;47(5):2150–2160.

[11] Wang H, Amini AA. Cardiac motion and deformation recovery from MRI: A review. IEEE Transactions on Medical Imaging. 2011 Oct 13;31(2):487–503.

[12] Ibrahim ES. Heart Mechanics: Magnetic Resonance Imaging – Advanced Techniques, Clinical Applications, and Future Trends. 2017 Sep 19; CRC Press.

[13] Cha J, Farhangi MM, Dunlap N, Amini AA. Segmentation and tracking of lung nodules via graph-cuts incorporating shape prior and motion from 4D CT. Medical Physics. 2018 Jan;45(1):297–306.

[14] Tavakoli V, Bhatia N, Longaker RA, Stoddard MF, Amini AA. Tissue Doppler imaging optical flow (TDIOF): A combined B-mode and tissue Doppler approach for cardiac motion estimation in echocardiographic images. IEEE Transactions on Biomedical Engineering. 2014 Mar 3;61 (8):2264–2277.

[15] Wang H, Stoeck CT, Kozerke S, Amini AA. Analysis of 3D cardiac deformations with 3D SinMod. In 2013 35th Annual International Conference of the IEEE Engineering in Medicine and Biology Society (EMBC) 2013 Jul 3 (pp. 4386–4389). IEEE.

[16] Dhasarathan C, Shanmugam M, Kumar M, Tripathi D, Khapre S, Shankar A. A nomadic multi-agent based privacy metrics for e-health care: A deep learning approach. Multimedia Tools and Applications. 2024 Jan;83(3):7249–7272.

[17] Yuan H. Overcoming computational resource limitations in deep learning for healthcare: Strategies targeting data, model, and computing. Medicine Advances. 2025 Mar;3(1):42–45.

[18] Panahi O. Deep Learning in Diagnostics. Journal of Medical Discoveries. 2025;2(1):1–6.

[19] Matheny ME, Goldsack JC, Saria S, Shah NH, Gerhart J, Cohen IG, Price WN, Patel B, Payne PR, Embí PJ, Anderson B. Artificial Intelligence In Health And Health Care: Priorities For Action: Article examines priorities for the uses of artificial intelligence in health and health care. Health Affairs. 2025 Feb 1;44(2):163–170.

[20] Prakash P, Prabu P, Sakthivel V, Sivakumar R. RETRACTED ARTICLE: An efficient deep learning based stress monitoring model through wearable devices for health care applications. Optical and Quantum Electronics. 2024 Feb;56(2):200.

Himani Bhatt, Saurabh Rana, Mandeep Mittal

15 Future of AI-ML and Their Potential Impact on Deep Learning Applications in Medical Diagnostics and Imaging

Abstract: The field of healthcare has been developing and changing rapidly. Healthcare organizations are now starting to develop management strategies to integrate these platforms, artificial intelligence (AI)-machine learning (ML) toolsets. All these platforms leverage the computational power of these sophisticated algorithms to analyze data and provide improved insights. Recent AI-ML systems are revolutionizing the industry by providing creative ways to improve clinical decision support, operational workflows, diagnostic accuracy, and clinical outcomes. Additionally, these tools are becoming more and more useful in healthcare research, where they facilitate automated image processing, drug development, clinical trials, biomarker discovery, and productive analytics. The use of ML-Operations to manage models in medical environments, multifunctional and multi-agent AI to leverage a variety of data sources, accelerated translational research, and virtualized education for training and simulation are additional connected themes. This chapter uses hypothetical studies of medical imaging data, recently developed DL interpretation data from electronic medical records or electronic health records, massive amounts of data in personalized medicine, and dataset shifts in AI to demonstrate the progress in data science.

Keywords: healthcare, machine learning, deep learning, ChatGPT, artificial intelligence

15.1 Introduction

The medical industry is undergoing rapid change in the modern period. Recent advances in artificial intelligence (AI) technology, such as deep learning (DL) technologies, have led to a paradigm shift in medicine in the areas of concentrate medication therapy, advanced treatment methods, advanced illness diagnostics, and surgery. The paradigm shift in medicine has been significantly impacted by the recent advancements in machine learning (ML). ML is a type of AI that makes accurate predictions using algorithms. These algorithms are used in medicine to produce novel results based on medical data. Currently, the medical industry makes extensive use of ML systems. There are potentials and challenges for healthcare research as a result of higher access to data. Developing a trustworthy medical system based on data-driven techniques, DL, and ML presents a significant challenge, which will figure out the relationship between all the data in large medical datasets. For instance, Google Deep-

 | https://doi.org/10.1515/9783112226223-015

Mind has proposed an approach to use its specialized knowledge to improve healthcare, Enclitic uses deep neural networks (DNNs) and X-rays to identify health issues on computed tomography (CT) scans, and International Business Machines (IBM) Watson Health uses this technology to communicate with patients and healthcare providers and concentrates on offering affordable healthcare to a large population. Using a development dataset of retinal pictures, Gulshan et al. [1] created a DL-based algorithm in 2016 to detect diabetic retinopathy. The DL algorithm generally shows high susceptibility and precision [1]. The next year, Esteva et al. [2] used a dataset of surgical photos to create an algorithm for the classification of skin cancer using DNNs. In the past, it is acknowledged that Alan Turing was the first to describe the idea of utilizing computers to simulate critical thinking and intelligent behavior in 1950. Six years later, John McCarthy came up with the term AI.

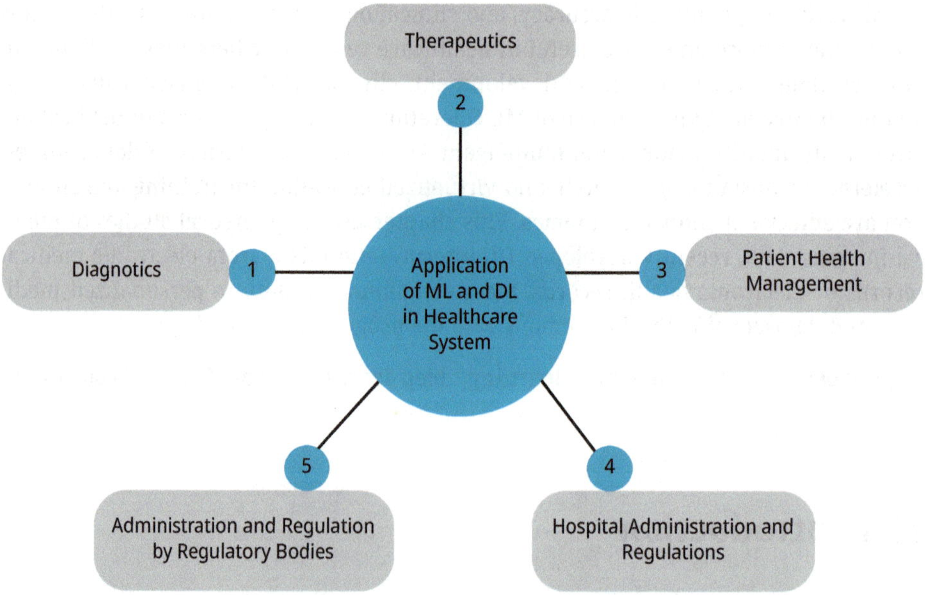

Figure 15.1: The numerous uses of ML and DL in medicine and healthcare.

AI is the scientific and technological method of building intelligent machines, according to computer scientists [3]. A number of advancements in AI took place during the 1970s and 2000s. Biosensors and AI have a big impact on the medical industry. This highlights the revolutionary effects of AI and biosensors on medical science and patient care while outlining the difficulties, possible problems, and opportunities in their application in the healthcare sector. In particular, Aisu et al. [4] used closed-loop control algorithms, insulin predictions, and assessments within constant glucose monitoring to explore the use of AI in diabetes management. The study looks at the opportunities and problems related to proactive and customized therapy. Even though ML-based algorithms have improved various medical specialties, data science and big data analytics have also significantly advanced in the field of medicine concurrently with ML-related algorithm improvements. Scholars are attempting to assess and identify the relationships between big data and actual medical datasets. Additionally, several ML or DL algorithms and healthcare equipment for the clinic have been approved by the US FDA and other developed nations, as shown in Figure 15.1. The ML and DL-based medical devices that have been approved in Japan are mentioned by Marshall et al. [5]. Eleven ML and DL-based devices were discovered to have obtained regulatory approval. Of these, six have to do with radiography and five with gastroenterology. The widespread use of computer-assisted decision-making and outcome evaluation in healthcare delivery emphasizes the technical importance of modeling knowledge and proficiency [6]. However, typical rule-based models cannot adequately replicate the complex structure of human brains since they mostly rely on feature representation of issue domains. Thus, this chapter used a deep model to try to solve this problem. The deep model combines learning and feature representation to produce a single model that can mimic human cognitive processes. Many medical problems that potentially benefit from DL and ML techniques haven't been thoroughly studied, though. The healthcare sector may benefit from the many benefits of DL and ML, including enhanced performance, a lateral learning model with integrated feature learning, and the ability to handle complex, varied data, among others. Better techniques and resources are also needed to incorporate DL into clinical decision-support operations.

15.2 Related Works

McCulloch and Pitts created a neural network model, which marked the beginning of ML [7]. Neural networks from earlier times might have proven more beneficial and perceptive. From their inception, ML models kept developing. Hinton's research established the foundation for data-driven ML. Data-driven ML has gained popularity recently as a particular use of AI. Data-driven ML uses algorithms to train and learn from knowledge and experiences [8]. A Chinese diabetes dataset was developed to create data-driven ML models by Rahmani et al. [9]. With the creation of innovative methods to address various issues, ML algorithms and models have advanced at a rapid pace. Graph ML (GML), a novel type of ML technique, is one example. It uses the structural properties of graphs and other irregular datasets to effectively describe nodes, edges, or even complete graphs with features [10]. Applications of ML can be found in a number of fields, such as cancer research, education, drought danger monitoring and forecasting, and medication discovery and development. Chakraborty et al. [11] investigated how blockchain, AI, and IoMT intersect in the healthcare industry. Through data-driven methods, the study explores how these technologies might be used to improve healthcare for patients. The study also carried out a comparison of several big data technologies in the field of data-driven caring. The enormous volumes of medical data needed for data-driven healthcare may be handled and processed with the help of this research. Leiserson et al. [12] offers a basis for implementing advanced approaches by recognizing the uses and boundaries of conventional ML in the healthcare industry. This opens the door for a data-driven approach to healthcare. Even with the significant advancements in ML research and instruction, there are still many issues that need to be addressed. Among the challenges to incorporating AI into education include privacy issues, ethical issues, and the necessity of excellent human-AI collaboration [13].

In their book chapter, Chahal and Byrne [3] emphasized the value of DL approaches in processing and analyzing large amounts of medical data. Mitchell et al. [14] also evaluated the importance of DL in the healthcare industry. They looked at new network topologies and their uses while studying DL models employed in healthcare. In their book chapter, Wu et al. [13] also emphasized the potential of ML and DL in health monitoring systems. They put out an effective model that would outperform conventional ones in terms of accuracy and credibility. They carried out a comparison of several big data tools in the field of data-driven care. The enormous volumes of medical data needed for data-driven patient care may be handled and processed with the help of this research. Antibiotic resistance is one of the global health issues that DL has been suggested to address. DL was introduced by Jackson et al. [15] as a solution to the antibiotic resistance issue. A method for explaining the potential of DL to address antibiotic resistance has been demonstrated. One of the most important questions is who owns healthcare data and who has the power to make use of it. Conflicts can be avoided, and responsible data use can be ensured by establishing explicit governance norms that cover data ownership, access, and sharing. Strong ethical frameworks, ethical supervision, and continual monitoring are necessary to handle these ethical issues and assure that AI systems are legitimate, accessible, and responsible. Lastly, training and education are prerequisites for DL integration in healthcare. In this sense, comprehensive training courses, seminars, and programs may close the knowledge gap and guarantee that the team is ready for the upcoming AI-centric healthcare age.

15.3 Progress in Medical Data Science Along the ML to DL Approach

The first invented computers were quite large and inefficient. With a storage capacity of less than 5 MB, it was quite little. In 1980, the first hard disk with a 1-GB capacity was launched. Every day, both the computing speed and the data storage capacity climbed. As a result, computer hardware and data storage systems have greatly improved in terms of recollection, speed, and efficiency. Computers nowadays are speedier and have a large processing capability when it comes to data [16].

15.3.1 ML to DL Interpretation of Medical Image Data

Medical image data identification has significantly improved because to the evolution of ML models. A new stage in the development of ML has been sparked by the recent development of DL. DL techniques for radiological image processing have demonstrated notable increases in accuracy in various medical imaging domains, including cardiac function evaluation, lung cancer detection, eye care detection, digestive health comprehension, and breast imaging comprehension. One crucial area of medical imaging data processing utilizing ML to DL models comprehends breast cancer (BC) images. More effective DL algorithms are particularly being used by scientists for picture data related to cancer and medicine. For BC screening, Wu et al. employed a DCNN model of DL. Here, more than 1,000,000 BC images were used by the researchers to train and assess the graphic data. Liu et al. [17] have created an enhanced mammography-based BC screening approach employing DL methods. The convolutional network approach was employed in this study to categorize mammograms. Here, they have employed the "end-to-end" training methodology, which effectively makes use of both the independent test set of Full-Field Digital Mammography and training datasets [17]. ML to DL techniques have been widely used for the detection of lung cancer. Scientists have recently been diagnosing cancer and other diseases utilizing three-dimensional (3D) DL approaches, in which a model has been created for the 3D perspective from an object using a series of 2D images that depict the scene. Researchers are using ML and DL to screen the image and other cardiac operation data. These models are advancing the science in crucial ways. Caldarini et al. [18] attempted to use DL to evaluate the echocardiography data in order to comprehend heart function. The local cardiac structures have been identified, and the researchers have assessed cardiovascular risk and cardiac performance. Numerous noteworthy models have already been widely applied in ophthalmology over the ML to DL transition. Several innovative algorithms have occasionally been used in this direction. DL systems have mostly been used by scientists to categorize and validate various ophthalmological components. Additionally, DL has improved gastroenterology, particularly colon and endoscopy procedures. Endoscopy and colonoscopy are now more reliable diagnostic procedures thanks to the assistance of ML and DL models. By using these simulations, scientists are attempting to enhance endoscopic procedures. Jovanović et al. [19] used a learning-based DL method to identify the area of interest in entire slide images. The study employed a dataset of 44,731 slide images from 15,186 participants to develop and evaluate classification algorithms.

15.3.2 Personalized Medical Care Using Big Data

The three main categories of data that make up big data, complicated datasets, or excessively huge datasets are semistructured, unstructured, and categorized information. Numerous experts in the field of biology have attempted to use big data to improve healthcare and personalized medicine. Dynamic simulation modeling (DSM) and big data have been combined by Rajput and Subasi [20] to provide personalized therapy. Occasionally, histological scans were used to try and identify lung tissue. Single-cell level pathology is crucial for identifying cancer cells in addition to ML or DL. Rajkomar et al. [21] have attempted to use a single-cell pathology landscape to identify intratumor phenotype heterogeneity. Rajpurkar et al. [22] enhanced healthcare quality and advanced personalized treatment by using DL models to predict several disease events from a sizable amount of data.

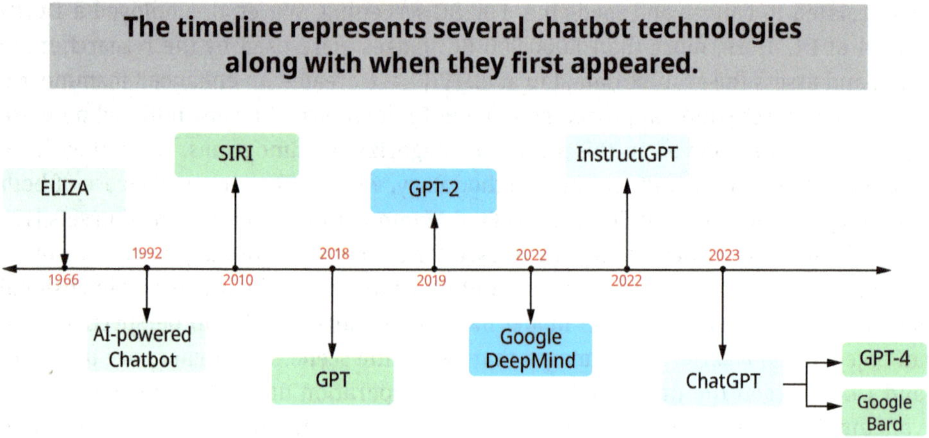

Figure 15.2: Timeline for various chatbots.

15.3.3 Medical Research Data

In addition to image categorization, ML and its incredibly effective DL-based systems are used in medication research and learning, antibiotic use and resistance, the structure and operation of protein prediction, and a number of other health science fields.

a) **Structure of Protein and Prediction of Function**
DL models were used recently in multiple fields of study related to protein structure as well as function prediction, including de novo protein modeling biochemical processes of proteins using self-supervised automatic encoders of DL models, and enhanced protein structure estimation using DL models.

b) **Drug Identification and Creation**
One significant area of medical research is developing drugs. DL models and AI have previously been used in various drug discovery and development domains, including the use of DL-based web screening for drug discovery called deep screening [23], DL-based medication toxicology testing, lead optimization model development using DL (lead optimization through the 3D ligand information using AI/ML methods), and optimization of hit preference in online screening utilizing machine intelligence.

c) **A Number of Additional Medical Science Fields**
Models based on ML and DL are useful in various other fields of medical study, including surgery [15]. Research has emerged as a significant field in biology in recent years. Here, ML to DL models have occasionally been created. To comprehend the anti-CRISPR peptide families, ML was employed. Likewise, SeqCor ML techniques were created to control the potential biases brought about by the organization of gRNA sequences.

15.4 Medical and Healthcare Chatbot Technology

Chatbots are AI-powered interpersonal assistants that can answer in natural language and assist users with a variety of activities. They have recently become more common in a number of areas, consisting of healthcare [24]. Patients who wish to examine their health issues before consulting a physician can utilize chatbots to self-diagnose. An additional chatbot application that provides patients with individualized guidance and assistance is health coaching [25]. Coach AI (a health coaching platform) makes it possible to provide both individual and group patients with health assistance. These chatbots can assist patients in maintaining their health and well-being by providing advice on stress management, exercise, and food choices. Chatbots may additionally be employed to make consent forms that are informed easier to read when viewed online. Another kind of chatbot that can identify how a user feels and modify discussions to support those undergoing psychological care is the sympathetic chatbot. Additionally, individuals can receive mental healthcare from chatbots. Wyse, an AI-powered bot that uses dialectical behavior therapy and cognitive behavioral therapy to assist people control their mental states, is one example. Another kind of chatbot that can identify an individual's psychological condition and modify discussions to support those undergoing psychological treatment is the sympathetic chatbot. Chatbots featuring healthcare design enable patients to make doctor's appointments and get medication reminders. The healthcare industry has been touched by these amazing, always-available health monitoring systems. Thus, by offering patients individualized and effective treatments, chatbots have the potential to revolutionize the healthcare industry. Self-monitoring health guidance, informed cooperation, psychological health support, and remote patient monitoring are just a few of the many medical services that are offered. As they advance, chatbots are anticipated to play an increasingly significant role in the healthcare sector.

15.4.1 ChatGPT: The Latest Chatbot Technology Using Various Models

Experimental teams worked diligently for several years to create DL-based chatbots, and chatbots with particular functions were released. OpenAI introduced ChatGPT, a novel chatbot that can produce conversation informally, which is strikingly comparable to human speech, in 2022. The upgraded ChatGPT version now operates on GPT-4, while the original ChatGPT model was built on GPT-3.5. ChatGPT generates text that resembles that of human using DNNs. The capabilities of ChatGPT have the potential to change research, teaching, and statistical process control (SPC) procedures. However, the technology is vulnerable to misuse or misconceptions because it is still in its early phases of development. The study examined ChatGPT's capacity to produce information relevant to SPC procedures, instruction, and research, as well as to deliver code and explain important ideas. Due to its ability to streamline radiological reports, ease patient-provider contact, and produce academic papers of a respectable caliber, ChatGPT has gained a lot of attention in the medical industry, as shown in Figure 15.2. ChatGPT fulfills a vital function by offering medical guidance and repositories of medical information, including symptom checks, fitness, and health counseling. To meet the requirements of patients or users, the system also offers a few special medical features like setting up appointments, notification alerts, and instructions for routine medication. Social networking sites like ChatGPT may additionally encourage communication and provide tools for the mental health of students in higher education. When talking about ChatGPT's advantages, it's important to note that its application in clinical settings is still difficult and that its results in the medical profession are on par with third-year medical graduates who receive grades that fall just short of the passing threshold. Notwithstanding ChatGPT's potential advantages for the healthcare system, there are serious ethical issues with its use. One of the major issues in this regard is the possibility of academic articles being misused when using ChatGPT. Additionally, because ChatGPT sometimes shows unpredictable answers and neglects or dismisses crucial data, there are concerns regarding the correctness of the translation results. The suggestion of doctors for in-person counseling, which they make after many years of practice in their various medical specialties, may clash with ChatGPT's response. Investigating and resolving each of these ethical concerns to a certain acceptable quality is essential prior to introducing ChatGPT in clinical settings.

15.5 The Wide Range of Applications of ML and DL in Healthcare Administration

In the medical industry, ML and DL are playing important roles in diagnosis, treatment, medical condition management, hospital governance and regulation, and national healthcare management operations [26, 27].

a) **Disease Identification**
Numerous diagnostic tasks are carried out by ML and DL, such as medical pictures identification for early cancer diagnosis, rapid infectious illness detection, and medical diagnosis in various gastrointestinal diseases, cardiac surgery, ophthalmology, etc. disciplines. They can be used for pathology-related data diagnosis. They might be employed to analyze data pertaining to pathologists. In addition to these diagnostic procedures, the Electronic Health Record or HER data is analyzed using ML and DL methods.

b) **Anti-dots**
When treating patients with HER, ML and DL can provide guidance to doctors and potentially aid in therapeutic methods. Additionally, it assists medical doctors doing surgery influenced by AI. It could support precise medicine based on data. Pharmacogenomics-based healthcare also benefits from these frameworks.

c) **Hospital Management and Governance**
Hospital management and governance can be guided by ML and DL algorithms on a daily basis. Additionally, it aids in the preservation and analysis of HER or EMR data. Subsequently, it could be useful for tracking illnesses. Regulatory agencies can use several ML metrics, including DL, to monitor and assess all health-related regulatory actions. Regulatory agencies like the FDA can use these frameworks to examine product-related applications first.

15.6 New Challenges to the Use of ML to DL in Healthcare

There are several challenges in the application of ML and DL models, which are also represented in Figure 15.3.

15.6.1 Standardization of Data

Results from ML to DL might be more precise. Thus, before it is used to diagnose the patient, data standardization, training of models, and testing are essential. The dataset shift, a significant issue for training and test data, has already been covered. The adoption of a sepsis-alerting approach at Michigan University Hospital was recently characterized as posing real-time challenges. As a result, before applying ML or DL models in the real healthcare sector, appropriate standards must be established.

15.6.2 Creating Policies, Standards, and the Regulatory Framework

In the actual world, ML will be used in the medical field. The technologies may occasionally be biased by the data. As a result, ML or DL technologies can be used while taking disease risk into account. Control by the regulatory agencies is necessary to guarantee efficacy and security. To guarantee efficacy and safety, each nation must have laws and regulations in such areas. As a result, a regulatory atmosphere will be established.

a) **Requirement of a Skilled and Knowledgeable Workforce**
The cost of these systems is high. As a result, the deployment of all technologies requires a significant financial investment. At the same time, the systems require constant upkeep, including hardware maintenance, software upgrades, and changes related to algorithms. These technologies, therefore, have significant financial ramifications. Therefore, appropriate budgetary allocation should be required prior to the technology implementation.

b) **Ethical Issue in Data Security**
Cybersecurity safeguards pertaining to patient safety are necessary for ML technologies. Misconduct may occur, and medical datasets may be compromised by hackers. It should be taken into account that it contains private data from patients in real time. Thus, protecting the patient's privacy is essential.

c) **Further Implementation Difficulties**
Large image sizes raise concerns since neural networks have the potential to enhance model complexity and consequently increase pixel count. The quantity of RAM needed has therefore grown. The possibility of noise in some weak supervision configurations is another drawback of the dataset. When data is gathered from many sources, a technical bias system is also required.

15.7 Conclusion

Significant advancements have been made in healthcare because of ML and DL methods. The recent paradigm change in healthcare is the result of ongoing research advancements from the fields of ML to DL. However, there are still many unanswered questions regarding the practical application of these technologies in the medical area. In order to enhance the standard of diagnoses and treatments, as well as for a variety

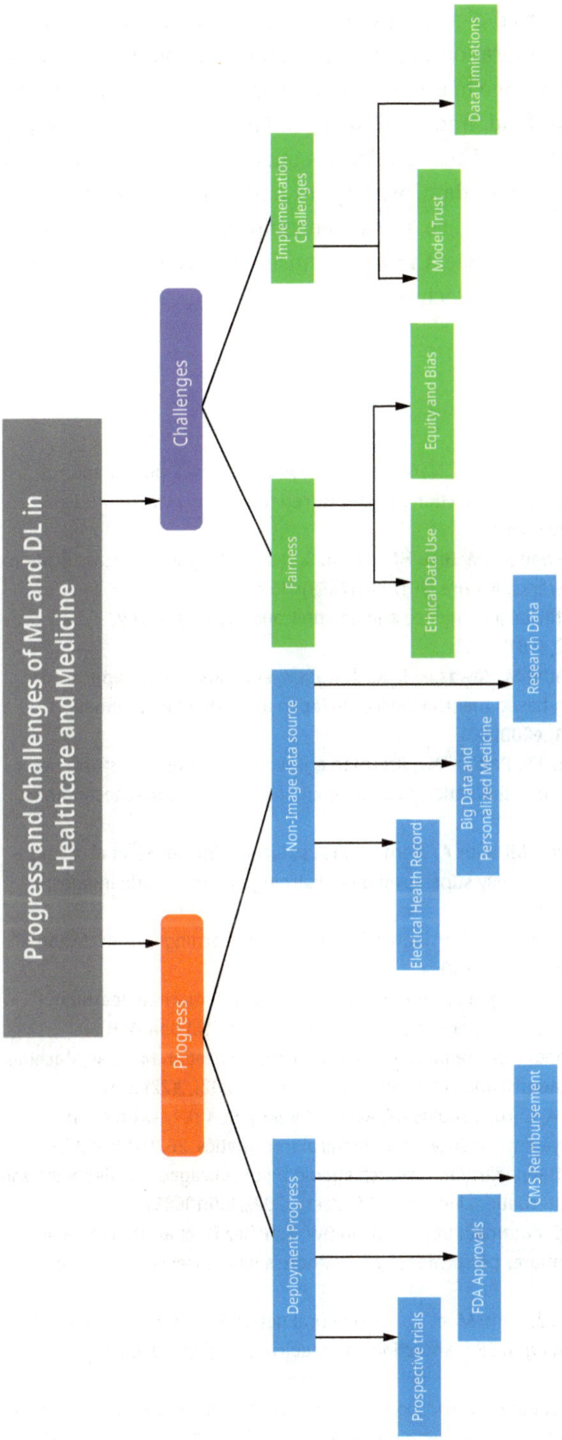

Figure 15.3: The challenges and progress in the development of ML and DL in healthcare system.

of medical applications, new and varied algorithms are required. The method is based on encouraging interdisciplinary cooperation between specialists in data science, engineering, medicine, and AI. It is imperative to establish norms and legal frameworks specifically suited to AI in the medical field, and worldwide collaboration is necessary to guarantee uniformity and thoroughness. These guidelines ought to cover crucial moral issues like accountability, openness, and data privacy. In addition to addressing the present issues with the use of AI in healthcare, this all-encompassing approach for ML to DL-based medical solutions will guarantee a more promising future for the industry as it brings in a new era driven by modern technologies.

References

[1] Gulshan V, Peng L, Coram M, Stumpe MC, Wu D, Narayanaswamy A, et al. Development and validation of a deep learning algorithm for detection of diabetic retinopathy in retinal fundus photographs. JAMA. 2016;316(22):2402–2410.

[2] Esteva A, Kuprel B, Novoa RA, Ko J, Swetter SM, Blau HM, Thrun S. Dermatologist-level classification of skin cancer with deep neural networks. Nature. 2017;542(7639):115–118.

[3] Chahal D, Byrne MF. A primer on artificial intelligence and its application to endoscopy. Gastrointestinal Endoscopy. 2020;92(4):813–820.

[4] Aisu N, Miyake M, Takeshita K, Akiyama M, Kawasaki R, Kashiwagi K, et al. Regulatory-approved deep learning and machine learning-based medical devices in Japan as of 2020: A systematic review. PLOS Digital Health. 2022;1(1):e0000001.

[5] Marshall DA, Burgos-Liz L, Pasupathy KS, Padula WV, IJzerman MJ, Wong PK, et al. Transforming healthcare delivery: Integrating dynamic simulation modelling and big data. Pharmacoeconomics. 2016;34:115–126.

[6] Campanella G, Hanna MG, Geneslaw L, Miraflor A, Werneck Krauss Silva V, Busam KJ, et al. Clinical-grade computational pathology using weakly supervised deep learning on whole slide images. Nature Medicine. 2019;25(8):1301–1309.

[7] Ghorbani A, Ouyang D, Abid A, He B, Chen JH, Harrington RA, et al. Deep learning interpretation of echocardiograms. NPJ Digital Medicine. 2020;3:10.

[8] Gaudelet T, Day B, Jamasb AR, Soman J, Regep C, Liu G, et al. Utilizing graph machine learning within drug discovery and development. Briefings in Bioinformatics. 2021;22(6):bbab159.

[9] Rahmani AM, Yousefpoor E, Yousefpoor MS, Mehmood Z, Haider A, Hosseinzadeh M, et al. Machine learning in medicine: Review, applications, and challenges. Mathematics. 2021;9(22):2970.

[10] Shamshirband S, Fathi M, Dehzangi A, Chronopoulos AT, Alinejad-Rokny H. A review on deep learning approaches in healthcare systems. Journal of Biomedical Informatics. 2021;113:103627.

[11] Chakraborty S, Nandi SS, Dhama K. Deep learning research should be encouraged for diagnosis and treatment of antibiotic resistance. International Journal of Surgery. 2022;105:106857.

[12] Leiserson CE, Thompson NC, Emer JS, Kuszmaul BC, Lampson BW, Sanchez D, et al. There's plenty of room at the top: What will drive computer performance after Moore's law?. Science. 2020;368 (6495):eaam9744.

[13] Wu N, Phang J, Park J, Shen Y, Huang Z, Zorin M, et al. Deep neural networks improve radiologists' performance in breast cancer screening. IEEE Transactions on Medical Imaging. 2020;39 (4):1184–1194.

[14] Mitchell EG, Maimone R, Cassells A, Tobin JN, Davidson P, Smaldone AM, et al. Automated vs human health coaching. Proceedings of the ACM on Human-Computer Interaction. 2021;5(CSCW1):1–37.

[15] Jackson HW, Fischer JR, Zanotelli VR, Ali HR, Mechera R, Soysal SD, et al. The single-cell pathology landscape of breast cancer. Nature. 2020;578(7796):615–620.

[16] Ng MF, Zhao J, Yan Q, Conduit GJ, Seh ZW. Predicting the state of charge and health of batteries using data-driven machine learning. Nature Machine Intelligence. 2020;2(3):161–170.

[17] Liu Z, Du J, Fang J, Yin Y, Xu G, Xie L. DeepScreening: A deep learning-based screening web server for accelerating drug discovery. Database. 2019;2019:baz104.

[18] Caldarini G, Jaf S, McGarry K. A literature survey of recent advances in chatbots. Information. 2022;13(1):41.

[19] Jovanović M, Baez M, Casati F. Chatbots as conversational healthcare services. IEEE Internet Computing. 2020;25(3):44–51.

[20] Rajput A, Subasi A. Lung cancer detection from histopathological lung tissue images using deep learning. In Applications of artificial intelligence in medical imaging 2023 (pp. 51–74). Elsevier.

[21] Rajkomar A, Dean J, Kohane I. Machine learning in medicine. The New England Journal of Medicine. 2019;380(14):1347–1358.

[22] Rajpurkar P, Irvin J, Ball RL, Zhu K, Yang B, Mehta H, et al. Deep learning for chest radiograph diagnosis. PLoS Medicine. 2018;15(11):e1002686.

[23] Green H, Durrant JD. DeepFrag: An open-source browser app for deep-learning lead optimization. Journal of Chemical Information and Modeling. 2021;61(6):2523–2529.

[24] He J, Baxter SL, Xu J, Zhou X, Zhang K. The practical implementation of artificial intelligence technologies in medicine. Nature Medicine. 2019;25(1):30–36.

[25] Yasnitsky L. Artificial intelligence and medicine: History, current state, and future forecasts. Current Hypertension Reviews. 2020;16(3):210–215.

[26] Wang H, Czerminski R, Jamieson AC. Neural networks and deep learning. In The machine age of customer insight 2021 (pp. 91–101). Emerald Publishing.

[27] Vornholt P, De Choudhury M. Understanding the role of social media-based mental health support among college students. JMIR Mental Health. 2021;8(7):e24512.

Index

 | https://doi.org/10.1515/9783112226223-016

Mathematical Methods in the Digital Age

Volume 1
Satya Prakash Yadav, Abhishek Kumar, Ashutosh Kumar Dubey, Victor Hugo C. de Albuquerque, Adrian Rodriguez-Aguiñaga, Angeles Quezada
Medical Image Processing - Multimodal Fusion Techniques, 2024
ISBN 978-3-11-139549-4, e-ISBN (PDF) 978-3-11-143597-8, e-ISBN (EPUB) 978-3-11-143599-2

Volume 2
Umesh Kumar Lilhore, Vishal Dutt, T. Ananth Kumar, Martin Margala, Kaamran Raahemifar
Math Optimization for Artificial Intelligence - Heuristic and Metaheuristic Methods for Robotics and Machine Learning
ISBN 978-3-11-143605-0, e-ISBN (PDF) 978-3-11-143618-0, e-ISBN (EPUB) 978-3-11-143624-1

Volume 3
Hemant Kumar Saini, S.B. Goyal, Sardar M.N. Islam and T. Ananth Kumar
Quantum Unmanned Aerial Vehicle - Achieving Cybersecurity and AI Strategies
ISBN 978-3-11-161930-9; e-ISBN (PDF) 978-3-11-161932-3; e-ISBN (EPUB) 978-3-11-161935-4

Volume 4
Shalli Rani, Syed Hassan Sha, Ayush Dogra
Mathematical Computing and Sustainability - Predictive Modeling and Ramifications of Intelligent Systems
ISBN 978-3-11-161111-2, e-ISBN (PDF) 978-3-11-161203-4; e-ISBN (EPUB) 978-3-11-161232-4

Volume 5
Satya Prakash Yadav, Abhay Bansal, Victor Hugo C. de Albuquerque, Fadi Al-Turjman, Houbing Herbert Song
Octonion-Sparse Based Image Processing - Multi Spectral Imaging and Feature Extraction
ISBN 978-3-11-914666-1, e-ISBN (PDF) 978-3-11-220807-6; e-ISBN (EPUB) 978-3-11-220835-9

Volume 6
Ganesh Gopal Devarajan, Daniel Arockiam, Senthil Murugan Nagarajan, Rashmi Agrawal
Healthcare 5.0 with Fuzzy Logic – Artificial Intelligence, Cyber-Physical Systems, and Medical Internet of Things
ISBN 978-3-11-914504-6, e-ISBN (PDF) 978-3-11-221572-2; e-ISBN (EPUB) 978-3-11-221592-0

Volume 7
Narinder Kaur, Bobbinpreet Kaur, Yonis Gulzar, Dac-Nhuong Le
Computational Optimization – Machine Learning and Fuzzy Systems
ISBN 978-3-11-914376-9, e-ISBN (PDF) 978-3-11-221943-0, e-ISBN (EPUB) 978-3-11-221944-7

Volume 8
Suresh Kumar Krishnadhas, Ananth Kumar Tamilarasan, Pramod Singh Rathore, Abhishek KumarArtificial Intelligence and Robotics - Transformative and Computational Algorithms for Enhanced Problem Solving
ISBN 978-3-11-914520-6, e-ISBN (PDF) 978-3-11-221437-4, e-ISBN (EPUB) 978-3-11-221466-4